Challenges in Physics Education

Series Editor

Marisa Michelini, Dipartimento di Fisica, University of Udine, Udine, Italy

This book series covers various aspects of physics teaching and learning across all educational levels and learning environments. The individual volumes cover a wide array of topics, including innovative approaches and pedagogical strategies for physics education, the development of effective methods to integrate multimedia into physics teaching and learning, pioneering lab experiments, and the utilization of web-based interactive activities. Selected research-based contributions will be prominently featured throughout.

The GIREP, the International Research Group on Physics Teaching, collaborates on this publication series, incorporating selected papers from globally recognized experts and monographs. Book proposals from other sources are also encouraged.

"Challenges in Physics Education" is targeted towards professionals, teachers, researchers, instructors, and curriculum developers, aiming to enhance physics teaching and learning and, consequently, the overall perception of physics within society.

Proposals for publication should be submitted to Marina Forlizzi (marina.forlizzi@springer.com).

Sergej Faletič · Jerneja Pavlin
Editors

Teaching and Learning Physics Effectively in Challenging Times

Springer

Editors
Sergej Faletič
Department of Physics
Faculty of Mathematics and Physics
University of Ljubljana
Ljubljana, Slovenia

Jerneja Pavlin
Department of Physics and Technical Studies
Faculty of Education
University of Ljubljana
Ljubljana, Slovenia

ISSN 2662-8422 ISSN 2662-8430 (electronic)
Challenges in Physics Education
ISBN 978-3-031-72540-1 ISBN 978-3-031-72541-8 (eBook)
https://doi.org/10.1007/978-3-031-72541-8

This Springer imprint is published by the registered company Springer Nature Switzerland AG
The registered company address is: Gewerbestrasse 11, 6330 Cham, Switzerland

Preface

The peer-reviewed book presents selected papers from the international conference—GIREP Conference 2022, Ljubljana, Slovenia.

The aim of the book is to familiarize readers with the discussion on physics education related to effective learning in physics, from contemporary physics to remote settings. The book is aimed at a wide range of readers, researchers in the field of physics education, university teachers, physics teachers at various levels of education and other interested parties.

The GIREP Conference 2022 of the International Research Group on Physics Teaching (GIREP, https://www.girep.org/) was organized by the University of Ljubljana, Faculty of Education and Faculty of Mathematics and Physics. It is one of a series of GIREP conferences that have been organized at regular intervals since 1966. The conference took place between 4 and 8 July 2022 in Ljubljana, Faculty of Education.

The main role in the organizing and supervising conference had the following members of the local organizing committee Sergej Faletič, Saša Ziherl, Katarina Susman, Anja Kranjc Horvat and Jerneja Pavlin in collaboration with other members of the local organizing committee and in scientific cooperation with the international advisory board. The event was financially sponsored by GIREP. The conference was attended by 204 participants from 37 countries from around the world.

The conference was organized in a hybrid form after 2 years of webinars. The majority of participants were present live. The presentations of the online participants were integrated into the regular sessions to allow for live feedback, comments, questions and answers.

A central theme of the conference was effective learning in physics from contemporary physics to remote settings. The aim was also to discuss the possibilities of contemporary topics to increase interest, context and scientific literacy in the context of modern technology, combined with the challenges and opportunities of distance teaching and learning. The conference programme enabled the discussion of various topics on contemporary physics, distance education and current aspects of research and practice in physics education.

The programme of the scientific conference offered plenary talks, parallel oral sessions, symposia, workshops and poster sessions. In total, 7 invited talks, 81 oral and 80 poster presentations were held. The conference categorized the contributions into nine categories:

1. Invited Talks: Plenary speakers invited to share their lifelong experience on teaching and learning.
2. Strategies and Methods to Improve Physics Learning and Teaching: From good practices to new tested approaches.
3. Physics Curriculum Materials: Development, Implementation, and Assessment: From the development of new materials to effective teaching of existing topics.
4. Lab work and Experiments in Physics Education: From productive use of experiments in learning and teaching to integrating laboratories and lessons.
5. Contemporary Physics and Modern Physics in School: From non-traditional topics to new approaches in learning and teaching modern physics topics.
6. Teaching and Learning in Remote Settings: From good practices to new findings that emerged during the time of the pandemic.
7. Informal Learning and Non-Formal Learning of Physics: From science centres and outreach to learning from everyday experience
8. Physics Teacher Education: From preparing future physics teachers to building physics teacher community.
9. Students' Identity and Wellbeing as Learners of Physics: From diversity, inclusion, and equity to special needs and gifted students.

The conference was a unique opportunity to provide most participants with an international space for a lively exchange of scientific ideas and best practice, to stimulate new research and projects, to network and to make new contacts for collaboration in physics education.

After the conference, the Local Organizing Committee received 86 submissions, which were subjected to a rigorous process of double-blind peer-review, many of which were of very high quality. Due to the diversity of the proposals and the wealth of topics suggested by the authors, the selection had to be made very carefully. The result was two publications—the Conference Proceedings published in the Journal of Physics: Conference Series and the Springer book in front of you entitled Effective teaching and learning of physics in challenging times. The book you are currently reading contains 21 contributions that were rated as particularly high quality and also includes the contributions of some invited speakers.

The content of the book in front of you is divided into five chapters:

- Lessons from experience.
- General aspects of effective teaching and learning.
- Experience-based teaching and learning.
- Contemporary and complex topics.
- Teacher education.

In the "Lessons from Experience" part, three contributions from selected plenary speakers are collected. These represent an overview of the plenary speakers' research and provide some background in three areas of physics education research. The first explores the study of learning through inquiry, not just for students but also for teachers. The second uses the example of wave optics to discuss the challenges associated with learning a topic, instruments to assess these challenges and suggests guidelines for addressing them. The third contribution addresses the study and importance of embodied cognition. Students sometimes do not know how to express themselves with words, but they may be able to show correct understanding of concepts with non-verbal communication, such as gestures and movements.

The part entitled "General Aspects of Effective Teaching and Learning" contains five contributions, which address topics of general interest for effective learning of physics. The first contribution is a summary of a symposium on recent developments in physics education research and innovative approaches to teaching and learning in the twenty-first century. The second proposes using human-centred design frameworks to engage students more effectively with the teaching materials. The third proposes teaching physics in the way that physicists do physics. The fourth and fifth address the role of self-efficacy assessments in learning practices, and the importance of a growth mindset for effective learning.

The third part, "Experience-Based Teaching and Learning," focuses on teaching and learning through experience. It comprises five contributions. These include a series of engaging activities with an evaluation for teaching physics in a swimming pool, a contribution focusing on inquiry-based tasks and online learning scenarios, along with their evaluation, and a presentation of laboratory exercises in mechanics using Arduino and smartphones, accompanied by an evaluation of these exercises. The fourth contribution addresses the connection between physics and mathematics and identifies students' difficulties in these areas. The final contribution describes a specific learning experience using game-based learning.

The "Contemporary and Complex Topics" part includes four contributions and addresses topics that may not be in all secondary school curricula but are increasingly important in present times. These include teaching quantum physics due to the increasing role of quantum technology, teaching climate change topics and an example of teaching a topic that students perceive as complex: electrical circuits in primary school.

The last part "Teacher Education" contains four contributions. They deal with the interplay between curricular knowledge and the perceived agency of pre-service physics teachers, the discovery of the relationship between teachers' self-efficacy and the teaching of physics, the interpretation of and reactions to students' explanations, and the development of a training programme for thermodynamics.

We would also like to thank the Scientific Advisory Board and the GIREP Board for the productive discussions and help in preparing the conference and its publications. The organizers would like to thank the authors for their enthusiasm and preparation of the manuscripts and the reviewers for their diligent work and the time they devoted to the evaluation process.

We anticipate that this book will make an important contribution to the development of physics education by providing insights into the various research and initiatives in this area. It is also intended to contribute to advances in the teaching and learning of physics and to address current challenges at different levels of education.

Ljubljana, Slovenia

Sergej Faletič
Jerneja Pavlin

About This Book

This book presents a selection of the best papers from the GIREP Conference 2022, the conference of the International Research Group on Physics Teaching (GIREP).

It introduces readers interested in the field of physics education after the COVID pandemic and discusses the content related to effective learning in physics from contemporary physics to remote areas.

Specifically, the conference covered different areas of physics education: from best practices to new tested approaches, from non-traditional topics to new approaches to learning and teaching modern physics topics, from best practices to new insights that have emerged during the time of the pandemic, from developing new materials to teaching existing topics effectively, from the productive use of experiments in learning and teaching to the integration of labs, from diversity, inclusion and equity to special needs and gifted students, from preparing future physics teachers to building a community of physics teachers, and from science centres and outreach to learning from everyday experiences.

The conference provided an opportunity for in-depth discussion on the above and other areas of physics education. Selected papers covering the above areas in some extents are included in this book.

Contents

Lessons from Experience

Inquiry for Meaningful Learning

Dagmara Sokołowska

Abstract Inborn curiosity and inquiry attitude are invaluable features of humans, rarely appreciated and maintained in today's education, overloaded by massive content knowledge in all subjects. Developing these features in the young generation is urgent, in times when, concededly, information can be easily accessed; nevertheless, profound reflection still needs to be appropriately cultivated since it is in high demand on personal and global levels. Teachers who want to maintain an inquiry stance and skills in students need to reinforce their own inquiry competencies in teaching and learning through active collaboration with other professionals. Teachers can become role models in inquiry for their students only when they get profoundly engaged in a shift from traditional to inquisitive education and the reflection on this transformation. For more than ten years, our group has been designing and implementing CPD strategies to promote the Inquiry-Based Learning method in science education. A sequence of CPD strategies developed in collaboration with other EU partners has led to a considerable rise in the sustainability of teacher inquiry stance and their engagement in reflective, collaborative learning.

Keywords Continuous professional development · Inquiry-based learning · Practitioner inquiry · In-service teacher · Collaborative learning · Science education · Sustainability of learning · Reflective teacher

1 Introduction

The concept of a human being born with an empty mind (*tabula rasa*, empty tablet), traced back to Aristotle (1936) and Stoic School, has been confronted for ages with innatism and ideas of rationalists (Plato and Descartes), who argued that humans are born with some already-formed (innate) ideas. Thanks to the studies in genetics and cognitive science developed over the last few decades, it has been concluded that newborns coming into the world possess some common for all humans attributes of

D. Sokołowska (✉)
M. Smoluchowski Institute of Physics, Jagiellonian University, Krakow, Poland
e-mail: dagmara.sokolowska@uj.edu.pl

S. Faletič and J. Pavlin (eds.), *Teaching and Learning Physics Effectively in Challenging Times*, Challenges in Physics Education, https://doi.org/10.1007/978-3-031-72541-8_1

the mind. Among others, they have a very basic sense of numbers, a sense of solid objects and some physics laws, the intuition of probability, and they exhibit basic logical thinking abilities (Dehaene 2021). On top of that, it was evidenced that young children are born with an innate curiosity (Carlton and Winsler 1998) and intrinsic motivation reflecting the natural human propensity to learn and assimilate (Ryan and Deci 2000) and associated with early academic achievements (Shah et al. 2018). We all experience being bombarded with never-ending questions, *Why?*, constantly asked by 2–5 year-olds to acquire more and more knowledge. It also seems that quite early, they realize gaps in their knowledge which is an indispensable metacognitive indication of a desire to learn (Dehaene 2021). We would not survive without this inbred, strong, inquisitive approach.

Nowadays, more than in any other period before, human beings face rapid and massive changes we already experience and can anticipate in the future—both in our lives and globally. Currently, in education, we are confronted with the situation in which we need to prepare learners for challenges in their lives and career that are not even yet defined. Thus we need to support and maintain the curiosity and inquisitive attitude of young children in order to extend it over their entire lives and thus promote in them a life-long-learning attitude. The modern world increasingly depends on exact, natural, and technical sciences. Thus there is a growing need to attract people who would associate their professional careers with developing these disciplines. And unlike in the past, we now need more than only geniuses for academic research. Common science and mathematics literacy is required to widen more equal accessibility to achievements of civilization and sustainable development of all societies.

When entering the education system, children are still curious and motivated. However, most of them lose their interest in learning quite soon. There are different hypotheses on why it happens (Dehaene 2021). Among them is the lack of an individual approach, which is harmful equally to gifted students and appearing-to-be low-achievers since it builds in both groups a perception of not being heard out with due attention and adjustment to their educational needs. Another hypothesis is that traditional, teacher-centered teaching methods dump down pupils' interest (Dehaene 2021). They can discourage asking questions and thinking independently when most classes are organized in a listening-to-a-teacher mode. In addition to that, overloaded curricula of all subjects overwhelm students' minds. Thus, their only way to survive the school period is to memorize most of the content, pass the exam, and forget it immediately in order to empty their minds for another dose of knowledge. What causes a specific, inherently limited requirement for learning by heart in the first period of education, that is to say, getting to know letters and basic mathematics, extends to all learning. It provides a fast-food feeding of students' brains—satisfying school demands on short notice and becoming unhealthy for their minds in the long term. Instead of reinforcing cohorts of motivated, self-reflective, and critically thinking young citizens, we produce bored people tired to death of any learning. The results from the SECURE project show a substantial drop in motivation and positive attitude towards STEM subjects between the ages of 8 and 13 across 10 European countries (Sokolowska et al. 2015). During the semi-structured interviews conducted in this project in Poland, we have already heard from eleven-year-olds: *We remember better if we do something; if we have time to inquire. We want challenges, something to be demanded from us! We do not like to learn by memorizing, however this is the most common way of learning at school.*

A transformation from the past format of teacher-centered learning to the modern and urgently required design of student-centered learning needs time and involvement of all players in education, starting from policy-makers, through academics and school principals, and ending up at the most critical group of education actors, the in-service teachers, who were proven in many studies to have a profound impact on student learning (Schroeder et al. 2007). The time is crucial and limited; thus, in-service teachers, who had been mostly educated traditionally (Sorge et al. 2023; OECD 2020a), do not experience sufficient induction programs (OECD 2020a) and feel isolated as a singular teacher of a particular subject at their school (Krakehl et al. 2020), are likely to leave their profession (up to half of novice teachers resign within the first five years, Stone 2015). To efficiently face today's educational challenges and succeed in their career, teachers urgently need Continuous Professional Development (CPD) in suitable active methods of teaching and learning, accompanied by constant, daily support from other educators.

2 Background

2.1 Strategies for Teacher Professional Learning

Teachers, in general, are socio-sensitive. The vast majority (approx. 90%), as reported in TALIS 2018 study (OECD 2020a), were moderately or highly motivated to become teachers to influence the development of youngsters and contribute to society. At the same time, approx. 75% of them acknowledged a desire to support socio-economically disadvantaged students (OECD 2020a). As such, they care about their students, but many do not report self-reflection as an important feature of a good teacher (De Winter and Airey 2019; Fragat et al. 2021). In particular, they do not associate the teacher profession with research literacy, although it was suggested as one of the main characteristics of a modern teacher (BERA-RSA 2014, Mezler 2011). Teachers worldwide agree that the most common collaborative activities among teachers at school are discussions on the learning development of specific students and exchanging teaching materials. Still, generally, "deep forms of collaboration are far less prevalent" (OECD 2020b).

McLoughlin and Sokolowska (McLoughlin and Sokołowska 2023) collected several evidence-based strategies for supporting thoughtful teacher professional development in collaborative groups based on the adaptation, design, or co-design of teaching modules, their implementation, and reflection on the resulting students' learning outcomes. Lesson study and learning study (Cheung and Wong 2014) promote planning and executing a lesson to make student learning visible and answer a shared question regarding students' learning. Professional Learning Communities (PLC) and Communities of Practice (CoP) endorse collaboration toward action, together with a dedication to continuous improvement of their teaching practice, but they have slightly different focuses. CoPs are more outcome-oriented, while PLCs are more process-oriented.

Action Research (AR) is a more advanced evidence-based strategy for teacher development. It is oriented on addressing a key problem in the organization or community (Reason and Bradbury 2008) in order to change practice while at the same time creating knowledge (Oosthuizen 2002). The strategy is based on a deep reflection on practice but can be implemented irregularly.

2.2 *Inquiry-Based Learning (IBL)*

In 2007, the European Commission published a document (Rocard et al. 2007) recommending Inquiry-Based Learning (IBL) as the best method for teaching science (Inquiry-Based Science Education, IBSE) in Europe. Soon, its promotion started, supported by numerous EU projects funded under the 7th Framework Programme (among others, Project Fibonacci 2010–2013, Project Establish 2010–2014, and Project SAILS 2012–2015), aiming to design IBSE teaching materials and provide teacher training. The document (Rocard et al. 2007) also forced adequate changes in all science-related EU curricula.

Inquiry-Based Learning is a strategy of learning in which students construct their knowledge and develop competencies through direct experience in authentic circumstances (McLoughlin and Sokołowska 2022) in a repetitive process usually based on the IBL cycle (Harlen 2013; Sokołowska 2020), Fig. 1, that mimics "acting like a scientist"—the idea launched by Bruner (1960) and argued by him to be like any other act of life, leading to acquiring knowledge with deepen understanding. He contrasted the IBL with other methods often used at school, which do not take the students through the process of developing learning but focus on providing them with conclusions, thus losing the inherent feature of science learning as an empirical endeavor.

Effective IBL starts with setting the scene and grounding the upcoming IBL activity in the context already known to the learners or in their interest (Sokołowska 2020), which was evidenced to have a very high impact on students' achievements (Hattie 2009). The IBL usually involves collaborative work, which has been proven to bring students considerable benefits of learning in different areas (social, psychological, academic) (Hattie 2009; Laal and Ghodsi 2012). Working together to formulate the inquiry question, putting forward reasoning-based hypotheses, planning investigation details, carrying out the experiments, and collecting data, stimulate students' joint creativity. Data analysis and drawing conclusions involve and develop students' critical thinking, and sharing their results with other groups builds their communication skills. Regarding those mentioned above, the IBL method gives the opportunity for the comprehensive development of different key competencies (EU 2018) in

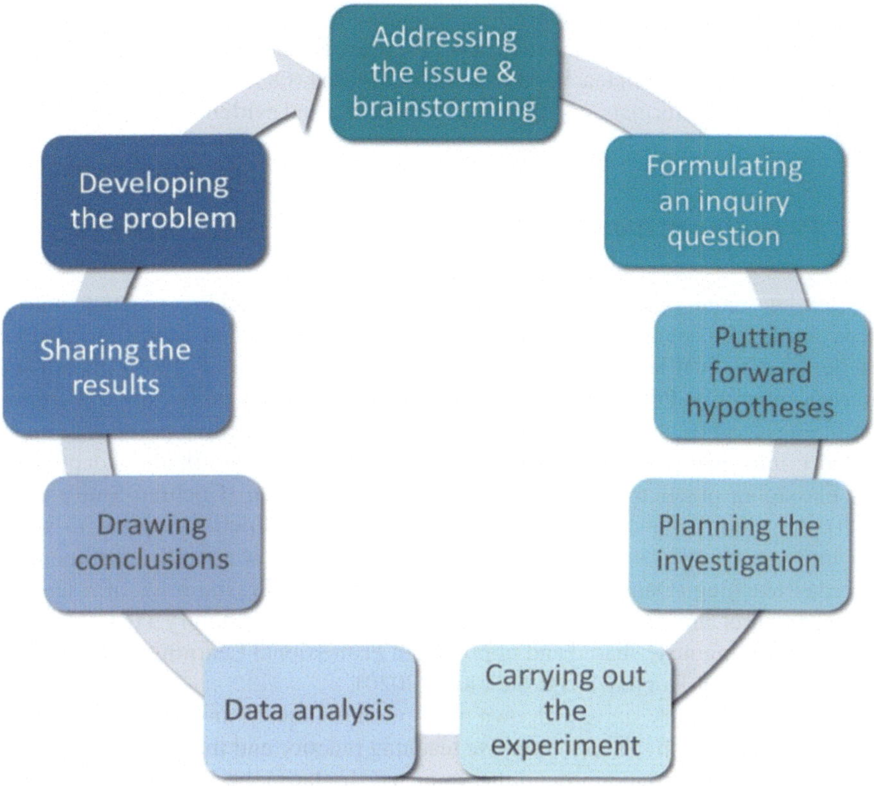

Fig. 1 The IBL cycle (Sokołowska 2020)

students. While inquiring, students are constantly challenged by the undiscovered (McLoughlin and Sokołowska 2022); thus, they become likely ready to adapt to unknown situations and cultivate a positive attitude toward lifelong learning.

The impact of IBL on students' learning outcomes depends on its mindful implementation. The IBL cycle is not rigid and should be adapted to learning purposes and class circumstances (Sokołowska 2020). The IBL process can be deployed on at least three levels (structured, guided, and open) (Sokołowska 2020), among which the guided IBL is reported as the most effective (Furtak et al. 2012; Lazonder and Harmsen 2016). Efficient management of the class working by inquiry requires careful planning of classroom spatial arrangement, adjustment of lesson timetable, selection of grouping strategies, collection of adequate equipment and resources, the adaptation of teaching materials (Sokołowska 2020), and on top of all that—consolidating in a teacher a positive attitude to the IBL philosophy of learning. All of the above implies that developing IBL teaching skills takes time and cannot result from listening to one or even several lectures on IBL. Such skills must be acquired in active repetitive participation in the IBL activities, preferably taking the role of a student and thus getting the chance to experience the spirit of inquiry. To teach by inquiry, teachers need to become inquirers. They need to change the passive, theoretical approach inscribed in teacher education in the past (OECD 2020a) and get used to more active CPD formats, collaboration with peers, and evidence-based reflection.

2.3 *Practitioner Inquiry (PI)*

The teacher inquiry approach, as a reflection on own teaching practice with a focus on the impact on student learning, was described as an essential pillar of teacher professional learning and development (Timperley et al. 2007), increasing their sustainability. Practitioner Inquiry is a form of teacher professional learning, combining rigorous, systematic, and intentional inquiry in teaching practice to make evidence-informed professional judgments and, resulting from them—adaptations of a classroom practice influencing their students' learning (Cochran-Smith et al. 2009). An inquiry stance is oriented on the process of repetitive cycles to study and improve practice while at the same time becoming a natural constituent of everyday teaching (Dana and Yendol-Hoppey 2015); in PI, the roles of a teacher and an inquirer are combined in one. As in the students' inquiry (IBL), teacher inquiry is continuously shared and improved in Professional Learning Communities (Dana and Yendol-Hoppey 2015; De Lange 2020).

Practitioner Inquiry starts with own motivation to inquiry, an issue bothering the teacher and selected by them to improve teaching practice and thus student learning. Then the process continues around the sequence similar to the IBL cycle (Fig. 2)—first, by posing an inquiry question, secondly, by planning the investigation (selecting particular students or groups of students, setting the time, selecting a place and a topic, and teaching–learning methods) and choosing the data collection tools. Each part of the PI cycle is profoundly discussed with other members of PLC who serve as critical friends to consolidate the teacher's thoughts and improve the process. When planning is finished, teachers conduct their investigations and collect data. They return to the PLC meetings with evidence for another round of thorough discussion on the results and conclusions derived from the investigations. Finally, the process is summarised in a presentation in front of the PLC or a broader audience outside the community. Each member of the PLC carries out their own PI, and all PIs get similar attention from other members of the PLC.

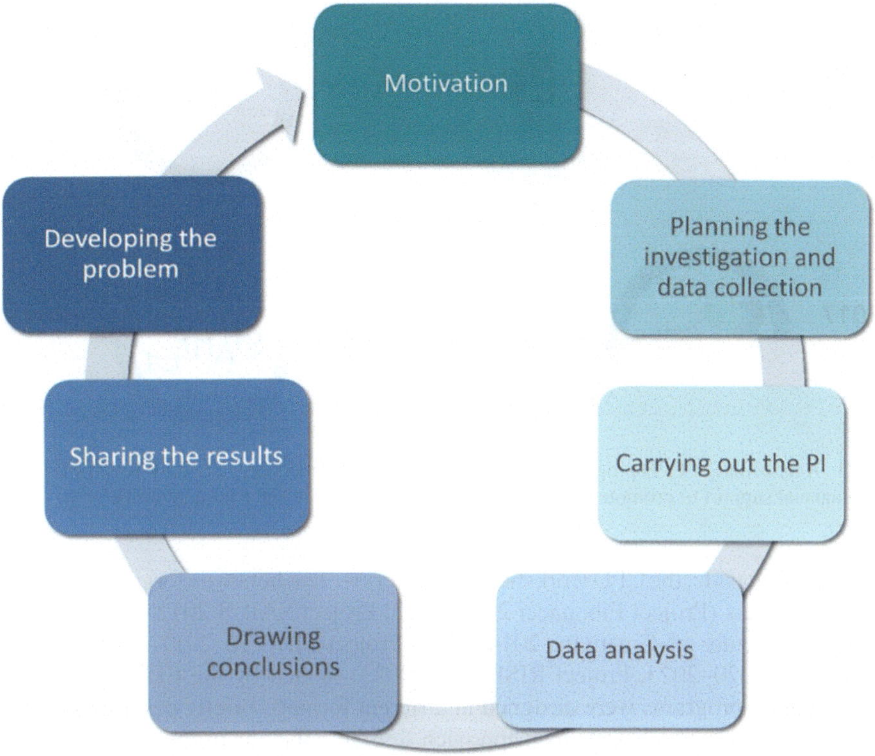

Fig. 2 The PI cycle (De Lange 2020)

3 Evolution of Strategies for Continuous Professional Development Training for Science Teachers

Darling-Hammond et al. (2017) proposed a comprehensive approach to effective teacher professional development, which secures the development of teacher understanding of the subject matter enhancing teacher collaboration by encouraging sharing teacher practice experiences and reflecting on them with a critical friend attitude. The model builds up teachers' ownership of learning and sets up a positive atmosphere for the growth of a positive mindset for life-long learning. In this approach to CPD, teachers work together with the use of effective practice models and get experts' coaching and support. They argue that effective teacher professional development should be of sustained duration.

Over the last thirteen years, a group of academic teachers and Ph.D. students (tutors and researchers) from the Faculty of Physics, Astronomy, and Applied Computer Science at the Jagiellonian University (WFAIS UJ) in Krakow have been developing didactic materials for teachers and learning units for students aged 6–18 to promote teacher inquiry skills and the IBL method of teaching and learning

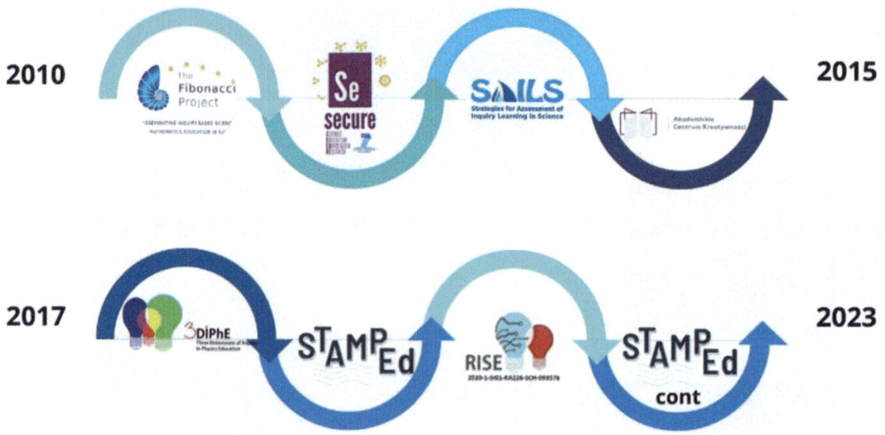

Fig. 3 A flowchart of CPD programs for in-service science teachers organized by WFAIS UJ with EU financial support to promote the IBL method in science and teacher inquiry skills

science. Alongside, the CPD workshops for in-service teachers were organized within several projects (Project Fibonacci 2010–2013, Project SAILS 2012–2015, Project Academic Center of Creativity 2014–2015, Project 3DIPHE 2017–2020, Project STAMPEd 2020–2023, Project RISE 2021–2023) funded by the EU (Fig. 3). The CPD training programs were designed in different formats, briefly described below. All of the training was based on an approach similar to the one proposed by Darling-Hammond et al. (2017), embracing most of its aspects to some extent. Our continuous concern was the sustainability of the teachers' lifelong learning attitude and their inquiry stance. Thus, we formulated a research question: *What formats of Continuous Professional Development for in-service teachers best support the sustainability of their long-term engagement in CPD and their inquiry attitude toward learning?*

Depending on the project, several data-collection tools were designed and implemented. Lecturers continuously documented their reflections and observations with the use of reflection tools and photos. Teachers were surveyed with pre- and post-questionnaires, asked to complete SWOT analyses on the CPD process, and selected ones were invited to the Focus Group semi-structured interviews. Based on data analysis, the format of inquiry workshops was evolving to increase the sustainability of the CPD results.

3.1 Launching Inquiry-Based Learning in Science in Poland—the Fibonacci Project

In 2010, our group was invited to participate in the Fibonacci project with the aim of disseminating the IBL method on a large scale. The science curriculum in Poland, similar to many other science curricula all over EU, was about to change and introduce

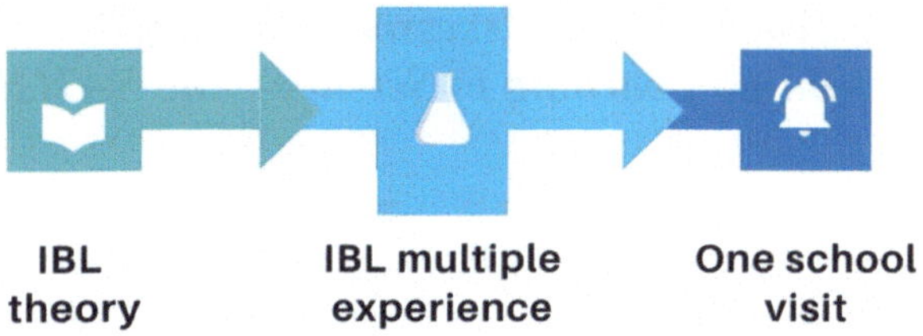

Fig. 4 The *Fibonacci* scheme of workshops. Training consisted of (1) a short introduction to the IBL method, (2) three meetings of experiencing the IBL units taking the role of learners, and (3) single field visit at each teacher's school to observe the teachers' IBL module implementations in their class

for the first time elements of the IBL method into the science classes, following the EU recommendation (Rocard et al. 2007) published three years earlier. At that time, it was perceived as a big challenge for teachers in Poland, who received pre-service training in a traditional format of teacher-centered pedagogy. The Fibonacci project strategy was to design the ready-to-use learning modules in the IBL and train as many in-service teachers as possible.

The teacher training took place in two rounds, each consisting of three meetings with teachers and tutors' field visits in all schools represented by the teachers in the project. From the very beginning, the training was oriented on giving the teachers genuine experience, and thus developing the IBL skills in teachers. It was planned to provide teachers only with an indispensable theoretical introduction to the IBL method and secure considerable time for them to experience the IBL units by taking the roles of learners. Each participant was asked to adapt the workshop materials to their school context and to organize a school visit to present tutors with participants' implementations of what they had learned during the workshops (Fig. 4).

3.2 Quick Check-Up on the Fibonacci Project Outcomes—the SECURE Project

During the second round of Project Fibonacci (2010–2013), the research in the SECURE project (Sokołowska et al. 2015) started. The aim of the project was to study mathematics, science, and technology (MST) curricula in ten EU countries, together with surveying approx. 9000 students at ages 5, 8, 11, and 13 and their 1500 teachers about the MST condition at schools (Sokołowska et al. 2015). No teacher training was planned for the project; however, in Poland, some of the teachers participating in the SECURE project previously took part in Project Fibonacci (2010–2013) meetings, which gave us an opportunity to study the short-term impact of that training on teachers and learners and draw conclusions for a redesign of the IBL training formats in the future.

3.3 *Equipping Teachers with Assessment Strategies in Inquiry-Based Learning in Science—the SAILS Project*

The SAILS project aimed to further disseminate the IBL method among teachers, upgrading the strategy by incorporating assessment strategies adequate to evaluate students' learning of science through inquiry. The format of teacher training was up to the project partners. Jagiellonian University decided to organize three week-long Teacher Schools.

The workshops followed the format proposed in Fibonacci, however, with two distinctions. One was "a letter to myself" written by every teacher at the end of the training, summarizing teachers' perceptions of the process and their hopes and plans for implementation. All teachers were supposed to seal their letters and they received them three months later to reflect individually on the mid-term CPD outcomes and training sustainability.

Another amendment proposed by our group and piloted only for physics teachers was an IBL design activity at the end of the third Teacher School (Fig. 5). Participants divided into groups of four were asked to prepare an IBL module together with the evaluation method and implement it during the last day of training in another group of participants. It was a very dynamic activity since all groups needed to design their own modules and follow their implementation, while at the same time, each group received a module from another group. So one day, all participants became the IBL module designers, and the other day different members of each group took alternative roles of students or teachers observing and evaluating the implementation of their own module by the others. Thus, in one extended activity, teachers had a chance to experience the IBL method from three different perspectives.

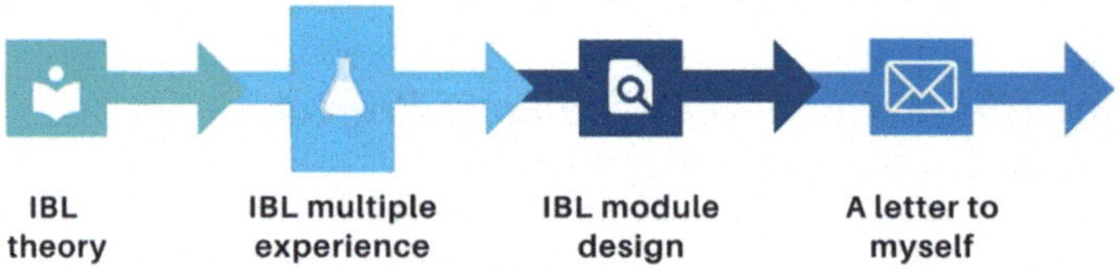

Fig. 5 The *SAILS* scheme of workshops for physics teachers (third cohort). Training consisted of (1) a short introduction to the IBL method, (2) three meetings of experiencing the IBL units taking the role of learners, (3) IBL module design and implementation (perspective of a designer, a student, and a teacher), and (4) a letter to myself

Fig. 6 The ACC scheme of workshops for in-service and pre-service teachers. Training consisted of (1) a short introduction to the IBL method, (2) five meetings of experiencing the IBL units taking the role of learners, (3) implementation of the IBL units by pre-service teachers in classes of in-service teachers

3.4 *Joint IBL Workshops for Pre-Service and In-Service Teachers—the Academic Center of Creativity (ACC) Project*

The Project Academic Center of Creativity (2014–2015) aimed at the implementation of the IBL method in ten classes of 10-13 yolds and the study of the IBL impact on learners' attitudes towards the IBL and sustainability of their learning over six months after the intervention in the classes. Before the implementation, five in-service teachers were paired with five pre-service teachers to participate in the IBL workshops and work on implementing ten IBL lessons in each class (Sokołowska 2018). During the IBL intervention, pre-service teachers took the role of regular teachers, while in-service teachers observed their students (Fig. 6). Pre-service teachers worked with the modules designed by our group, took part in CPD training and collaborated with our group on the research.

3.5 *Building up Teachers' Inquiry Skills—the Introduction of the Practitioner Inquiry—the 3DIPhE Project*

The Project 3DIPhE (2017–2020) brought extra dimensions to the science teacher CPD meetings organized by our group. In addition to regular training in the IBL, similar to the Fibonacci approach, teachers formed a Professional Learning Community facilitated by the academic teachers. They were coached in the Practitioner Inquiry method—an evidence-based reflection concerning their IBL implementation at schools. The CPD strategy for participants was founded on three pillars: (1) building the PLC group, 2) the development of practical active teaching skills (the IBL skills), and 3) the development of Practitioner Inquiry competencies through repetitive group discussions with other participants and dissemination of the PI results to the public (Sokołowska and Kniaź 2023), Fig. 7.

Up to the launch of the 3DIPhE project, teachers participated in group work, conducted experiments in small groups, and, in some projects, also designed new IBL

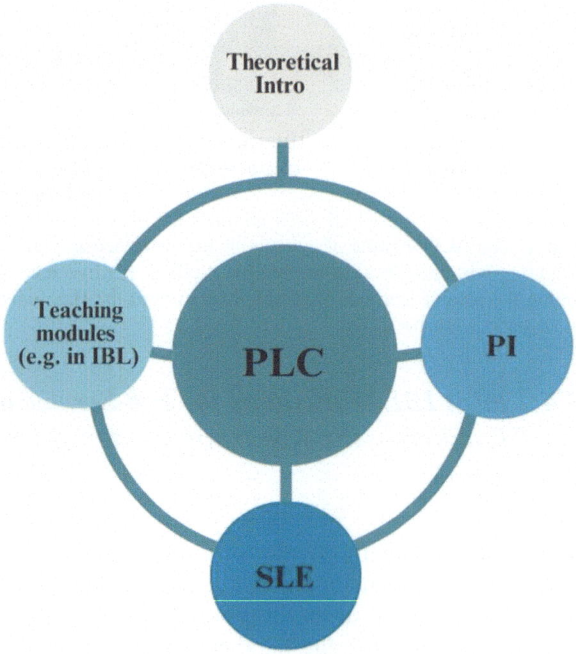

Fig. 7 The 3DIPhE and STAMPEd projects scheme of workshops for in-service teachers. Training consisted of (1) a short introduction to the IBL method, (2) seven meetings based on three pillars: building PLC, enhancing the active teaching methods (IBL in 3DIPhE project and math/ physics transition modules in STAMPEd project), and developing PI competencies by repetitive group discussions with other participants and public presentation of the PI results. All participants presented the results of their PLs during-a small local event (SLE).

modules. In the 3DIPhE format, they formed a PLC focusing on inquiring about their individual practice and helping each other improve each step of this inquiry. Teachers started to collaborating to enhance their teaching practice in a rigorous manner of investigation and a trustful atmosphere of PLCs. Each of the three cohorts of in-service teachers ended their training by preparing presentations of their individual PI investigations. All participants got the opportunity to present their evidence-based results to the external public during a small local event (SLE).

3.6 Developing Practitioner Inquiry Skills Out of the Context of the IBL—the STAMPEd Project

Three pillars of the CPD workshop strategy implemented in the 3DIPhe project were the foundation of the subsequent in-service teacher training in the STAMPEd project. The only alteration (Fig. 7) was the focus of practical activities—the IBL was replaced by workshops addressing the transitions across mathematics and physics education, in particular issues with noncompatible curricula of both subjects and problems related to transfer between schools at two different levels (lower and upper secondary).

Teachers participated in basic and advanced training, taking the role of students while experiencing the active learning modules. They also went together through the entire PI cycle, discussing each other's motivations, inquiry questions, data collection tools, data analysis, and the design of presentations summarizing the PI investigations at schools. Some teachers from another (RISE) project took part in the STAMPED small local event.

3.7 Remote format of the IBL—RISE Project

During the COVID-19 pandemic, our group got involved in Project RISE (2021–2013) promoting Inquiry-Based Learning in online activities. The project aimed to transfer the existing IBL units from in-person to remote formats and develop other teaching and learning aids. The entire CPD training was organized online in two groups of novice teachers. We used the structure of meetings based primarily on experiencing the IBL modules while taking the role of learners. Most teachers adapted or designed their own IBL modules and implemented them in at least one class. Teachers got invitations to the STAMPEd small local event, where they learned about the Practitioner Inquiry strategy as a possible extension of inquiry at the level of teachers and saw genuine examples of PI. Most teachers perceived their IBL activities at schools as an excellent opportunity to reflect on what benefits the IBL brings to their students based on some evidence they collected (observations, surveys administered to students).

4 Results

Development of CPD strategies for conducting in-service science teacher training by the group of academic teachers and Ph.D. students at our faculty emerged from the desire to address the gap between the traditional preparation of teachers and the emergent need for the transition from teacher-centered learning to student-centered learning. So-called Rocard's report (Rocard et al. 2007), recommending the IBL method for science education across Europe, triggered adaptations of science curricula at all school levels in most countries and launched huge EU funds to disseminate the IBL among teachers. The Fibonacci project gave us the first opportunity to organize in-service teacher training for and beyond the local community.

From the very beginning, our group was designing teacher training in the format of practicing and actively responding to challenges regarding both the student-centered approach and teacher constant reflection. We were convinced that to get involved profoundly, teachers needed enough time in each training slot, so all the CPD meetings in Fibonacci and subsequent projects (except SAILS) were organized on Saturdays when teachers were not busy with school obligations. Unlike most teacher workshops available at that time, according to our Fibonacci participants, *the workshops were very engaging and demanding*, nevertheless *giving a lot of satisfaction* and *eye-opening*. As a group of coaches, we cared not only about the immediate and visible teacher learning and gaining competencies but also about the long-term sustainability of their initial enthusiasm and attitude toward the paradigm change in teaching science.

Thus, although satisfied with the excellent beginning of the IBL dissemination in the Fibonacci project among altogether 50 teachers from 24 schools, we realized that this quite basic strategy of the Fibonacci workshops allowed teachers to encounter active learning experiences and maybe even understand the spirit of IBL; however, teachers did not develop their ownership of learning by inquiry. We observed they willingly participated in the training, and most of them implemented the ready-to-use IBL units in their classes, however, with little follow-up to taking their own initiative in the future. Only four teachers took part in both rounds of Fibonacci training. Admittedly, many Fibonacci teachers agreed to engage with their students in the next SECURE project, researching the current situation of MST education in a broad sense. Still, they never contacted us again for future training. This situation raised a big concern about the sustainability of our CPD format.

The study of the SECURE project in Poland (also with the involvement of some Fibonacci teachers) revealed that students still did not do experiments regularly, although they love learning by experience. Pupils admitted t*here is exciting science out of school, but at school, it is boring; we learn far too much by heart.*

In the subsequent SAILS project funded for teacher training in the IBL supplied with adequate assessment methods of learning, the Polish coordinator of the project decided to organize three week-long Teacher Schools in which we followed previously created Fibonacci CDP format, separately in three groups of subject teachers (of biology, chemistry, and physics). Altogether, more than 80 physics teachers participated in the project over three years. During the last Teacher School, our group proposed that physics teachers would participate in a modified program and, after experiencing several ready-to-use IBL units, would design their own IBL units, implemented the next day by another group of teachers. Such a dynamic situation released much teacher creativity and boosted their ownership of learning. Teachers commonly admitted that they had never participated in such demanding CPD activity and expressed gratitude for giving them space for such an experience. One teacher said: *Now I see I can do that* [design and implement the IBL unit] *by myself*. The other added: *It helped me that we designed a new teaching unit together in a group.* This newly implemented approach made a huge change. After the project, many teachers reported the implementation of the IBL regularly. Many teachers expressed their will to stay in contact for future projects. Six of them returned for ERASMUS + projects (Project 3DIPHE 2017–2020, Project STAMPEd 2020–2023, Project RISE 2021–2023).

In the Academic Center of Creativity, we had this unique opportunity to pair the in-service and pre-service science to work on implementing a series of lessons in the in-service teachers' classes. Practitioners who were educated traditionally were very skeptical about effective class management during the IBL classes and students' outcomes afterward. One of them said before the ACC training: *Academic teachers design units that do not fit our everyday situation at school. Here it is not university but reality.* After the training, the same teacher admitted: *The IBL method would be engaging and maybe even motivating for my students, but I do not have time for it in my lessons.* A significant change in her opinion was noticed after implementing ten IBL lessons in her class: *Now I can see it is feasible. Even a pre-service teacher, who hadn't known my students before, could manage the class and finish lessons on time.* Teachers could not believe that six months after the IBL intervention, the median of the normalized change in test results of their students was only -9%, independently of student gender or their ability level (Sokołowska 2018). Three of the five pre-service teachers continued working with us in other educational events. Among them, one continued the collaboration with us in all three subsequent projects (Project 3DIPhE 2017–2020, Project STAMPEd 2020–2023, Project RISE 2021–2023), already as an in-service teacher.

The 3DIPhE and STAMPEd projects allowed teachers to further develop their inquiry skills by conducting their Practitioner Inquiries and sharing their design elements in profound discussions with other members of the PLC groups. All six participants of the first cohort of the 3DIPhE project continued, forming a group of experienced teachers in the second round. From two cohorts of the 3DIPhE project, nine out of twelve teachers continued their work in STAMPEd project (however, due to the COVID-19 pandemic situation at that moment, three of them needed to resign right after the beginning). Two teachers staying in the STAMPEd project invited their school colleagues to get involved.

Teachers engaged in the 3DIPhE CPD developed their ownership of learning and strong dedication to continuing the evidence-based reflection on their teaching practice (Sokołowska and Kniaź 2023). Repeated IBL activities, investigated through Practitioner Inquiries, helped them to increase their students' motivation and interest in physics, even in classes with basic physics programs (McLoughlin and Sokołowska 2022). Most students learning by inquiry achieved better results in typical physics tests than their colleagues from control groups, learning in a traditional format. PLC groups of teachers built up during CPD training consolidated as groups of professional friends trusting each other and taking the joint endeavor. On top of all that, the in-service teachers appreciated the bonds established with the university as partnership relationships (Sokołowska and Kniaź 2023).

Teachers presenting their implementation and own PIs to the broader audience during the small local events built up their confidence in teaching and the perception of the usefulness of their PIs for other educators. They also developed their public presentation skills (Sokołowska and Kniaź 2023). Seeing the results of teachers' genuine PIs convinced some other teachers to get involved in the RISE project and, as the follow-up, also in the second cohort of the STAMPEd project (five out of nine novice participants recruited from the RISE project). The examples from their colleagues help some teachers respond to the challenge of implementing active methods and the rigorous inquiry of the student learning outcomes resulting from such innovations in their teaching practice.

5 Conclusions

The era of education for industrial workers, who were bonded for life to one job and repeated the same actions all their lives, is gone. The future of education is learning competencies, not being fed with tons of (still growing) encyclopedic knowledge only because "it may come in useful." The education of upcoming generations is likely to become "on-demand," not so much learning in advance; learning what is needed and interesting to address current, personal, and global challenges and to maintain the motivation for learning. Generations Z and Alpha, with easy access to information and AI, concerned about their future in a fast-developing world and fast-changing circumstances, are more likely than ever to question dogmas and the existing world order. It is their vast potential, but only if equipped with reflection and mindful inquisitiveness.

Learning by inquiry is not just another method of learning. It is an immanent human being future, an attitude that is easy to lose if not cherished and not assisted. Thus one of the main challenges in education now is promoting and developing the inquiry stance. To be able to do that, teachers urgently need to regain their own inquiry skills and attitude toward their teaching practice and development as human beings and thus retrieve their passion for what they do to become role models for their students.

Reinforcement of the inquiry stance takes time, both for learners and teachers. As we observed at the beginning of our CPD training more than ten years ago, even experiential engagement of teachers in the IBL activities, taking the role of learners, did not guarantee a sustainable transfer of the IBL to their students. Such a strategy is good for the teacher induction to inquiry, especially when aiming at large-scale dissemination of inquiry, but should be followed by training on the IBL design, tailored to teachers' own context to strengthen their ownership of learning and their self-confidence in teaching by inquiry.

Supplementing the CPD training with an IBL-design element allowed us to fulfill most of the criteria for effective teacher professional development (Darling-Hammond et al. 2017) and launched learning sustainability in some teachers. But yet another aspect was missing in our CPD design—a deepen reflection of the participants on their practice. To secure this learning element, we went even beyond by proposing an evidence-based reflection in terms of Practitioner Inquiry in collaboration with other EU partners (Project 3DIPhE 2017–2020, Project STAMPEd 2020–2023, Project RISE 2021–2023). This method of classroom investigation of learning issues and implementations with a focus on student learning, while conducted by teachers individually, is constantly put under the reflection of other teachers—participants of the CPD, forming together a Professional Learning Community of teachers. The strategy initiated in the 3DIPhE project enlarged the sustainability of teacher learning, their inquiry stance, and subsequent involvement in CPD (at 50–75%), as evidenced in the following training meetings organized in the same format. In this group, two teachers developed their inquiry skills to the level of the IBL trainer.

References

Aristotle [c. 350 BC] (1936) De Anima. English edition: On the Soul. In: Aristotle 8 Loeb Classical Library. William Heineman, London

BERA-RSA (2014) Research and the teaching profession. Building the capacity for a self-improving education system. https://www.bera.ac.uk/wp-content/uploads/2013/12/BERA-RSA-Research-Teaching-Profession-FULL-REPORT-for-web.pdf. Accessed 14 June 2024

Bruner J (1960) The process of education. Harvard University Press, Cambridge Massachusetts, London, p 14

Carlton MP, Winsler A (1998) Fostering intrinsic motivation in early childhood classrooms. Early Child Educ J 25:159–166. https://doi.org/10.1023/A:1025601110383

Cheung WM, Wong WY (2014) Does lesson study work? A systematic review on the effects of lesson study and learning study on teachers and students. Int J Lesson Learn S 3(2):137–149. https://doi.org/10.1108/IJLLS-05-2013-0024

Cochran-Smith M, Barnatt J, Friedman A et al (2009) Inquiry on Inquiry: practitioner research and student learning. Action Teach Educ 31(2):17–32. https://doi.org/10.1080/01626620.2009.10463515

Dana NF, Yendol-Hoppey D (2015) The PLC book. Corwin, USA

Darling-Hammond L, Hyler ME, Gardner M et al. (2017) Effective teachers professional development. Learning Policy Institute, Palo Alto. https://bibliotecadigital.mineduc.cl/bitstream/handle/20.500.12365/17357/46%20Effective_Teacher_Professional_Development_REPORT.pdf?sequence=1. Accessed 14 June 2024

Dehaene S (2021) How we learn. Why Brains Learn Better than Any Machine… for Now. Penguin Publishing Group, pp 53–68

De Lange J (2020) Practitioner inquiry in the context of inquiry based learning. In: M. Čepič, De Lange J (eds) 3DIPhE, vol. 2. University of Ljubljana, Faculty of Education, Ljubljana. https://www.3diphe.si/files/2021/12/3D_volume2_v1_MC.pdf. Accessed 14 June 2024

De Winter J, Airey J (2019) What makes a good physics teacher? Views from the English stakeholder community. Phys Educ 55(1):015017–1–015017–12. https://doi.org/10.1088/1361-6552/ab5215

EU (2028) EU key competences for lifelong learning. https://op.europa.eu/en/publication-detail/-/publication/297a33c8-a1f3-11e9-9d01-01aa75ed71a1/language-en. Accessed 14 June 2024

Fragat T, Henriksen EK, Tellefsen CW (2021) Pre-service science teachers' and in-service physics teachers' views on the knowledge and skills of a good teacher. Nord Stud Sci Educ 17:277–292

Furtak EM, Seidel T, Iverson H et al (2012) Experimental and quasi-experimental studies of inquiry-based science teaching: a meta-analysis. Rev Educ Res 82(3):300–329. https://doi.org/10.3102/0034654312457206

Harlen W (2013) Inquiry-based learning in science and mathematics. Rev Sci Math ICT Educ 7:9–33. https://efe.library.upatras.gr/index.php/review/article/viewFile/2042/2085

Hattie J (2009) Visible learning. Routledge Taylor & Francis Group, London and New York, p 148

Krakehl R, Kelly AM, Sheppard K et al (2020) Physics teacher isolation, contextual characteristics, and student performance. Phys Rev Phys Educ Res 16:020117–1–020117–17. https://doi.org/10.1103/PhysRevPhysEducRes.16.020117

Laal M, Ghodsi SM (2012) Benefits of collaborative learning. Procedia Soc Behav Sci 31:486–490. https://doi.org/10.1016/j.sbspro.2011.12.091

Lazonder AW, Harmsen R (2016) Meta-analysis of inquiry-based learning: effects of guidance. Rev Educ Res 86:681–718. https://doi.org/10.3102/0034654315627366

McLoughlin E, Sokołowska D (2022) Inquiry approaches in physics education. In: Guisasola J, McLoughlin E (eds) Connecting Research in Physics Education with Teacher Education, vol.3, International Commission on Physics Education, pp 209–221

McLoughlin E, Sokołowska D (2023) Physics teacher professional learning . In: Tasar MF and Heron PRL (eds) The International Handbook of Physics Education Research: Teaching Physics. AIP Publishing Melville, New York, pp 15–1–14–21

Mezler DE, Shaffer PS (eds) (2011) Teacher education in physics: research, curriculum, and practice. American Physical Society, College Park, MD, pp 3–14
OECD (2020a) Teachers' guide to TALIS 2018, vol I. OECD Publishing, Paris
OECD (2020b) Teachers' guide to TALIS 2018 results, vol II. OECD Publishing, Paris
Oosthuizen MJH (2002) Action research methods for students, academics and professionals. In: Williamson K (ed) Information management and systems, 2nd edn. Chandos Publishing, Wagga Wagga NSW
Project 3DIPhE (2017–2020) Three dimensions of inquiry in physics education (ERASMUS+ KA2). https://www.3diphe.si/. Accessed 14 June 2024
Project Academic Center of Creativity (2014–2015) Akademickie Cemtrum Kreatywności Jagiellonian University (7FP) http://www.ack.fais.uj.edu.pl/. Accessed 14 June 2024
Project Establish (2010–2014) European science and technology in action: building links with industry, *Schools and Home* (7FP). http://www.establish-fp7.eu/project.html. Accessed 14 June 2024
Project Fibonacci (2010–2013) Disseminating inquiry-based science and mathematics education in Europe, (7FP). http://www.fibonacci-project.eu/. Accessed 14 June 2024
Project RISE (2021–2023) Remote inquiry in science education (ERASMUS+ KA2 2021–2023). https://castel.ie/research_project/remote-inquiries-in-science-education-rise/. Accessed 14 June 2024
Project SAILS (2012–2015) Strategies for assessment of inquiry learning in science (7FP). http://www.sails-project.eu/project.html. Accessed 14 June 2024
Project STAMPEd (2020–2023) Supporting transitions across mathematics and physics education (ERASMUS+ KA2) https://stampedproject.eu/. Accessed 14 June 2024
Reason P, Bradbury H (eds) (2008) The SAGE handbook of action research. SAGE Publications, Los Angeles. Oxford. New Delhi. Singapore, p 1
Rocard M, Csermely P, Jorde D et al (2007) Science education now: a renewed pedagogy for the future of Europe. European Commission, Brussels
Ryan RM, Deci EL (2000) Intrinsic and extrinsic motivations: classic definitions and new directions. Contemp Educ Psychol 25:54–67. https://doi.org/10.1006/ceps.1999.1020
Schroeder CM, Scott TP, Tolson H et al (2007) A meta-analysis of national research: effects of teaching strategies on student achievement in science in the United States. J Res Sci Teach 44:1436–1460. https://doi.org/10.1002/tea.20212
Shah PE, Weeks HM, Richards N et al (2018) Early childhood curiosity and kindergarten reading and math academic achievement. Pediatr Res 84:380–386. https://doi.org/10.1038/s41390-018-0039-3
Sokołowska D (2018) Effectiveness of learning through guided inquiry. In: Sokołowska D, Michelini M (eds) The Role of Laboratory Work in Improving Physics Teaching and Learning. Springer Nature Switzerland AG, pp 243–255
Sokołowska D (2020) Inquiry based learning to enhance teaching. In: Čepič M, Sokołowska D (eds) 3DIPhE, vol. 1. University of Ljubljana, Faculty of Education, Ljubljana. https://www.3diphe.si/files/2021/12/3D_volume1_v1_MC.pdf. Accessed 14 June 2024
Sokołowska D, Kniaź J (2023) Building the professional learning communities of in-service science teachers in collaboration with university lecturers: opportunities and benefits for both players. In: Bugaj JM, Budzanowska-Drzewiecka M, Mikołajczyk B (eds) Building an Academic Community. Brill, Leiden, pp 86–105. https://doi.org/10.1163/9789004689060_006
Sokołowska D, de Meyer J, Wojtaszek M et al (2015) Attitude, motivation and self-esteem in mathematics, science and technology researched by the SECURE project—polish results against the background in Europe. Edukacja Biologiczna i Środowiskowa (Eng.) 1(2015):26–37

Sorge S, Otero VK, Lavonen J (2023) Pre-service physics education. In: Tasar MF, Heron PRL (eds) The international handbook of physics education research: teaching physics. AIP Publishing Melville, New York, pp 14–1–14–20

Stone E (2015) If not us, then who? KDP Record 51:126–130. https://doi.org/10.1080/00228958.2015.1056664

Timperley H, Wilson A, Barrar H et al (2007) Teacher professional learning and development: best evidence synthesis iteration (BES). Wellington, New Zeland Ministry of Education, pp x, xxi-xliii

What is Difficult in Learning Physics and What We Can Do About It: The Case of Wave Optics

M. Planinić, K. Matejak Cvenić, A. Sušac, L. Ivanjek, and K. Jeličić

Abstract The questions of what students find difficult in physics and how physics teachers and faculty can help them learn easier and deeper underlie the whole discipline of PER. These important questions will be addressed in the light of the recent findings of our research group on high school students' understanding of wave optics. Wave optics, a good representative of a difficult topic for high school students, was investigated since 2018 in a research project conducted at the Faculty of Science, University of Zagreb. Secondary school students' understanding of wave optics was investigated first through 27 demonstration interviews and later with a newly constructed instrument, the Conceptual Survey of Wave Optics (CSWO), administered to 224 students. The results of the interviews and of the CSWO suggested some possible origins of student difficulties, and the existence of several distinct levels of difficulties. The implications of the findings for the teaching of wave optics and physics in general are discussed.

1 Introduction

The question of what is difficult in learning physics and what physics instructors and teachers at all levels can do about it is the question that underlies the whole discipline of physics education research (PER). It is also the question that motivates the work of the Physics Education Group at the Faculty of Science, University of Zagreb, with

M. Planinić (✉) · K. Matejak Cvenić · K. Jeličić
Faculty of Science, Department of Physics, University of Zagreb, Bijenička c. 32, 10000 Zagreb, Croatia
e-mail: maja.planinic@phy.hr

A. Sušac
Faculty of Electrical Engineering and Computing, University of Zagreb, Unska 3, 10000 Zagreb, Croatia

L. Ivanjek
Physics Education Group, School of Education, JKU Linz, Altenberger Straße 69, 4040 Linz, Austria

S. Faletič and J. Pavlin (eds.), *Teaching and Learning Physics Effectively in Challenging Times*, Challenges in Physics Education, https://doi.org/10.1007/978-3-031-72541-8_2

its collaborators from other institutions, and which led us to explore many physics topics in that sense, most recently wave optics.

From 2018 we are conducting a research project (INVESTIGATE n. d.) which has the goal to investigate the effect of including students' investigative experiments in the secondary school wave optics teaching unit on students' conceptual understanding of wave optics and their scientific reasoning. The reason for starting the project was the fact that in Croatia the secondary school physics teaching is still mostly of the lecturing type (including some demonstrations), although there are constant efforts to transition to a more inquiry-based teaching. One step in that direction is also our research project (whose acronym INVESTIGATE reflects well its intention), in which we have developed a new inquiry-based teaching sequence for secondary school wave optics in duration of 8 teaching periods of 45 min each. The teaching sequence includes the topics of the double-slit interference of light (Young's experiment), interference on optical grating, single-slit diffraction and polarization of light. Another problem for introducing inquiry-based teaching in Croatian secondary schools is that not all schools are well equipped for experiments That is why, in addition to developing a new teaching sequence, we have also, as a part of our project, selected, purchased, and partly developed adequate experimental wave optics equipment for the participating schools. A new diagnostic test on wave optics, the Conceptual Survey on Wave Optics (CSWO) (Matejak Cvenic et al. 2022), was also constructed and will serve to evaluate the conceptual understanding of wave optics of secondary school students after the planned teaching intervention. But, at the start of the project, it was necessary to first investigate the level of student knowledge after standard instruction so a series of interviews was conducted with secondary school students in Croatia, after they had completed their regular instruction on wave optics at schools, which yielded new insights in the difficulties that students have with this difficult physics topic. These findings also guided the construction of the CSWO. The sequence of the activities in the project is presented in Table 1.

During the interviews and after the testing of a larger sample of students with the CSWO (224 secondary school students in Zagreb) we have gained some additional insights in the levels of difficulty that physics topics may present to students and to some new perspectives on student learning, which will be discussed in this paper.

Table 1 The sequence of activities within the research project INVESTIGATE

Activity
Investigation of students' difficulties through semi-structured demonstration interviews
Construction and validation of a new diagnostic instrument (Conceptual Survey on Wave Optics)
Development and preliminary testing of the new teaching sequence on wave optics
Training of the participating physics teachers
Implementation and evaluation of the new teaching sequence

2 Investigating Student Difficulties with Wave Optics Through Demonstration Interviews

Student difficulties with wave optics were previously investigated in PER, but mostly on university students (e.g., Ambrose et al. (1999), Wosilait et al. (1999)). The studies showed that university students had significant difficulties with conceptualizing and understanding interference, diffraction, and polarization of light. However, it was important to extend the research to secondary school students as well.

2.1 Interference and Diffraction of Light

We chose wave optics for our project because it is a difficult topic which relies very much on experiments. Croatian students are usually shown some demonstrations of basic wave optics phenomena in standard teaching, such as Young's experiment, optical grating pattern or a single-slit diffraction pattern, but the focus on the instruction is mostly on mathematical description of the phenomena and on solving numerical problems.

We started our investigation with the series of 27 semi-structured demonstration interviews with secondary school students in Zagreb (aged 17–18 years) that lasted ca. 45 min each. In those interviews, students were asked about their predictions of the patterns in typical wave optics school experiments (in words and drawing), then they were shown the experiment and asked to describe their observations (in words and drawing) and later to provide the explanation of the observed patterns. The experiments that were used were the double-slit interference with monochromatic (laser) light, optical grating interference with monochromatic and white light, and the single-slit diffraction with monochromatic light. The interviews were audiotaped and transcribed. The results are described in detail in some other publications (Matejak Cvenić et al. 2021b; Matejak Cvenić et al. 2023), but here we bring only a brief summary of the most important findings.

2.1.1 Predictions

Generally, it was noticed that students were unable to predict the patterns that they had studied at school and which they saw in demonstrations or textbooks. Most students were not able to predict the double-slit interference pattern, but they usually recognized it when they saw it. Common predictions for the double-slit pattern were one or two dots, and sometimes a vertical pattern, because of the vertical orientation of the slit. With optical grating some students expected to see as many maxima as there were slits, and many expected a denser pattern (closer maxima) than in the double-slit experiment. On single-slit diffraction some students expected that no light would pass through the slit, and others predicted a small bright spot or, less often, a double-slit interference pattern.

2.1.2 Observations

Most students had difficulties describing their observations. The observations were usually focused on only one (the most prominent) feature of the pattern, most often the intensity of maxima, and less often their spacing.

Quite often students described the interference pattern as a "*broken line of light*", and not as a series of maxima and minima. They usually did not comment on the equal spacings of the maxima. On the single-slit pattern, they mostly noticed the wide and bright central maximum and the decreasing intensity of side maxima. On optical grating, on the other hand, they noticed that the spacing of the maxima increased in comparison with the double-slit pattern. With white light on the optical grating, they focused on the colours that appeared.

2.1.3 Explanations

The explanations were generally lacking, and those that were given seemed to be formed on the spot. The students did not seem to possess previously formed models of wave optics phenomena. Their explanations often contained elements of geometrical optics—students tried to use reflection or refraction of the light to explain the patterns.

Some students who described the interference pattern as a "*broken line of light*" formed different explanations for how that happened, usually attributing the effect of "*breaking of the light*" (not in the sense of refraction, but of splitting one beam of light into many) to the obstacle it encountered between the two slits, which caused it to split into many parts. On single-slit diffraction, some students concluded, after observing the pattern, which reminded them of interference, that there had to be two sources of light for interference to occur and ascribed the role of the two sources to the two edges of the slit. For the middle part of the light beam they used geometrical optics as explanation and concluded that the middle part just passes through undisturbed and forms the central maximum as a geometrical image of the slit. When observing colours with white light incident on a grating, some students explained the effect as the refraction of light and likened the grating to a glass prism or a lens. There were very few correct explanations.

2.2 *Polarization*

For polarization of light, students were given one experimental task: to find out, with the help of a known polarizer, which of the two equally looking slides was also a polarizer. After that, students were asked to explain what polarization of light was. The detailed analysis of the interview results on polarization can be found in (Matejak Cvenić et al. 2021a).

2.2.1 Experimental Task

A third of the students used the correct strategy to find out which slide was a polarizer (looking through two slides, the known polarizer and one investigated slide, and rotating one of them to see whether there would be a change in the intensity of the transmitted light with rotation). Other students used incorrect strategies, such as looking through one of the slides at the time or through two, but without rotation.

2.2.2 Explanations

It was noticed that students relied very much on the schematic representations and mechanical analogies presented in class in their explanations. They remembered, but often misinterpreted, them. While they were explaining polarization, it was additionally discovered that students' models of light were mostly vague and incomplete, and sometimes non-existent.

Students tended to confuse oscillation direction with propagation direction ("*polarized light propagates in only one direction*"). Some spoke vaguely of light oscillating, conceptualizing the light wave similarly to a mechanical wave on a string. Only one student gave the correct explanation of polarization, that included the electric field oscillations. Some students described polarization as *the weakening of the light* (focusing only on the reduction of light intensity). Almost half of the students tried to explain polarization using geometrical optics (stating that light is reflected or refracted by the polarizer) and some described polarizer as a slit or a grating, referring probably to the mechanical analogy of a polarized wave on a rope and vertical or horizontal slits through which it can or cannot pass.

3 What Do We Get After Standard Instruction on Wave Optics?

The results of the interviews suggested that students had not formed an adequate understanding of the wave optics phenomena during standard teaching of the topic. The interviewed students were volunteers who had good, very good or excellent grades in physics (equally represented), and physics is a compulsory subject in their schools, so the implications were disturbing. In summary, the interviews showed that most students:

- could not predict the patterns which they had seen in class,
- had trouble describing observations systematically—they tended to focus on only one prominent feature,
- could not explain the observed patterns,

- in their attempts to explain the patterns often used fragments of knowledge and simplified reasoning to construct the explanation on the spot,
- seemed to lack any previous models of the phenomena,
- often tried to explain wave optics phenomena with geometrical optics,
- remembered, but often misinterpreted, schematic representations and analogies that were presented in class,
- tended to reduce complex concepts to simpler ones in explanations.

4 The Development and the Results of the CSWO

The development of the CSWO was motivated by the lack of the available diagnostic instruments on wave optics, suitable for secondary school students. The expected learning outcomes for the secondary school wave optics instruction (Matejak Cvenic et al. 2022) were identified as the following:

- Demonstrate knowledge of basic wave concepts and the wave model of light.
- Apply mathematical conditions for the interference of light from two sources.
- Reason about school experiments in wave optics (interference of light on two slits, interference on an optical grating, diffraction of light on a single slit, and polarization of light with polarizers or by reflection on different media).
- Differentiate patterns of basic interference and diffraction phenomena introduced in high school physics.
- Explain wave optics phenomena and apply them to real-life situations.

We constructed the CSWO, intended primarily as a posttest for high school students, to probe those learning outcomes. After several cycles of item testing and Rasch analysis, the last version of the test consisted of 26 multiple-choice items and was administered to 224 high school students in Croatia after their regular instruction on wave optics. The Rasch analysis showed good structure and targeting of the test, as well as good functioning of the individual items and the test as a whole (Matejak Cvenic et al. 2022).

During the test construction, an important question emerged for us of how to organize the content of a conceptual survey. The Rasch model requires that the construct underlying the instrument is organized hierarchically, with different levels of difficulty. However, it is not easy to form a hierarchy in a conceptual test. We decided therefore to use the Webb's depth of knowledge (WDK) model (Webb 2002) and organize the construct around the cognitive complexity of items, expecting that the increase in cognitive complexity will also be followed by the increase in the difficulty. We chose the WDK model because it is simpler and better attuned to the needs of physics education, than for example, the more widely used Bloom's cognitive taxonomy. It is also, in our opinion, easier to use in teaching and research practice, since it links cognitive levels to the complexity of the items (number of steps required for its solving), and complexity is easier to recognize than the cognitive level.

Table 2 The three levels of the Webb's Depth of Knowledge model that can be probed with a test, and their application to the CSWO

Level	Content in the CSWO
Level 1. Recall and reproduction	• knowledge about basic wave concepts, such as e.g., wavelength of light or path difference • knowledge about the wave model of light and its links with wave optics phenomena (e.g., polarization suggests that light is a transverse wave)
Level 2. Concepts and skills	• applying the interference condition • reasoning about typical school experiments in wave optics (using skills like proportional reasoning or control of variables) • recognizing and using key features and variables of each phenomenon
Level 3. Strategic thinking	• explaining wave optics phenomena • applying phenomena to real-life situations

The WDK model has three levels, that can be probed by a written test. The levels are presented in Table 2. The fourth level, Extended thinking, requires other ways of probing, e.g. through students' projects or independent open inquiry.

Figure 1 shows how average difficulties of items, probing different learning outcomes, confirmed the theoretically predicted construct, based on the WDK model. Recognition of wave optics patterns turned out to be a relatively difficult and cognitively demanding task for students, which was confirmed also by our additional eye-tracking studies on wave optics pattern recognition (Susac et al. 2021, 2020). Students probably need to know the key features of the requested pattern first, and then look for them in the presented images—the obtained images of patterns can vary with the equipment used (e.g. the type of light source, color, shape of maxima, slit width, degree of diffraction effects superimposed to pure interference etc.), so it is not just a task of simple recall, it seems to be cognitively much more demanding.

5 What is Difficult in Learning Wave Optics and What Can We Do About It?

As we can see from Fig. 1 students have the least difficulty with the recall and reproduction part of the CSWO, which tests the basic knowledge about wave optics phenomena. Students have more difficulties with reasoning about typical experiments, but mostly they have problems with explanations, real-life applications of phenomena and with pattern recognition and differentiation.

The result concerning patterns may seem surprising, but we have reasons to believe (as explained above) that pattern differentiation in wave optics is a cognitively demanding task. On the other hand, its difficulty may stem, in addition, from the fact

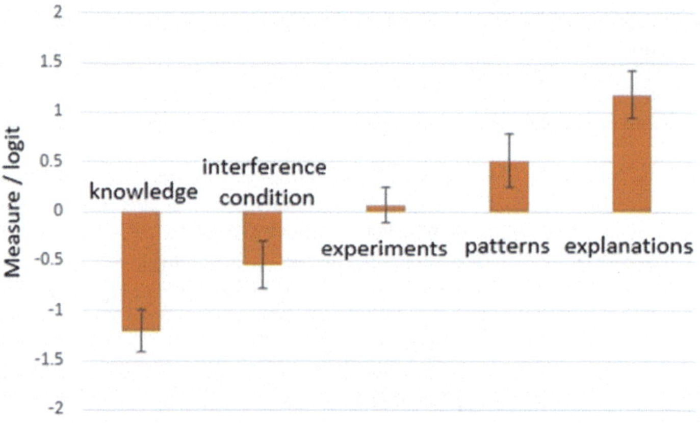

Fig. 1 Comparison of difficulties of the CSWO items (in logit) representing different learning outcomes (Matejak Cvenic et al. 2022). The zero level represents the average difficulty of items in the test. Error bars are 1 SEM (standard error of the mean)

that students have not engaged enough with these patterns, e.g., through hands-on experiments, to really notice their key features and differences.

Generally, as cognitive demand increases, so does the difficulty of the respective items. The CSWO results, but also the results of the interviews, as well as our teaching experience, suggest: **complex reasoning** is difficult for students. This is a problem that transcends wave optics alone, since there is a lot of complex reasoning in all parts of physics. Problem solving (which was not probed by the CSWO), for example, also involves complex multi-step reasoning and procedures that students find difficult.

However, there is one other important conclusion from the results: it is not possible to develop complex strategic thinking (Level 3) unless students have a good foundation of the relevant underlying concepts and reasoning skills (Level 2). Level 3 relies heavily on good and solid Level 2, but this level seems to be neglected in typical physics teaching and assessment practice in Croatia, and probably in many other countries as well. Teaching and assessment are often mostly directed either to Level 1 (especially in elementary school) or to Level 3 (in secondary school, e.g., through complex numerical problems).

Instead, the major part of the elementary and secondary school physics teaching should be directed towards fundamental concepts and skills—building in that way a strong basis for future learning and gradual development of complex reasoning. Experiments may play a very significant role in this process, since it is through experiments, particularly investigative experiments, that students really come to know the studied physics phenomena and then investigate their key aspects, building and using many important skills in the process, such as the skills of systematic observation of physics phenomena, proportional reasoning, control of variables, designing simple experiments, performing measurements, interpreting measurement results or testing hypotheses.

The good news is that Level 2 is attainable for most students, and is far easier than Level 3, but still very important. What physics teachers in secondary school can do to improve the current situation is to first build a strong Level 2, and then attempt Level 3. Teaching should rely more on experiments and inquiry—on building systematic observation skills, acquainting students with phenomena, testing hypotheses, developing conceptual understanding and representation skills.

This was also our plan in designing our new teaching sequence for wave optics. It was constructed in a way to foster as much as possible (within the limits of the short time typically allocated to wave optics in physics teaching) the development of the basic concepts and scientific reasoning skills needed for wave optics, but also to acquaint students well with the phenomena, their respective experimental patterns and to help them form the explanations of phenomena. The teaching sequence has the following structure and content in eight teaching periods:

1. Introduction to the historical dilemma of whether light could be described by the wave or the particle model, discussion of how models are evaluated, recognizing the fundamental difference in behavior of particles and waves, and the interference as a crucial phenomenon that sets them apart. Observing the double slit experiment with laser light and investigating with the mechanical model (transparencies with spherical wave fronts) whether the observed pattern is consistent with the interference of mechanical waves.
2. Students' investigative experiment of the double-slit pattern: how is the spacing of the maxima affected by different variables, finding the mathematical description.
3. Describing and applying to light waves the mathematical conditions for constructive and destructive interference, investigating, in light of those conditions, the double-slit pattern to obtain the explanation of how it is formed.
4. Investigating the interference on optical grating: students' investigative experiments of optical grating pattern with optical gratings of different constants and with monochromatic and white light. Deriving mathematical description of interference on optical grating.
5. Investigating what happens when light is incident on only one slit. Demonstration of the single—slit diffraction. Students' investigative experiment: What happens to the width of the central maximum as the slit is being narrowed? Forming the explanation of the diffraction with the Huygens- Fresnel principle (without mathematical description).
6. Systematization of the different observed patterns—finding in which aspects they are similar or different. Conceptual questions on interference and diffraction of light and some numerical problem solving in small groups related to these topics.
7. Introduction to polarization of light through demonstration and mechanical analogy. Explaining polarization by describing light as a transverse electromagnetic wave.
8. Investigation of why polarizing sunglasses reduce glare. Students' investigative experiment: testing the hypothesis that light is polarized by reflection from some surfaces. Introduction of Brewster's law.

The sequence was piloted by the research team in one secondary school in Zagreb (which was later not included in the experimental group) and found appropriate for the secondary school students and the allocated teaching time. The necessary equipment was selected and purchased for the participating schools. In cooperation with one firm from Croatia, we designed and produced relatively cheap printed single and double slits and optical gratings, which showed good performance in wave optics school experiments.

6 Continuing the Project: The Implementation and Evaluation of the New Teaching Sequence on Wave Optics

The next step in our project was the training of the involved physics teachers from six Zagreb secondary schools, who volunteered to implement the new teaching sequence on wave optics in their classes. We first held a two-day workshop for these teachers alone, performing for them the whole sequence and discussing all the details. The teachers were in the role of students in that workshop.

After that, they obtained all the necessary equipment, lesson plans, and teaching materials. In the next three-day workshop, we invited ca. 50 participants, physics teachers from secondary schools in Croatia, to learn about the new teaching sequence, this time performed by the six teachers who participated in the project. In this way, the six teachers had a chance to try out some parts of the sequence and receive feedback from us and the participants.

In the fall of 2022, the teaching sequence was implemented in six Zagreb secondary schools, under monitoring of the research team. Students were pre- and post-tested with the Lawsons' Classroom Test of Scientific Reasoning (LCTSR) and post-tested with the CSWO.

The collection and analysis of the data and the comparison with the control group results (students of the same teachers who were taught in previous years by the traditional method and tested in the same way as the experimental group) is still ongoing, but the first results suggest an improvement in the conceptual understanding of students who were taught with the new teaching sequence.

7 Conclusion

Wave optics is a difficult topic for secondary school students—in a short time, they are exposed to a number of similar and yet different patterns, and a number of challenging phenomena. Students are often unable after standard teaching to distinguish, let alone explain, different phenomena of wave optics.

The results of our project so far suggest that complex reasoning is difficult for students, but that they also lack some important skills and conceptual understanding to engage in such reasoning, so more emphasis in teaching on building first the necessary skills and concepts (Level 2) is needed.

We hope to find out soon to what degree our new teaching sequence may have helped students to build a stronger and more coherent knowledge of wave optics.

Acknowledgements This work has been fully supported by Croatian Science Foundation's funding of the project IP-2018-01-9085.

References

Ambrose BS, Shaffer PS, Steinberg RN, McDermott LC (1999) An investigation of student understanding of single-slit diffraction and double-slit interference. Am J Phys 67:146. https://doi.org/10.1119/1.19210

INVESTIGATE project website in English, https://www.pmf.unizg.hr/phy/en/zavodi/zef/physics_education

Matejak Cvenic K, Planinic M, Susac A, Ivanjek L, Jelicic K, Hopf M (2022) Development and validation of the conceptual survey on wave optics. Phys Rev Phys Educ Res 18:010103. https://doi.org/10.1103/PhysRevPhysEducRes.18.010103

Matejak Cvenić K, Ivanjek L, Planinić M, Sušac A, Hopf M, Cindrić M (2023) Probing high school students' understanding of interference and diffraction of light using standard wave optics experiments. Phys Rev Phys Educ Res 19:020118. https://doi.org/10.1103/PhysRevPhysEducRes.19.020118

Matejak Cvenic K, Ivanjek L, Planinic M, Jelicic K, Susac A, Hopf M (2021) Analyzing high school students' reasoning about polarization of light. Phys Rev Phys Educ Res 17:010136. https://doi.org/10.1103/PhysRevPhysEducRes.17.010136

Matejak Cvenić K, Planinić M, Jeličić K, Hopf M, Ivanjek L, Srnka C, Sušac A (2021b) Exploring secondary school students' understanding of basic phenomena relating to wave optics. J Phys Conf Ser 1929:12007. https://https://doi.org/10.1088/1742-6596/1929/1/012007

Susac A, Planinic M, Bubic A, Jelicic K, Ivanjek L, Matejak Cvenic K, Palmovic M (2021) Effect of students' investigative experiments on students' recognition of interference and diffraction patterns: an eye-tracking study. Phys Rev Phys Educ Res 17:010110. https://doi.org/10.1103/PhysRevPhysEducRes.17.010110

Susac A, Planinic M, Bubic A, Ivanjek L, Palmovic M (2020) Student recognition of interference and diffraction patterns: an eye-tracking study. Phys Rev Phys Educ Res 16:020133. https://doi.org/10.1103/PhysRevPhysEducRes.16.020133

Wosilait K, Heron PRL, Shaffer PS, McDermott LC (1999) Addressing student difficulties in applying a wave model to the interference and diffraction of light. Phys Educ Res Suppl Am J Phys 67(7):5. https://doi.org/10.1119/1.19083

Webb N (2002) Depth-of-knowledge levels for four content areas. http://ossucurr.pbworks.com/w/file/fetch/49691156/norm%20web%20dok%20by%20subject%20area.pdf

Students' Embodied Interactions During Collaborative Learning Activities in Physics

Bor Gregorcic

Abstract In this paper, I outline some qualitatively different ways in which students can use their bodies when learning physics. I pay special attention to student embodiment in the presence of digital technologies and provide illustrative examples of students using their bodies in range of ways while engaging in different phases of the physics learning process.

Keywords Embodiment · Embodied cognition · Student interactions · Physics education · Collaborative learning · Gestures

1 Introduction

Learning physics entails developing an understanding of physical phenomena, as well as learning to speak the "language" of physics and to participate in physics disciplinary discourse (Lemke 1982; Airey and Linder 2009). Physics learning thus necessarily involves learning to communicate and work with relevant representations or semiotic resources, such as written and spoken language, mathematics, graphical representations such as diagrams, graphs, and even physics equipment (Airey and Linder 2017).

The study of communication between learners or between the teacher and the learners is often seen as a crucial aspect of studying educational processes. This is especially characteristic of an interactionist perspective on learning. From this perspective, learning to "talk physics" (Lemke 1982), or to participate in physics discourse, is in essence considered as the equivalent of learning physics. Yet, today, in the time of the ever-growing presence of artificial intelligence and large language model chatbots (Gregorcic and Pendrill 2023; Kortemeyer 2023), it is more relevant than ever to recognize that discourse proficiency does not always correspond to an underlying conceptual understanding (Mahowald et al. 2023); it has become possible

B. Gregorcic (✉)
Division of Physics Education Research, Department of Physics and Astronomy, Uppsala University, Box 516, 75120 Uppsala, Sweden
e-mail: bor.gregorcic@physics.uu.se

S. Faletič and J. Pavlin (eds.), *Teaching and Learning Physics Effectively in Challenging Times*, Challenges in Physics Education, https://doi.org/10.1007/978-3-031-72541-8_3

for AI-based algorithms to successfully imitate (DiSessa 1993; Airey 2009) scientific and physics discourse in many contexts.

I would like to argue that paying attention to forms of communication that more directly stem from our students' bodies, such as gestures and physical interaction with people and equipment in their surroundings, may give educators an alternative, and arguably more "direct" and genuine window into their students' thinking and meaning making; the study of communication between physically co-present learners allows us to investigate aspects of learning that often escape the analytic eye when the focus is predominantly put on language and graphical representations. Embodied forms of communication are still outside of the domain of large language models, such as ChatGPT, and may attract ever more interest in the future, when human and machine-generated language and graphical outputs become harder to differentiate.

2 Embodiment in Physics Teaching and Learning

Embodiment is a collection of related theoretical perspectives on human cognition ad communication with implications for the learning sciences. The idea that learning and cognition should not be understood merely as the product of information processing within the brain, but also involve the body of the learner, has a long history in philosophy, education and education research (Dewey 1916; Merleau-Ponty 1945). Similarly, communication does not only take place through language, but also involves other modalities, some of which rely on the use of our bodies, such as performing hand gestures and actions (Nevile 2015).

2.1 *Embodied Cognition and Physics Learning*

In late twentieth century, the Cartesian mind–body divide has loosened in the field of psychology, giving rise to a field of study often referred to as *embodied cognition* (Shapiro and Spaulding 2024). The psychological and educational communities have since seen an increasing interest the role of the body in cognition and learning. The notion of *embodiment*, an umbrella term for processes of thought, learning and communication involving the body, has also gained traction in the science and physics education research communities (Kersting et al. 2023). Notably, the terms *embodiment* and *embodied cognition* do not have a single universally defined meaning and are used in slightly different ways by different philosophical, psychological and educational schools of thought (Kersting et al. 2021, 2023). Yet, for the purpose of this paper, it will be enough to begin with a broad understanding of the main idea of embodied cognition—that we can gain significant insights into the learning processes by paying attention to the students' use of their bodies.

In this paper, I provide illustrative examples of students engaging in learning activities by actively using their bodies. I use these examples to outline a qualitative framework for understanding the ways in which students may involve their bodies when learning physics with the assistance of digital technology. The examples are taken from my own research activities that I pursued over the last decade in collaboration with colleagues and pre-service physics teachers.

2.2 Kinaesthetic and Embodied Learning Activities

Perhaps the most familiar sense in which students can use their bodies when learning physics is through kinaesthetic learning activities. While one could see any student activity in which they move their body as a kinaesthetic activity, I believe it is beneficial to make a distinction, at least temporarily, between *kinaesthetic* learning activities and *embodied* learning activities. This distinction will help us as physics educators better appreciate some qualitatively distinct ways in which the involvement of students' bodies can enhance learning.

Kinaesthetic learning activities (KLAs) have been described as activities which physically engage students in the learning process (Bernhard and Bernhard 1999; Whitworth et al. 2014; Oliveira and Fischer 2017; Ruiz 2017; Coletta et al. 2019). They are often used in the context of mechanics, where students can use their bodies as sensors for physical interactions and motion. This way, the students are participating as part of the actual physical phenomenon under study. The concept of embodied learning activities (ELAs) has been suggested (Scherr et al. 2012) to differentiate a subset of kinaesthetic learning activities in which students enact physical phenomena or concepts, using their "bodies, or parts of their bodies, as metaphorical substitutes for physical entities in a role playing of physical phenomena" (Euler et al. 2019, p. 5). ELAs therefore engage students' bodies in analogical activities meant to mirror the phenomena of interest, when such phenomena are removed from everyday experience because of the spatial or temporal scale at which they take place (*e.g.* microscopic (Richards and Etkina 2013; Richards 2019), astronomical (Morrow 2000; Gregorcic 2015)) or their level of abstraction (*e.g.* energy (Close and Scherr 2015), spin (Hahn and Gire 2022)).

It is worth noting that in some cases, ELAs can facilitate sensorimotor experiences where students' bodies also become sensors for physical quantities, which may be directly relevant for the studied phenomenon. For example, students can sense forces acting on their bodies when enacting an astronomical scenario and make conceptual links, connecting their embodied experiences to forces between astronomical bodies (Euler et al. 2019). Therefore, as my colleagues and I have argued, ELAs and KLAs are not always clearly separable categories of learning activities (Euler et al. 2019) and may flow into each other quite naturally.

2.3 *Embodied Communication in Physics Learning*

So far, I have presented a rough overview of activities which are designed for helping students learn or develop physics concepts, either by having them experience phenomena through their bodies, or having them enact abstract ideas into concrete, embodied representations. In this section, I will instead focus on a different instantiation of the idea of embodiment—embodied communication.

Similarly to the way in which the body gained traction in psychology research, resulting in the idea of embodied cognition, there happened a so-called "embodied turn" in research on language and social interaction (Nevile 2015). A shift and widening of attention from just language to embodied modalities such as gesture, gaze, haptic touch and body positioning, carries with it quite naturally also an increased sensitivity to physical features of the environment where social interactions take place. In many ways, the field of embodied communication (Streeck et al. 2011) shares areas of interest with the field of distributed cognition (Hutchins 1995) in that they consider the relevant unit of analysis to also contain physical artefacts in the surroundings. For physicists and physics education researchers, such a unit of analysis can be interesting when trying to better understand physics disciplinary and educational practices in the presence of measuring equipment (Volkwyn et al. 2019a) or in multimodal contexts involving several non-linguistic representations of physics ideas (Fredlund et al. 2014).

2.3.1 Gestures: An Overview

Gestures are bodily actions, usually involving the hands, that typically accompany speech and carry implicit or explicit meaning. The study of gestures is an academic field in itself (McNeill 1992, 2005; Kendon 1997, 2004). In this paper, I limit the review of the literature to publications whose findings are most relevant for the interpretation of examples presented further below.

Gestures have been suggested to be deeply linked to our cognitive processes (Iverson and Goldin-Meadow 1998; Goldin-Meadow 2005). For example, studies show that the mismatch content of student's spoken language and gestures is indicative of learning taking place (Alibali and Goldin-Meadow 1993; Goldin-Meadow and Wagner 2005) and that seeing and performing gestures helps students think and learn (Goldin-Meadow 2005; Cook and Goldin-Meadow 2006). Furthermore, it has been shown that gestures can pre-shadow students' verbal articulation of scientific concepts during learning (Roth and Lawless 2002a). A likely explanation for this observation is that gestures represent a separate, though not completely independent channel of communication that is much less codified than language, which enables the communication of ideas that learners are not yet capable of formulating into words. This means that gestures have the potential to offer us an alternative window into our students' thinking (Goldin-Meadow 2006).

Gestures can roughly be categorized into beats (rhythmical movement of hands during speech that do not carry independent meaning and help in pacing speech), deictic gestures (pointing gestures, indicating a position in space), iconic gestures (gestures that represent visual or spatial features of an object or phenomena, such as a shape of a trajectory) and metaphorical gestures (similar to iconic, but communicating an abstraction that is not directly perceivable, such as momentum or energy) (Roth and Lawless 2002a).

In recent decades, the study of gesture has found its way into the field of education (Roth 2001), including science (Crowder 1996; Chue et al. 2015), mathematics (Edwards 2009; Alibali and Nathan 2012), chemistry (Flood et al. 2014) and physics education research (Scherr 2008; Gregorcic et al. 2017). While some studies have focused on the role of teachers' gestures during lectures (Hwang and Roth 2011), this paper will focus on *students'* use of gestures and other embodied means of communication during physics learning activities using digital learning tools. Against this background, I will mostly focus on deictic and iconic gestures, as they seem to play an especially central role in students' communication.

2.3.2 Symbiotic Gestures

The studies that I will present in this paper investigate student engagement and interaction in learning environments with a significant digital component. In most cases, the students are working in small groups in front of a large digital screen or an interactive whiteboard (IWB). The learning activities involved the students interacting with each other, as well as with the content on the IWB screen. It comes as no surprise that this is also reflected in their use of gestures. For example, students often point to and illustrate actions against the backdrop of digital content. A term that is useful for describing this behaviour was coined by Goodwin as *symbiotic* (Goodwin 2003) or *environmentally-coupled* (Goodwin 2007) gestures (I will use the term symbiotic gestures). These are gestures that do not carry their intended meaning if we look at them in isolation from the physical surroundings or environment in which they occur, even when they are coupled to accompanying speech. These kinds of gestures are not unusual in our everyday lives. The television weather forecast, for example, is typically abound with symbiotic gestures. Symbiotic gestures are also common in physics learning contexts, even though they are not often referred to explicitly as symbiotic gestures. In the literature, we can see examples of students and teachers gesturing against computer screens and physics equipment to communicate physics content (Roth and Welzel 2001; Roth and Lawless 2002b; Adams 2004; Volkwyn et al. 2020).

Below I present the first illustrative example taken from a study of student communication in a technology-supported learning environment (Fig. 1). In it, three students explore for the first time the shapes of planetary orbits around a massive star. From the interpretation of their gestures, we can see that they have successfully identified some critical features of elliptical orbits, even if they did not express these findings in disciplinary language.

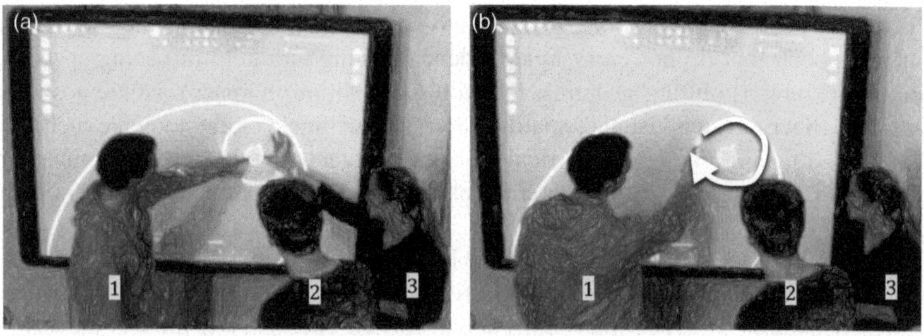

Fig. 1 Two students using pointing and symbiotic gestures for describing and highlighting features of a simulated movement of a planet around the Sun, displayed on the interactive whiteboard screen. Student 1 first points to the location of the Sun (**a**) and then outlines the movement of the planet around it (**b**). Student 2 uses a distance-measuring gesture to point out that the Sun is positioned eccentrically in regards to the orbit (**a**). Image is reproduced from (Gregorcic et al. 2017) under the Creative Commons 4.0 International License

2.4 Equipment and Other Resources in the Environment

While gestures represent an important component of embodied communication in physics learning contexts, students' physical interaction with equipment may also play an important role (Tang 2022). In a research project led by Trevor S. Volkwyn, we examined the role that physics devices play in the process of making meaning in a laboratory setting (Volkwyn et al. 2019b). We were particularly interested in the way in which student interaction with measuring devices and their physical manipulation can be an important source of information for different participants of the lesson. We saw that student manipulation of a 3-dimensional magnetometer (IoLab) was important not only for the student holding it, but also allowed others (including the teacher) to infer the reasoning of the student holding it.

The task for the students was to find the direction of the Earth's magnetic field in the room by using a 3D magnetometer embedded into a multiple-sensor educational device called the IoLab. The values of magnetic field components along three perpendicular axes were displayed in a time-dependent graph on a computer screen. As the graph was being drawn in real time, the wirelessly-connected sensor could be freely translated and rotated by students. The axes names and directions were also inscribed onto the sensor housing, allowing students to relate the graphs to the physical position of the sensor box.

Students came to explore what the values of the different components depended on. The manipulating student was able to hold the device, look at the screen and slowly learn to coordinate their movements to reliably achieve the desired outcome. While one could argue that this sort of hands-on engagement was likely beneficial for the student holding the device, the benefits extend beyond the manipulating student.

Fig. 2 Two students working in a pair, manipulating a wireless magnetometer connected to a graphing software to explore how moving the device in space impacts the graphs on the screen. The coordinate axes are labelled on the device (exaggerated in the picture). The student who is not holding the device attentively observes the other students' movements and the graphs as they change in real time. Throughout the activity, both students took turns in manipulating the sensor box. Images are reproduced from (Volkwyn et al. 2020) under the Creative Commons 4.0 International License

When a student attempted to make all but one component of the magnetic field zero, the teacher could observe the student's embodied actions—the manipulation of the magnetometer—and infer the student's reasoning. Other students were also able to comment and suggest alternative ways of translating and rotating the magnetometer by giving feedback based on their observations of the student and the graphs being displayed in real time (Fig. 2). The manipulation of the device thus also played a communicative role. In addition to gestures, paying attention to the students' use of their bodies during physical manipulation of equipment can also give important information to other learners and valuable insight into student thinking to teachers.

2.5 *Interpersonal Touch and Body Position*

The third way in which the body can be involved in the learning process is perhaps less familiar to physics educators, but is in fact a common mode of instruction in disciplines such as dance, yoga, and martial arts. It is in essence a kinaesthetic approach, one in which the input to the learner and feedback to the teacher is provided through direct bodily contact with another person—interpersonal touch (Gallace and Spence 2010)—or by means of a spatial relationship to another person.

In collaboration with a colleague Elias Euler, leading the project, and a former student Elmer Rådahl, we have studied an example of two students engaging in a dance-like activity while making sense of binary star dynamics (Euler et al. 2019). While the range of embodied resources that the students used during the activity includes aforementioned gestures (both "ordinary" and symbiotic) and interactions

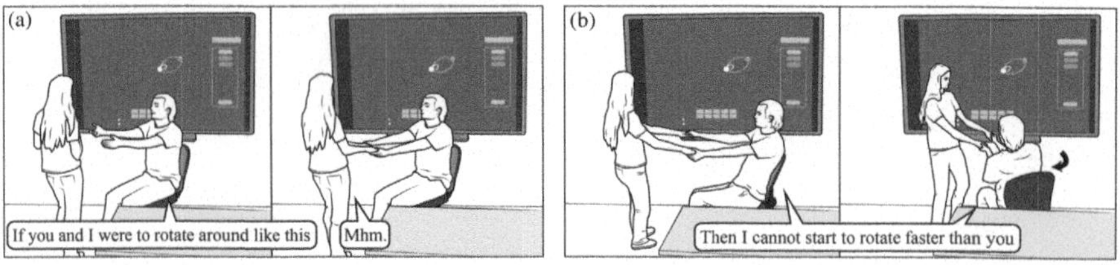

Fig. 3 Two students using their bodies to enact a scenario analogical to a binary star system. Their bodies represent the two stars and them holding hands represents the attractive forces between the stars. The question that they tried to answer was why the two stars in a binary star system always move synchronously, having the same orbital period. Image is reproduced from (Euler et al. 2019) under the Creative Commons 4.0 International License

with equipment (manipulating the surface of the IWB), the students also enacted a movement resembling the iconic dancing scene of the movie characters Jack and Rose from the movie Titanic (1997) in order to make sense of the dynamics of a binary-star system. They used their bodies to collaboratively explore and communicate salient features of a physics phenomenon (Fig. 3). The sense of haptic (pushing and pulling) touch and the relative positions of their bodies in space served a central role in their communication and reasoning.

3 Mapping Embodiment onto Physics Learning Processes

In this section, I would like to more explicitly exemplify how the above-described types of embodied reasoning and embodied communication can be used by students in the process of learning physics. For this purpose, I divide the physics learning process into three rough categories of activity (see below). These categories are not meant to be an exhaustive list of all possible types of activities within the domain of physics learning, but rather represent *some* qualitatively distinct ways in which I have, throughout my research and teaching experience, come to see students engage in physics learning. My views on this topic have been heavily influenced by the Investigative Science Learning Environment (ISLE) (Etkina et al. 2019; Brookes et al. 2020), as a framework for instruction and interpretation of student activity. Note that my intention is to give examples of students *productively* engaging in these activities in order to demonstrate the potential of embodiment and possibilities it affords for physics learning, and to encourage teachers and researchers to pay more attention to students' use of embodied resources. The three types of embodied engagement presented in the previous section can map onto different categories of physics learning activities.

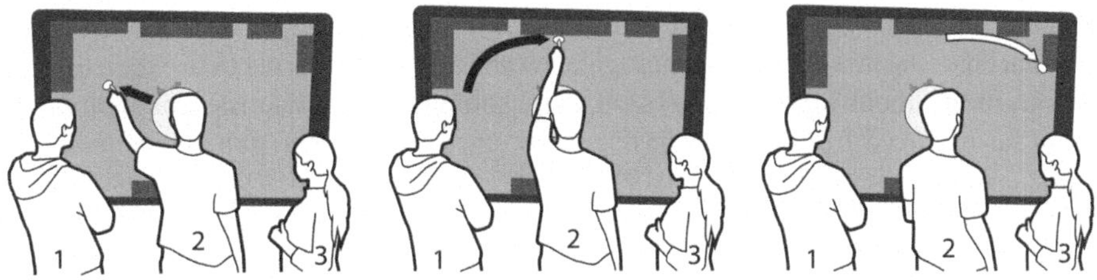

Fig. 4 Students create (draw), grab (with a touch screen action) and throw virtual planets into orbit around the "Sun". Through repeating this action several times, they experience a range of different possible trajectories that planets can take, including open and closed orbits of different eccentricity. It is important to note that the term eccentricity is not familiar to most students, but that they can formulate some aspects of the differences between the shapes of orbits using symbiotic hand gestures (see Fig. 1a). Image is reproduced from (Gregorcic and Haglund 2021) under the Creative Commons 4.0 International License

3.1 Students Familiarizing Themselves with Physics Phenomena

The first category of activity involves students familiarizing themselves with physics phenomena. Typically, the familiarization takes place through observation and description of observed patterns (in ISLE referred to as observational experiments). However, the body can also play a direct role through kinaesthetic engagement. It can serve as a tool for generating novel observations (almost all physics experiments require some type of physical manipulation). This can be seen in most hands-on learning environments. In the example below, I show how the idea of kinaesthetic engagement can be fruitfully expanded into the digital domain. Students can use an IWB, a type of large touch screen, to engage in playful manipulation of virtual content much like playing with physical objects. What makes this example particularly interesting is that they can engage in a kinaesthetic activity on a topic which is typically far-removed from our embodied experiences—orbital motion of planets (Fig. 4).

3.2 Students Discussing Physics Concepts and Developing Models

In this category of activity, I include students reasoning about, proposing, developing and finding ramifications of models of physical phenomena. This includes theoretical and conceptual discussions about models and connecting them to real-world phenomena that they are intended to represent. One example of such an activity is presented in Fig. 3, where students try to make sense of the dynamics of binary stars by enacting an embodied dance as a model of a pair of stars orbiting each other. While this example contains many features that can be analysed from a perspective of embodiment (Euler et al. 2019; Kersting et al. 2023), it is perhaps most interesting to note that enacting this embodied analogical model allowed students to use their everyday intuitions about forces and motion to reason about and discuss a phenomenon that is far-removed from their everyday experience. For a more exhaustive analysis of this episode, see (Euler et al. 2019).

3.3 Students Engaging in Collaborative Science-Like Activities

This category is particularly strongly inspired by the ISLE framework. Activities in this category are those that reflect an authentic scientific epistemology (Brookes et al. 2020). By this I mean that students engage in activities that resemble an authentic approach to science in terms of the development of new knowledge. To illustrate this, I present the example shown in Figs. 5 and 6.

Student 5 first presents a possible explanation of why planets have closed orbits in the form of a kinematic argument, relating the displacement of a planet during interval in the tangential and radial directions, relative to the Sun. To this, Student 4 replies with what he sees as a limitation of Student 5's proposal; he suggests that if a planet moves too fast in the tangential direction, that it may "escape" to never return again. Student 5 objects to this and suggests that no matter how fast they "throw" the planet, it will always return back, but allows for the possibility that the orbit may not necessarily be perfectly circular. Throughout this first part of the interaction,

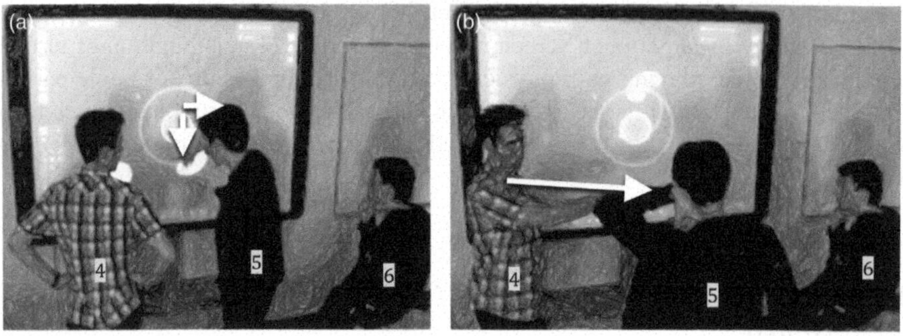

5: If it makes this distance [moving the hand horizontally] before it makes this distance [moving the hand vertically], then it will start circling.

4: Yeah, but it shouldn't go too fast, so it doesn't get thrown out [moving his hand].

5: But it is circling, no matter how far you throw it. How to make it really circular is another question...

Fig. 5 Student 5 is proposing a kinematic model for orbiting planets. A discussion develops about whether a planet with a very high initial velocity will ever come back to where it was thrown from. Image is reproduced from (Gregorcic et al. 2017) under the Creative Commons 4.0 International License

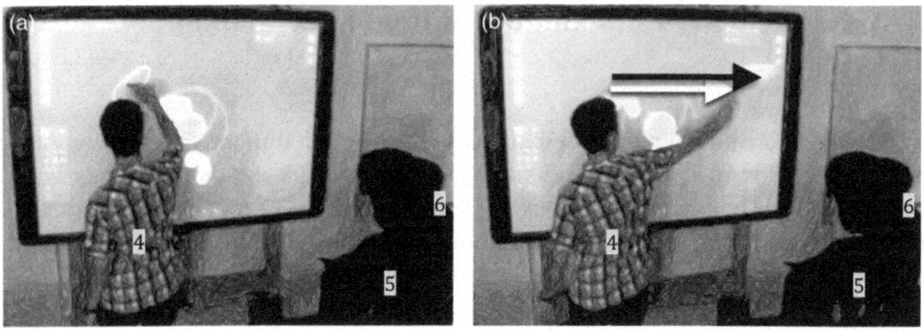

Fig. 6 A student draws a new object (planet)and grabs it (**a**), then throws it towards the right (**b**). The black arrow represents the displacement of the planet during a short time interval, as it was being thrown. The white arrow represents the motion of the student's hand while performing the throw. The student's intention was to test the other students' proposal that planets always return back to their original place, no matter how fast one throws them. Image is reproduced from (Gregorcic et al. 2017) under the Creative Commons 4.0 International License

we can see students not only talking, but also gesturing against the screen (*i.e.*, using symbiotic gestures), their gestures representing distances, directions and even reflecting the speed of hypothetical motion of planets. The aspect of their interaction that is crucial for the purpose of this paper's analysis, is that they are engaged in a science-like discussion. By science-like, I mean an exchange of arguments about what can be expected of the behaviour of the simulated physical phenomenon at hand. As we can see from the short analysis presented above, the students' bodies play a central role in this discussion.

Immediately following the exchange shown in Fig. 5, Student 4 creates a new planet by drawing it on the IWB, then grabs it and throws it with a large velocity towards the right (Fig. 6). His intention is to perform an experiment, a test of the idea proposed by Student 5, that a planet should always return back. In Fig. 6 Student 4 thus used his body in conjunction with the IWB to perform a virtual testing experiment. Their discussion continued after this excerpt.

The above presented exchange between the two students resembles an authentic scientific discussion, where their bodies are first employed as means of communicating ideas (Fig. 5). Following this, Student 4 used his body to perform experiments to test those ideas. These are central components of authentic knowledge-building practices in physics and other sciences (Brookes et al. 2020).

4 Going Remote: Embodiment in Remote-Learning Activities

In 2020, the educational reality quickly shifted to a predominantly remote-learning setup. One could expect that as a side-effect of this transition, students' bodies would no longer play a significant role in physics learning activities. Yet, I have experienced and documented in an ongoing research project, that embodied aspects of student activities did not disappear, but rather took on new and sometimes unexpected forms.

Here I will illustrate with a brief example, how the body can "make its way" through a video-conferencing interface. Teachers who were committed to following the pedagogical practices associated to active learning approaches found ways to engage students even in remote learning. The practice of "whiteboarding" (MacIsaac)—students working together to share content on a whiteboard—turned out to be one of the techniques, which could be transferred into remote education with relative ease. For example, a study led by Gregor Brumec (Brumec et al. 2022), used ISLE-based materials (Etkina et al. 2020) to engage students in "remote whiteboarding" using virtual whiteboards in a widespread video-conferencing tool. We paid special attention to the ways in which students divide the work and how they communicate in the online environment. I myself performed a similar study at Uppsala University already during the first year of the Covid-19 pandemic.

Most interesting for the paper at hand is the finding that students employ new and sometimes unexpected modalities of communication to compensate for the inability to produce symbiotic gestures. For example, in in-person whiteboarding, students normally draw and write on the whiteboard and routinely point to and gesture against the content of the board. In remote settings, we saw that this function of symbiotic and pointing gestures was taken over by drawing. Students used their pens (often of different colours), to circle, draw arrows and lines, in order to highlight features of whiteboard content, in parallel with their speech (see Fig. 7). I refer to this phenomenon as *semiotic drift*—a process where the communicational (semiotic) function of one modality (gesture) *drifts* into a different modality (drawing in this case). This implies that the function that gestures, particularly symbiotic and pointing gestures, serve in student communication, are relatively important for them, allowing them to communicate in ways not allowed by language and diagrams alone.

5 Implications for Education

My hope is that the theoretical ideas and empirical examples provided in this paper will inspire interested physics educators to increase their sensibility to students' use of their bodies when learning physics. Physics teachers who would like to consider the ideas of embodiment in the educational context can do this by:

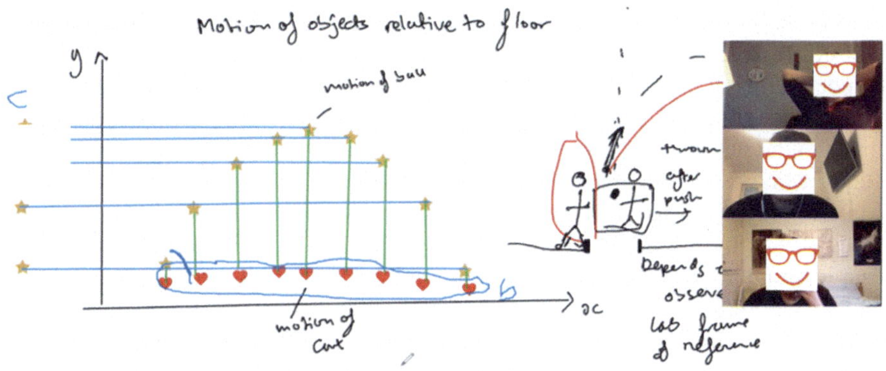

Fig. 7 The image shows a screenshot from a video-conferencing tool with a virtual whiteboard tool. The task involved students observing a video of an experiment of a ball shot vertically from a moving low-friction cart and landing back on it. The hearts represent the position of the cart at equally spaced time intervals and the stars represent the position of the ball at equally spaced time intervals. The red free-hand-drawn lines on the right side of the image and the blue free-hand-drawn line circling the hearts were drawn when students were referring to different parts of the whiteboard, but were not able to point to it using their hands

- *Listening to the students by paying attention to their use of their bodies.* This is true both in contexts where students can be expected to already have significant knowledge of the physics content and relevant vocabulary, but especially in contexts where students are still learning the concepts within a topic and perhaps still struggling to find words for expressing their ideas.
- *Learning to appreciate the richness of communication that happens.* It is easy to not give credit to students who do not express themselves scientifically through language. Yet, as I have shown, students may actually be engaging in science-like discussions without using scientific language. By knowing how to "listen" to their bodies, we can better recognize these moments and build on them.
- *Creating space for embodied engagement and encourage it.* When students feel that they are free to move, it is more likely that they will use their bodies in productive ways. Having students stand up and write on a shared surface is one practical way of achieving this. Kinaesthetic learning activities are another option. In remote learning settings, choose tasks that require the use of multiple representations and have them use an online whiteboard to work together. Home-based kinaesthetic activities may also be an option.

6 Discussion and Conclusion

The ideas presented in this paper emerged through the last 10 years of me studying students' use of their bodies while learning physics. The presented literature background on embodied learning activities and embodied communication may serve as a starting point for the exploration of the role of the body in physics education for interested researchers and teachers. The selection of episodes from empirical studies of student embodiment should be seen as illustrative examples from which readers can glean the utility of using an embodiment perspective to examine student interactions in the physics classroom. The list of references will provide further reading to those who wish to delve deeper into the topic.

In summary, students can use their bodies in several different ways and in different phases of the physics learning process. Ways of using the body range from kinaesthetic learning activities to embodied communication, as well as embodied interaction with laboratory equipment. Students can use their bodies throughout the physics learning process, from experiencing new phenomena and discovering regularities and patterns, modelling physics phenomena though analogical reasoning, and engaging in science-like activities, such as scientific discussions and proposing and performing testing experiments.

Acknowledgements I would like to express my gratitude to my PhD advisors Gorazd Planinsic and Eugenia Etkina for encouraging me to pursue this exciting line of research and introducing me to ISLE. My deepest thanks also go to Cedric and Anne Linder, for their support and guidance in the years after my PhD. The work presented in this paper has been the product of close collaboration with PhD students Elias Euler and Trevor S. Volkwyn, and undergraduate students Gregor Brumec and Elmer Rådahl. Many thanks also to the "embodiment crew," Jesper Haglund, Magdalena Kersting, Tamer Amin, Liv Kundrup Hardahl and Rolf Steier for expanding my knowledge and perspective on the topic of embodiment.

Competing Interests The author has no competing interests to declare.

Ethics Approval All the data for the original studies reported in this work has been collected, processed and stored in accordance with local regulations at Uppsala University, Swedish law and GDPR. Further ethics approval was not required by law, as we did not collect or process sensitive personal data.

References

Adams W (2004) Gesture with Interactive Computer Simulations. http://phet.colorado.edu/publications/Gesture.pdf

Airey J (2009) Science, Language and Literacy: Case Studies of Learning in Swedish University Physics. Doctoral dissertation, Uppsala University

Airey J, Linder C (2009) A disciplinary discourse perspective on university science learning: achieving fluency in a critical constellation of modes. J Res Sci Teach 46:27–49. https://doi.org/10.1002/tea.20265

Airey J, Linder C (2017) Social semiotics in university physics education. In: Treagust DF, Duit R, Fischer HE (eds) Multiple representations in physics education. Springer International Publishing, Cham, pp 95–122

Alibali MW, Goldin-Meadow S (1993) Gesture-speech mismatch and mechanisms of learning: what the hands reveal about a child's state of mind. Cogn Psychol 25:468–523. https://doi.org/10.1006/cogp.1993.1012

Alibali MW, Nathan MJ (2012) Embodiment in mathematics teaching and learning: evidence from learners' and teachers' gestures. J Learn Sci 21:247–286. https://doi.org/10.1080/10508406.2011.611446

Bernhard K, Bernhard J (1999) Mechanics in a wheelchair. Phys Teac 37:555–556. https://doi.org/10.1119/1.880405

Brookes DT, Ektina E, Planinsic G (2020) Implementing an epistemologically authentic approach to student-centered inquiry learning. Phys Rev Phy Educ Res 16:20148–20148. https://doi.org/10.1103/PhysRevPhysEducRes.16.020148

Brumec G, Gregorcic B, Planinsic G (2022) Students' use of different representations while communicating in a remote setting. In: GIREP 2022 Conference Book of Abstracts. Ljubljana, Slovenia

Chue S, Lee Y-J, Tan KCD, Boylan M (2015) Iconic gestures as undervalued representations during science teaching. Cogent Education 2:1021554–1021554. https://doi.org/10.1080/2331186X.2015.1021554

Close HG, Scherr RE (2015) Enacting conceptual metaphor through blending: learning activities embodying the substance metaphor for energy. Int J Sci Educ 37:839–866. https://doi.org/10.1080/09500693.2015.1025307

Coletta VP, Bernardin J, Pascoe D, Hoemke A (2019) Feeling Newton's second law. Phys Teac 57:88–90. https://doi.org/10.1119/1.5088467

Cook SW, Goldin-Meadow S (2006) The role of gesture in learning: Do children use their hands to change their minds? J Cogn Dev 7:211–232. https://doi.org/10.1207/s15327647jcd0702_4

Crowder EM (1996) Gestures at work in sense-making science talk. J Learn Sci 5:173–208. https://doi.org/10.1207/s15327809jls0503_2

Dewey J (1916) Democracy and education: an introduction to the philosophy of education. Macmillan, New York, NY

DiSessa A (1993) Toward an epistemology of physics. Cogn Instr 10:105–225

Edwards LD (2009) Gestures and conceptual integration in mathematical talk. Educ Stud Math 70:127–141. https://doi.org/10.1007/s10649-008-9124-6

Etkina E, Brookes DT, Planinsic G (2019) Investigative science learning environment: when learning physics mirrors doing physics. Morgan & Claypool Publishers

Etkina E, Brookes DT, Planinsic G, Van Heuvelen A (2020) The physics active learning guide for teaching on-line, ISBN: 9780137381319. Pearson

Euler E, Rådahl E, Gregorcic B (2019) Embodiment in physics learning: a social-semiotic look. Phys Rev Phys Educ Res 15:010134–010134. https://doi.org/10.1103/PhysRevPhysEducRes.15.010134

Flood VJ, Amar FG, Nemirovsky R et al (2014) Paying attention to gesture when students talk chemistry: interactional resources for responsive teaching. J Chem Educ 92:11–22. https://doi.org/10.1021/ed400477b

Fredlund T, Linder C, Airey J, Linder A (2014) Unpacking physics representations: towards an appreciation of disciplinary affordance. Phys Rev Spec Topics Phys Educ Res 10:1–13. https://doi.org/10.1103/PhysRevSTPER.10.020129

Gallace A, Spence C (2010) The science of interpersonal touch: an overview. Neurosci Biobehav Rev 34:246–259. https://doi.org/10.1016/j.neubiorev.2008.10.004

Goldin-Meadow S (2005) Hearing gesture: how our hands help us think. Harvard University Press, Cambridge, MA

Goldin-Meadow S (2006) Talking and thinking with our hands. Curr Dir Psychol Sci 15:34–39. https://doi.org/10.1111/j.0963-7214.2006.00402.x

Goldin-Meadow S, Wagner SM (2005) How our hands help us learn. Trends Cogn Sci 9:234–241. https://doi.org/10.1016/j.tics.2005.03.006

Goodwin C (2003) The body in action. In: Coupland J, Gwyn R (eds) Discourse, the body, and identity. Palgrave Macmillan, New York, NY, pp 19–42

Goodwin C (2007) Environmentally coupled gestures. In: Duncan SD, Cassell J, Levy ET (eds) Gesture and the dynamic dimension of language. John Benjamins Publishing Company, Amsterdam, pp 195–212
Gregorcic B (2015) Exploring Kepler's laws using an interactive whiteboard and Algodoo. Phys Educ 50:511–515. https://doi.org/10.1088/0031-9120/50/5/511
Gregorcic B, Haglund J (2021) Conceptual blending as an interpretive lens for student engagement with technology: exploring celestial motion on an interactive whiteboard. Res Sci Educ 51:235–275. https://doi.org/10.1007/s11165-018-9794-8
Gregorcic B, Pendrill A-M (2023) ChatGPT and the frustrated Socrates. Phys Educ 58:035021. https://doi.org/10.1088/1361-6552/acc299
Gregorcic B, Planinsic G, Etkina E (2017) Doing science by waving hands: talk, symbiotic gesture, and interaction with digital content as resources in student inquiry. Phys Rev Phys Educ Res 13:020104–020104. https://doi.org/10.1103/PhysRevPhysEducRes.13.020104
Hahn KT, Gire E (2022) Waving arms around to teach quantum mechanics. Am J Phys 90:778–786. https://doi.org/10.1119/5.0073946
Hutchins E (1995) Cognition in the Wild. MIT Press, Cambridge, MA
Hwang S, Roth W-M (2011) The (Embodied) performance of physics concepts in lectures. Res Sci Educ 41:461–477. https://doi.org/10.1007/s11165-010-9175-4
Iverson J, Goldin-Meadow S (1998) Why people gesture when they speak. Nature 396:228–228
Kendon A (2004) Gesture: visible action as utterance. Cambridge University Press, Cambridge
Kendon A (1997) Gesture. Annu Rev Anthropol 26:109–128
Kersting M, Amin TG, Euler E, et al (2023) What is the role of the body in science education? A conversation between traditions. Sci & Educ. https://doi.org/10.1007/s11191-023-00434-7
Kersting M, Haglund J, Steier R (2021) A growing body of knowledge: on four different senses of embodiment in science education. Sci Educ 30:1183–1210. https://doi.org/10.1007/s11191-021-00232-z
Kortemeyer G (2023) Could an artificial-intelligence agent pass an introductory physics course? Phys Rev Phys Educ Res 19:010132. https://doi.org/10.1103/PhysRevPhysEducRes.19.010132
Lemke JL (1982) Talking physics. Phys Educ 17:263–267
MacIsaac D, Whiteboarding in the physics classroom. http://physicsed.buffalostate.edu/AZTEC/BP_WB/index.html
Mahowald K, Ivanova AA, Blank IA, et al (2023) Dissociating language and thought in large language models: a cognitive perspective. https://doi.org/10.48550/arXiv.2301.06627
McNeill D (2005) Gesture and thought. University of Chicago Press, Chicago, IL
McNeill D (1992) Hand and mind: what gestures reveal about thought. University of Chicago Press, Chicago
Merleau-Ponty M (1945) Phenomenologie de La preception. Gallimard, Paris
Morrow CA (2000) Kinesthetic astronomy: the sky time lesson. Phys Teac 38:252–253. https://doi.org/10.1119/1.880520
Nevile M (2015) The embodied turn in research on language and social interaction. Res Lang Soc Interact 48:121–151. https://doi.org/10.1080/08351813.2015.1025499
de Oliveira MHA, Fischer R (2017) Ciênsação: gaining a feeling for sciences. Phys Educ 52:025011. https://doi.org/10.1088/1361-6552/aa5430
Richards A (2019) Teaching mechanics using Kinesthetic learning activities. Phys Teac 57:35–38. https://doi.org/10.1119/1.5084926
Richards AJ, Etkina E (2013) Kinaesthetic learning activities and learning about solar cells. Phys Educ 48:578. https://doi.org/10.1088/0031-9120/48/5/578
Roth W-M (2001) Gestures: their role in teaching and learning. Rev Educ Res 71:365–392. https://doi.org/10.3102/00346543071003365
Roth W-M, Lawless DV (2002a) Scientific investigations, metaphorical gestures, and the emergence of abstract scientific concepts. Learn Instr 12:285–304. https://doi.org/10.1016/S0959-4752(01)00023-8

Roth W-M, Lawless DV (2002b) How does the body get into the mind? Hum Stud 25:333–358. https://doi.org/10.1023/A:1020127419047
Roth W-M, Welzel M (2001) From activity to gestures and scientific language. J Res Sci Teach 38:103–136
Ruiz MJ (2017) Students dance longitudinal standing waves. Phys Educ 52:033006. https://doi.org/10.1088/1361-6552/aa648e
Scherr RE (2008) Gesture analysis for physics education researchers. Phys Rev Spec Topics—Phys Educ Res 4:010101–010101. https://doi.org/10.1103/PhysRevSTPER.4.010101
Scherr RE, Close HG, Close EW, Vokos S (2012) Representing energy. II. Energy tracking representations. Phys Rev Spec Topics—Phys Educ Res 8:020115–020115. https://doi.org/10.1103/PhysRevSTPER.8.020115
Shapiro L, Spaulding S (2024) Embodied cognition. In: Zalta EN, Nodelman U (eds) The stanford encyclopedia of philosophy, Summer 2024. Metaphysics Research Lab, Stanford University
Streeck J, Goodwin C, Lebaron C (2011) Embodied interaction in the material world: an introduction. In: Streeck J, Goodwin C, Lebaron C (eds) Embodied interaction: language and body in the material world. Cambridge University Press, Cambridge, pp 1–26
Tang K-S (2022) Material inquiry and transformation as prerequisite processes of scientific argumentation: toward a social-material theory of argumentation. J Res Sci Teach 59:969–1009. https://doi.org/10.1002/tea.21749
Volkwyn TS, Airey J, Gregorcic B, Heijkenskjöld F (2019a) Transduction and science learning: multimodality in the physics laboratory. Des Learn 11:16–29. https://doi.org/10.16993/dfl.118
Volkwyn TS, Airey J, Gregorcic B, Heijkenskjöld F (2019b) Transduction and science learning: multimodality in the physics laboratory. Des Learn 11:16–29. https://doi.org/10.16993/dfl.118
Volkwyn TS, Gregorcic B, Airey J, Linder C (2020) Learning to use cartesian coordinate systems to solve physics problems: the case of 'movability.' Eur J Phys 41:045701–045701. https://doi.org/10.1088/1361-6404/ab8b54
Whitworth BA, Chiu JL, Bell RL (2014) Kinesthetic investigations in the physics classroom. Phy Teac 52:91–93. https://doi.org/10.1119/1.4862112
(1997) Titanic. Paramount Pictures & 20th Century Fox

General Aspects of Effective Teaching and Learning

Connecting Research in Physics Education, Curriculum Decisions and Teaching Practices

Symposium Organized by GTG Physics Education Research at University-PERU

J. Guisasola, J. Ametller, D. Baccino, M. Carli, S. Kapon, A. C. Marti, M. Monteiro, O. Pantano, T. Peer, P. Sarriugarte, M. Schvartzer, A. Suárez, and K. Zuza

Abstract In the symposium contributions we discuss research in physics education and the consequences of its results for physics teaching. The symposium presents four different aspects of physics teaching and learning, but all of them have research-based problem analysis in common. The problems analysed cover different aspects of the physics teaching–learning process. Innovative aspects such as the effect on learning of the integration of engineering projects in the science teaching process, the influence on the learning process of conceptions about science and attitudes, and aspects related to teaching contents and students' learning difficulties. Its conclusions

J. Guisasola (✉) · P. Sarriugarte · K. Zuza
Department of Applied Physics, University of the Basque Country (UPV-EHU) and Donostia Physics Education Research Group (DoPER), Bilbao, Spain
e-mail: jenaro.guisasola@ehu.e

J. Ametller
Department of Specific Didactics, University of Girona, 17004 Girona, Spain

D. Baccino · A. Suárez
Consejo de Formación en Educación, ANEP, Montevideo, Uruguay

M. Carli · O. Pantano
Department of Physics and Astronomy, University of Padova, Padova, Italy

S. Kapon · T. Peer · M. Schvartzer
Faculty of Education in Science and Technology, Technion –Israel Institute of Technology, Haifa, Israel

A. C. Marti
Instituto de Física, Universidad de La República, Montevideo, Uruguay

M. Monteiro
Facultad de Ingeniería, Universidad ORT Uruguay, Montevideo, Uruguay

T. Peer
Acheret Center—Multicultural Research Fellowship, Kibbutz Cabri, Israel

S. Faletič and J. Pavlin (eds.), *Teaching and Learning Physics Effectively in Challenging Times*, Challenges in Physics Education, https://doi.org/10.1007/978-3-031-72541-8_4

are not merely intuitive proposals based on teaching experience, but on a careful planning of data collection, analysis of results and empirical basis.

1 Introduction

Regarding research results in any field, their transfer into practice is not necessarily straightforward. In the Girep Thematic Group PERU we work in an international cooperation effort to transfer the results of research in physics education to teachers and curriculum designers (Guisasola 2019). One of the principal objectives of the Girep Thematic Group PERU is to share Physics Education Research results and its relevance to physics education and classroom practice. In particular, to suggest ways in which the facilities for the study of physics at introductory physics courses might be improved.

Since 2012 PERU thematic group has organized symposia in Conferences and Seminars with the general objective of bringing together significant experiences and points of view from different areas of the world that are expressed in simple language, with the aim also of encouraging the application of innovative classroom practices and the implementation of research-based teaching initiatives. This chapter presents four different aspects of physics teaching and learning, but which have in common research-based problem analysis. The results of the research suggest the presence of very different factors that influence the teaching of physics and that makes this task complex. This rejects a simplistic conception of physics education that considers it a simple task that would consist of mastering the contents and having 'diplomacy' to deal with the students. On the contrary, as we will see in the symposium, the results accepted by the international community of physics teachers indicate that the task to be developed and the problems to be faced are sufficiently complex to constitute a field of research with multiple dimensions.

The four problems analysed offer a broad overview of the lines of research in PER such as students' attitudes towards the study of physics, students' epistemological beliefs and difficulties in the use of mathematical concepts, formal reasoning and conceptual learning. The common aim of the symposium is to show that the different lines of research in PER are complementary and that they analyse the different aspects that need to be taken into account in the complex task of teaching physics.

In the study developed by Kapon et al. (Sect. 2) students' interest in learning physics in a STEM context is analysed. The research shows the possibilities and, the challenges of involving students in engineering projects as a place for learning physics at the advanced high school level are discussed. The study developed by Arturo C. Marti and colleagues (Sect. 3), compares the attitudes and beliefs about science at the beginning of their university careers of two groups of students: physical sciences and life sciences. Some of the possible causes of the differences found and their implications for teaching are discussed. The research carried out by Ornella Pantano et al. (Sect. 4) and P. Sarriugarte et al. (Sect. 5) is in the more traditional line of PER research on student learning difficulties. Pantano et al. compare students' ability to answer questions about derivatives, integrals and vectors in a purely mathematical context and in the context of physics. The usefulness of the tool for both students and teachers is discussed. Sarriugarte et al. show the difficulties encountered by first-year university students in understanding the moment of inertia in the phenomenon of rotation of a rigid body around a fixed axis.

2 Learning Physics While Engaging in an Engineering Project

Several studies have suggested that the integration of engineering projects into the instruction of science can enhance students' attendance and engagement and support the learning of the traditional content of science. However, some studies have indicated that the learning of scientific content and practices through engagement in engineering projects is not straightforward and requires the additional design of various scaffolds. For example, some studies have reported students' tendency to engage in fulfilling the goals of their engineering design challenges, and only inconsistently engaging with the related math and science content (Berland and Steingut 2016). Others have cautioned that such activities can turn into arts and crafts activities, in which students and teachers focus on getting to a working solution by trial and error that is disconnected from the targeted science.

The problem described above is considered in the literature to reflect the "mixed success" of integrated STEM education (Berland and Steingut 2016). Berland and Steingut argued that students' perception of the value of math and science content for engineering predicts their efforts to integrate math and science content into their engineering coursework. The instructional implications they derived from this conclusion were that teachers and curriculum designers should mediate and emphasize the value of math and science to meeting these engineering challenges (Berland and Steingut 2016).

While we agree that 'value' is an influencing factor, in our view the reasons for the "mixed success" go beyond students' sense of 'value'. We suggest a complementary explanation situated in the very essence of the activity the students engage in. We will present findings from a recently published study (Kapon et al. 2021) that illustrate how learning physics by engaging in an engineering challenge generates tensions related to fundamental incongruencies between doing engineering and doing physics, and how these tensions affect students' learning. This explanation entails somewhat different instructional implications, which will be discussed in the concluding section.

2.1 Conceptual Framework and Goals

We examine the nature of learning physics through engagement in an engineering-maker project as a case of participating in a particular figured world. Figured worlds are "socially and culturally constructed realms of interpretation, in which particular characters and actors are recognized, significance is assigned to certain acts, and particular outcomes are valued over others. Each is a simplified world populated by a set of agents /…/ who engage in a limited range of meaningful acts or changes of state /…/ as moved by a specific set of forces." (Holland et al. 1998). We argue that the figured world of Engineering Maker-Based Inquiry (EMBI) in physics differs from the figured world of authentic scientific inquiry in physics and the figured world of traditional learning of physics in school, and that these differences have implications that affect students' learning of physics.

2.2 The Study

The data were drawn from an extended ethnographic case study (18 months, 3 h a week) of the authentic working sessions of two 11th grade students who were working on a long term engineering challenge as part of their mandatory learning of physics at the advanced high school level (Kapon et al. 2021). The physics teachers in the school in which the study took place are part of the Acheret Center (ACHERET 2005). Acheret is a community of physics educators that aims to incorporate authentic physics inquiry in schools. The project represented 40% of the requirements for matriculation in advanced level physics. The two students who worked together on the project were very different in terms of their academic achievement. The male was a low achieving student and the female was a high achieving, ambitious student. The students worked on the project in the school makerspace, and were mentored by a physics teacher (3rd author) who was trained in Acheret to serve as a research mentor in this context. The students and the mentor decided together that the goals of the project would be to design and build a working model of a solar panel that tracks and follows the movement of the sun and evaluate its performance by comparing the power efficiency of a stationary and tracking solar panel. Figure 1 presents an overview of the project.

The participants in the study were the two students and the educational staff: the teacher who mentored the students' inquiry project (a senior member in the Acheret community), the maker-space coordinator who assisted the teacher and served as the laboratory technician, the physics department head of the school (a senior member of Acheret) who helped the students to wrap up the project and write the final report in the second year when their mentor was on sabbatical, and the students' "regular" classroom physics teacher (a junior member of Acheret), who also stepped in to help during the second year. The data included videotaped observations of the authentic working sessions every second or third week over a full school year, including the

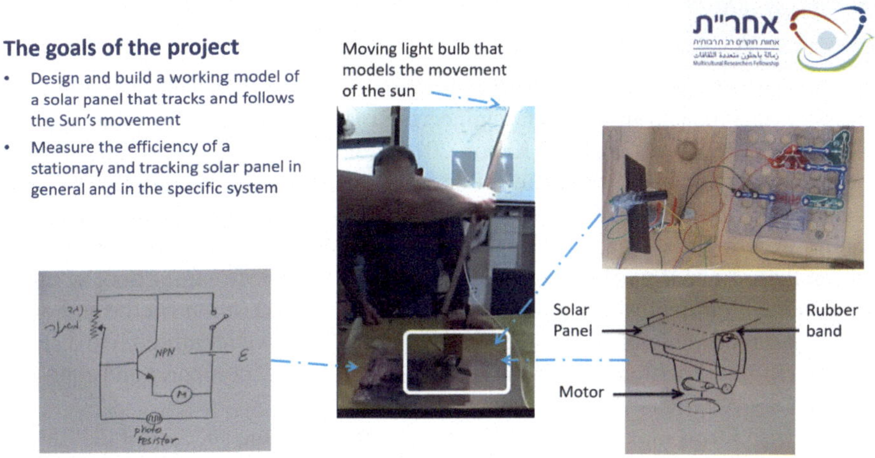

Fig. 1 Learning physics while engaging in an engineering project

oral exam in the second year, interviews with the students and the educational staff, and the artefacts that were generated during the inquiry (notes, final research report, etc.).

The analysis combined ethnographic accounts and fine-grained socio-linguistic discourse analysis of selected episodes. The discourse analysis of mentor–mentee interactions in the authentic working sessions was complemented by interviews and other ethnographic accounts. In a full paper that summarizes the analysis of this case study [8] we (1) characterized the figured world of EMBI, (2) identified the central legitimate forms of participation that were enacted by the students and influenced their learning, (3) articulated how these forms of participation were socially communicated, constructed, and enforced over time in the interaction between the two students and the educational staff, and (4) examined how these forms of participation facilitated (or impaired) the learning of the content and practices of the related physics.

In this chapter we use this analysis (Kapon et al. 2021) and other case studies we have conducted in physics classrooms at the Acheret center (Kapon 2016) to compare and contrast the figured world of EMBI with the figured world of authentic inquiry in physics in schools associated with Acheret Center, and with the figured world of physics in school which also shaped the students' experience. Then we compare and contrast two legitimate forms of participation in EMBI that we identified in the case study [8] and discuss the implications for the teaching and learning of physics.

2.3 Analysis

Table 1 compares the goals and the meaningful acts in the three laminated figured worlds we documented in Acheret: engineering maker-based inquiry (Kapon et al. 2021), authentic physics inquiry (Kapon 2016), and traditional school physics. As Table 1 shows, while there are some overlaps between the meaningful acts of EMBI and authentic scientific inquiry, there are also many differences in the goals and meaningful actions that can generate tensions, and there is a considerable difference between these figured worlds and the figured world of mainstream school science. Based on the sociolinguistic discourse analysis of the conversations in the authentic working sessions of the EMBI project (Kapon et al. 2021), we identified two distinct legitimate forms of participation in the EMBI project. Table 2 compares and contrasts them.

While both forms of participation contributed greatly to the success of the EMBI, only the participation as an engineer contributed to the students' learning of the

Table 1 Laminated figured world

	Engineering maker-based inquiry in physics (Acheret)	Authentic scientific inquiry in physics (Acheret version)	Learning physics in school
Goal	Design and build a working engineering device, evaluate its performance and improve it accordingly	Produce reliable knowledge and explanations about the physical world	Pass the external matriculation exam in advanced level physics, for which the project counts 40%
Meaningful acts	• Characterize what the device should do and translate these characteristics into scientifically measurable features • Design the device by employing scientific principles as well as incorporating practical considerations such as the availability of materials, how easy it is to work with them, etc • Build a working porotype, engaging in its construction and troubleshooting Design and perform experiments that test the performance of the device and improve the design if necessary	• Generate theory and quantitative models that can explain and produce testable hypotheses, and iteratively improve the models • Design experimental set up, build and improve it • Learn to use and choose wisely between a variety of measurement and data processing tools, by applying considerations such as uncertainty, direct vs. indirect measures, etc • Deal with ambiguity and gaps between the experimental results and theory Present the findings in a way that peers and experts can understand and evaluate them	• Attend working sessions in the laboratory • Submit a research proposal and final research report for external evaluation • Defend the work in an externally evaluated oral exam Present conceptual proficiency and understanding

Table 2 Legitimate forms of participation in EMBI

Participate as an engineer	Participate as a technician
Engage in the following actions while employing scientific as well as practical considerations: • system characterization and design • troubleshooting the system • testing system performance (designing and conducting experiments, interpreting results) • achieve a robust understanding of the underlying scientific principles and employ them in the design and evaluation of the system	The engagement is limited solely to practical and functional considerations • construction • repair of technical faults (disconnected wiring, mechanical jamming and the like) • receive rather than generate explanations • understand the scientific principles underlying the system solely at a functional level

related physics in terms of the figured world of learning physics in school, as well the figured world of authentic scientific inquiry. Our longitudinal analysis [3] shows that at the beginning of the project both students were participating as engineers as well as technicians. However, as the project progressed and developed, the high achieving student mainly participated as an engineer, while the participation of the low achieving student was limited to participation as a technician. As the year unfolded, the low achieving student gradually became an apprentice to the laboratory technician instead of to the research mentor (Table 3).

Our findings show that the low achieving student developed a deep sense of ownership and pride about the project. He contributed significantly to it by constructing and stabilizing the physical model of moving "sun" (see Fig. 1), and contributed to the mechanical construction of the moving the solar panel. However, our data also demonstrate that he did not understand some of the very basic physics principles that underpinned the engineering design. The student referred to this dissonance in an

Table 3 Categorization of question Q1. Name and description of each category are shown in the first and second columns from the left respectively. The column on the right show the percentages of the answers classified in each category (N = 97)

Cat	Explanation	% of answers
A1	They understand that the moment of inertia is related to the difficulty/ease of rotating a system and calculate it correctly	34
A2	They understand that the moment of inertia is related to the difficulty/ease of rotating a system but they do not compute it correctly	11
B	Arguments based on the idea that the difficulty of rotation depends on the mass and/or the distance to the axis	22
C	Affirmations based on rote resources	24
D	Incoherent	4
E	No answer	4

interview after the oral exam: "He (the external examiner) did not ask about the practical side. I was really disappointed /.../ because we worked on it, and prepared the axes, and the angles, and we cut the wood. This is the part I loved. The part I enjoyed most, that was most mine. So I was little disappointed." His partner also brought it up in the last interview: "/.../ The truth is that in this project, ah, Ram was more, he was doing the basic technical work, because... Well, now, in a retrospect, this does not seem fair. I knew electricity, so we ((the student and the research mentor)) kind of did this part quickly, and left the explanations to Ram to the end. So.../It was much more fun doing it with Ram, although he did not do much in the physics part." We interpret this as an unequitable opportunity to engage in doing physics.

2.4 Conclusion

Engineering maker-based inquiry offers inclusive pathways into physics that can nurture an emotional connection to physics for a more diverse body of learners. However, while supporting students' sense of belonging, which is a worthy goal in itself, it does not necessarily facilitate meaningful learning of physics. To achieve this, its implementation in physics classrooms requires careful pedagogical design that engages *all* the students as engineers and makes sure that no student will participate solely as a technician.

3 What Are the Differences in the Attitudes and Beliefs About Science of Students in the Physics-Mathematics and Life Sciences Areas and What Are Their Impact on Teaching?

It has long been accepted in the physics-teaching community that, in addition to the specific contents, the set of ideas, assumptions and previous conceptions about science, in general, the epistemological beliefs, strong influence the transformation that occurs as students move through the educational system. This specific field of epistemological attitudes and beliefs does not escape the claim of researchers to obtain quantitative information. In that sense, in the last twenty-five years a series of questionnaires or tests standardized to assess the epistemological attitudes and beliefs of the students have been proposed in the literature. There are many open questions in relation to the attitudes and beliefs of students, especially those related to the comparison between different groups, the causes of possible differences and their impact on learning. Several quantitative studies have been performed in recent years on the changes that occur in students' attitudes and beliefs depending on their previous training, the types of courses, and the teaching strategies used, as well as on the relationship between the pre-test and academic performance, among other aspects. In this section we address results recently published in the specialized literature about this topic.

3.1 *Attitudes and Beliefs of Physical Sciences Students in Comparison with Other Groups*

One aspect regarding attitudes and beliefs about science that has received considerable attention is the comparison of the performance of different student groups as a function of the academic options adopted in the early stages of their careers. In a recent study (Suarez et al. 2022), the attitudes and beliefs about science of students of physical sciences (physics and mathematics) were compared with those of life sciences (biology and biochemistry) students at the beginning of their university degrees using the CLASS toorl (Adams et al. 2006). It is worth noting, that both groups received similar physics courses during their high-school education. The differences in performance in each of the areas that make up the questionnaire were examined and, perhaps the flashiest result, is that a larger percentage of life science students (higher than that of physical science students) adopted a "novice" behavior in problem solving.

This research was carried out with university students from the School of Sciences of Universidad de la República (Montevideo, Uruguay) who were taking General Physics I in the first semester of the undergraduate courses in Physics and Mathematics and Biology and Biochemistry. It is relevant to point out that in Uruguay, primary, secondary and pre-university schools are characterized by a common curriculum framework for all the educational institutions in the country. The usual bibliography includes algebra-based "College Physics" textbooks such as those widely used worldwide (Serway, Resnick among others). Upon completing high school, students can opt for different university degrees. The focus of this work was on two sets of recently admitted students in the first year of the aforementioned School of Science. One set is comprised by those who pursue bachelor's degrees in Physics and Mathematics (hereinafter, "physical sciences" or "PhS") and the other by those who pursue bachelor's degrees in Biology and Biochemistry (hereinafter, "life sciences" or "LS").

The authors proposed to all the students to report online their degree of agreement or disagreement with the 42 statements of the CLASS test according to a Likert scale. Some examples of the statements are the following:

> A significant problem in learning physics is being able to memorize all the information I need to know.
>
> It is useful for me to do lots and lots of problems when learning physics
>
> After I study a physics topic and feel I understand it, I have difficulty solving problems on the same topic.
>
> When solving a physics problem, I look for an equation that uses the variables given in the problem and I substitute the values.

I enjoy solving physics problems.

Of a total of 42 statements, 27 of them are grouped into 8 categories: (a) Real world connection, (b) Personal interest, (c) Sense making/Effort, (d) Conceptual comprehension, (e) Applied conceptual understanding, (f) Problem solving/general, (g) Problem solving/confidence, (h) Problem solving/sophistication. The remaining 15 questions are not categorized. There is no agreement among the experts on some of the latter, while one in particular is used to rule out inconsistent responses. A question can be categorized in more than one category. For each category, and for the set of questions where there is agreement among the experts (36 of the 42 statements), the percentage of student responses that agree with that of the experts (favorable responses) were reported and evaluated using statistical tools as the non-parametric Mann–Whitney U test, the Cohen effect size and contrasted with a null hypothesis to be rejected if the probability is smaller than a given threshold.

The differences in performance in each of the areas of the questionnaire reveals some interesting aspects. Among other findings stands out that a considerable percentage of life science students (higher than that of physical science students) adopted a novice type of behavior in problem solving. From the analysis of the responses in the different categories, the authors highlight that a significant percentage of life sciences students try to solve physics problems following the strategy of "finding the right equation and substituting", known as "plug and chug". Being aware of this type of thinking allows us, as teachers, to anticipate the problems of our students, to stand differently in the classroom, and to develop actions aimed at changing this type of thinking, which will result in a better-quality science education. In this sense, it is important to keep in mind that in every action or omission that we make in the classroom (and that is part of the hidden curriculum), we are directly or indirectly affecting the epistemological beliefs of our students.

3.2 Physics (and Future) Teachers' Attitudes and Beliefs About Science

The epistemological conceptions of the Uruguayan Physics teachers and future Physics teachers were also analyzed through the application of the CLASS test (Suarez et al. 2021). This study was focused on identifying categories in which there are significant similarities or differences between the two groups and also in comparison with the experts' opinions taken as reference. The importance of this study is related to the fact that aspects which exhibit positive or negative variations between the opinions of future teachers and present teachers suggest that these specific training resulted favorable or unfavorable. Moreover, the areas where the divergences with the opinions of the experts are notorious indicate that they should be reinforced.

The survey was open online to physics teachers and students from all over Uruguay for a month in the first half of the (pandemic) 2020. In relation to the teachers an electronic announcement was sent to almost all active Physics teacher in the country. The responses received were 143, nearly equally distributed by genre. The age of the respondents was distributed as 31% between 31 and 40 years, 22% declare in the age groups contiguous to the previous one (18–30 and 41–50), 25% register over 51 years of age. Regarding future physics teaching, first-year students of the Physics schools were contacted directly by their teachers in several of the schools. 138 students participated, approximately two thirds women. All the responses were analyzed using the electronic spreadsheets available in the well-known site *Physport*.

Several points can be outlined from this study. The first one is that, not surprisingly, there is a significant difference in the averages of teacher and student responses (80% for teachers and 67% for students). The categories that present the most agreement with the reference opinion are the same for teachers and future teachers. These categories are Sense making/Effort and Confidence in problem solving. In contrast, the greatest observed differences can be found in the categories Applied conceptual understanding and Sophistication in problem solving. Assuming that it is possible to consider the two groups of surveyed individuals as part of a pseudo-longitudinal approach, these differences could be hypothetically attributed to the systematic training processes and the teaching–learning experiences.

Although this study provides interesting details several open questions about the attitudes and beliefs about science of Physics teachers remain open. Some of these questions could be at what stage do student teachers transform their epistemological beliefs, approaching that of experts. It would be also interesting to know how do these attitudes impact in the teacher training processes and how do they transmit this grounding to their students.

3.3 A Comparison with Other Studies Recently Published

As mentioned before several studies about attitudes and belief about sciences were published in the literature. A meta-analysis published in 2015 summarized 24 results available until this date based on the application of the CLASS and MPEX tests in several universities, mainly from U.S.A, Canada and United Kingdom, to students who took calculus-based physics courses. Among the main points highlighted in this meta-analysis, we can mention that students with expert-like opinions tend to choose physics majors more probably than other options. Also, there is some correlation between incoming beliefs and their gains on conceptual tests. Regarding the overall punctuation obtained in the CLASS test in these international studies and those mentioned in the previous section the comparison does not suggest large differences.

Another study in relation to the attitudes and beliefs of physics teachers was conducted in Thailand, a country with very different demographic, social, economic and cultural characteristics from those of Uruguay. In the study the responses of 196 physics teachers and 211 secondary students were analyzed. Despite the differences mentioned between the contexts of both teaching groups, the answers were very similar in the majority of the questions and categories of the CLASS. Only in two of the 42 questions of the test, significant differences are found between teachers from Thailand and from Uruguay. A similar situation was found in relation to physics teacher students in these countries (Suwonjandeen et al. 2018) who also show great differences with the experts precisely in the same questions which, in turn, present the highest degree of discrepancy between the responses of physics teachers and physics teacher students.

Very recently, a study about the epistemological beliefs and attitudes about science among Malaysian students was published (Ibrahim et al. 2022). This study was motivated by the poor student's performance and it discuss the role of the epistemological beliefs in this phenomenon. With this objective a specific tool was developed adapting the MPEX and CLASS tests. These findings suggest that positive epistemological beliefs correlate with a better performance.

To conclude this section, a couple of remarks can be stated. First, that all the previous works suggest that attitudes and beliefs about sciences play a significant role in the teaching and learning of physics. Although this role is often neglected ("the hidden curriculum"), it is manifested in several stages. Secondly, the attitudes and beliefs of physics teachers strong influence those of the students. Finally, it is clear that there are many open questions about this topic that deserve to be addressed.

4 Test of Calculus and Vector in Mathematics and Physics: A Research-Based Tool for Improving the Teaching and Learning of Physics in First-Year Courses

4.1 Background

Mathematics has a paramount importance in physics education, not only because it provides the language and the technical tools for describing and making predictions about phenomena, but also due to its structural role (Uhden et al. 2012). However, in many Science and Engineering courses, mathematics is seen merely as a prerequisite for physics. Courses like calculus and algebra are supposed to provide students with mathematical tools, which students are then expected to proficiently use in their physics courses in the following semester. This belief contrasts with findings in physics education research, which show quite clearly that using mathematics in physics is different from using it in a purely mathematical context. Even students who complete a calculus course successfully may have difficulties in using the same mathematical tools in a physics context (Redish and Kuo 2015).

With the aim of raising awareness on the issue and supporting students in overcoming these difficulties, in 2018 we designed an assessment tool, the *Test of Calculus and Vectors in Mathematics and Physics* (TCV-MP). Its development and validation were described in detail in a previously published paper (Carli et al 2020). In this contribution, after retracing the history of the instrument, we describe how its use has evolved at our University and beyond through the years.

4.2 Test Design and Validation

The TCV-MP was developed in the framework of a bigger initiative called "FisicaMente", aimed at supporting first-year Science and Engineering students as they approached their "Physics 1" course. The test focuses on three mathematical tools—derivatives, integrals, and vectors—which are pivotal for any university-level physics course, and with which students have recognized difficulties. These challenges were confirmed by a survey conducted in the initial phase of the project, involving physics instructors.

Given the large target population (students from 23 courses), the instrument was designed as a multiple-choice test. Physics Education Research (PER) informed the design of the items, in combination with an analysis of students' exams. From this analysis we identified the main students' difficulties related to the use of derivatives, integrals and vectors, and we related them to those documented in the literature to identify the specific subtopics and distractors for each item. We drew upon both existing concept inventories and topic-specific literature (for further details see Carli et al. 2020).

For each subtopic, different representational formats (words, graphs, formal language, and numbers) were considered, as students may adopt different problem-solving strategies depending on the representational format used in the problem (Ibrahim and Rebello 2012). Multiple versions of each item were developed, involving different combinations of the representations used in the question and answers. Finally, for each subtopic and combination of representations two parallel items were constructed, one in the context of mathematics and one in the context of physics, in order to compare students' answers in the two contexts. The chosen physics context was mechanics, as it is the common core of all Physics 1 courses. Through this process we constructed an initial pool of 78 items, which were administered to a pilot group of 70 students. Following item analysis and student interviews, 17 pairs of parallel items were selected and included in the final version of the test. The test was administered to first-year students in Science or Engineering at the beginning of the Spring semester in 2018. Participation was voluntary and physics instructors were in charge of students' involvement. In total, 1252 students completed the test.

4.3 Results from the First Administration of the Test

We analyzed students' answers both at the whole test level and for individual items. As an example of the results that we obtained, here we discuss students' answers on items 2 and 3, regarding the topic of derivatives. In item 2M (context of mathematics), students were given the graph of a function and were asked to select the correct graph displaying its derivative; in its parallel physics item (2P), a position-time graph was provided, and students had to select the corresponding velocity–time graph. In item

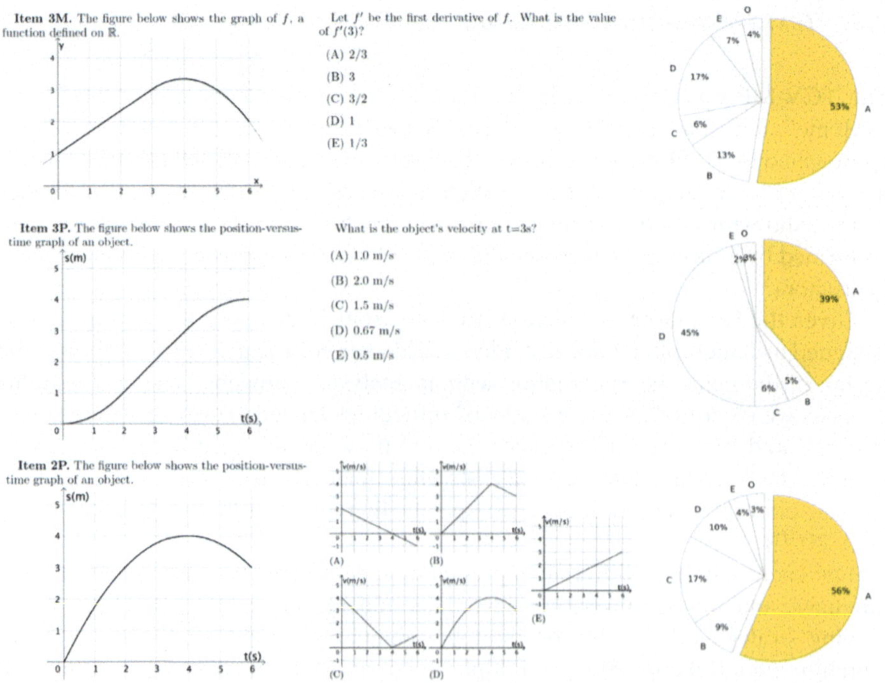

Fig. 2 Comparison of students' answers across two parallel mathematics/physics items (items 3M and 3P) and across two items regarding the same topic, but with different representations (items 3P and 2P)

3M, students were tasked with calculating the first derivative of a function at a specific point based on its graph; in the parallel physics item (3P), they had to calculate an object's velocity at a given instant based on its position-time graph. In Fig. 2 we compare students' answers in items 3M, 3P and 2P.

The answer profile was quite different between items 3M and 3P, with many more students choosing the most common incorrect answer (option D, corresponding to calculating y/x or s/t) in physics than in mathematics. This result suggests that the mathematics content of the item was only part of the problem; instead, the common incorrect conceptualization of velocity as "space over time" probably played a crucial role in students' selection of distractor D in physics. On the other hand, the comparison between items 3P and 2P offers some insights into the role of different representations. The percent of correct answers in item 2P, which did not involve calculation, was higher than in answer 3P and more aligned to that of item 3M. It was also comparable with item 2M (not shown, 64%).

4.4 Evolution Beyond the Initial Administration

Our results support the need, advocated by PER, for the explicit training of students' mathematization skills in physics. As a possible means of addressing this need, in 2019 we developed online learning modules, available to students upon completing the test. Organized as "chapters" within a Geogebra book, each module corresponds to one of the topics and encompasses an introduction, a series of interactive exercises similar to the test items, and supplementary exercises. An illustrative example is presented in Fig. 3. Here, students can move a slider on a position-time graph, observe the tangent line at that point, visualize its slope, and see how the velocity–time graph is constructed as the point is moved on the graph. They can then use the graph to answer a set of multiple-choice questions (not shown). A parallel item in a purely mathematical context was provided just before.

Since 2018, the test is made available for students every year at the beginning of their Physics 1 courses. For students, the test serves as a self-assessment tool, based on which they can identify the areas requiring more practice. The online modules are also provided as self-training tools. The results of the test are shared with the instructors of each course, so that they can gain insights into the situation of their class and possibly devise interventions. As an example of the information shared with the instructors, Fig. 4 shows the scores distribution for the degree courses that participated in 2022 and a single-item analysis for one of the degree courses (Mechanical Engineering).

The TCV-MP has become an established tool offered to Science and Engineering students and instructors at our university. However, the number of participants has

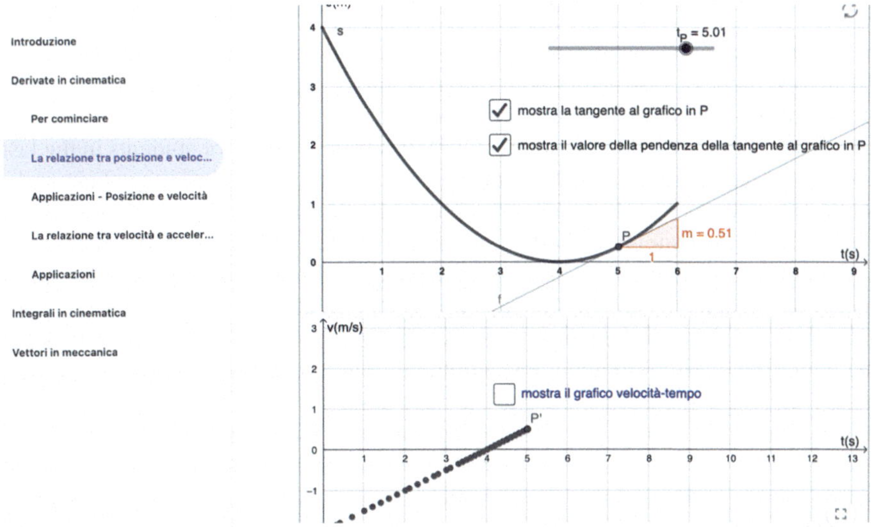

Fig. 3 An example from the interactive online learning modules developed from the TCV-MP test

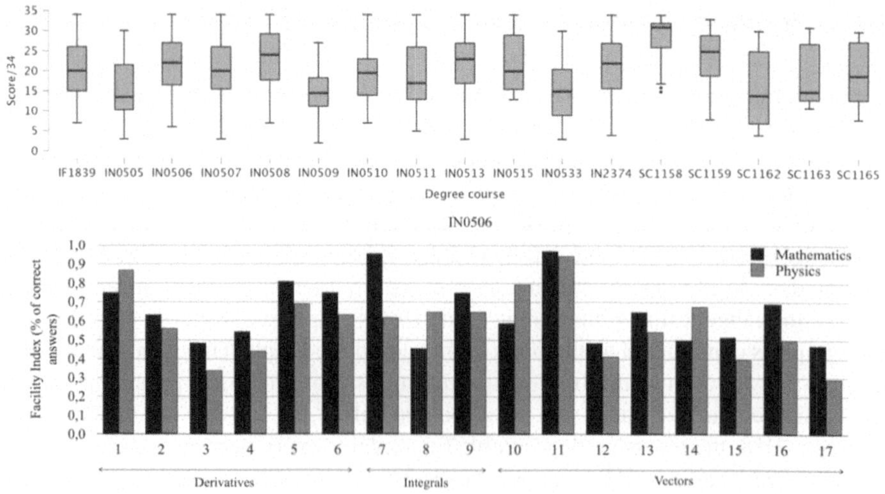

Fig. 4 Boxplots representing the scores distribution for the different degree courses that participated in the 2022 administration of the TCV-MP. Codes beginning with 'IN' identify courses belonging to the school of Engineering (e.g. IN0506 is Mechanical Engineering), while codes beginning with 'SC' identify courses belonging to the school of Science (e.g. SC1158 is Physics). 'IF' identifies an interfaculty course

lowered since the first administration, stabilizing at approximately 700 students per year. The main factor affecting participation is the presence or absence of entire large-enrollment courses, which strongly depends on instructor engagement. While some colleagues consistently propose the test and encourage student participation in different ways, others are only superficially interested. The instructors are continuously changing and many of the new ones have not participated in the initial process. The involvement of instructors with different ideas about physic teaching and the role of physics education research remains a challenge.

In 2019, a modified version of the TCV-MP was developed for students in the last two years of secondary school (12-13th grade), with the aim of reinforcing their mathematization skills before they enter university (Lipiello et al. 2022). While the section about vectors was maintained with minor modifications both in the mathematics and in the physics context, the section on derivatives and integrals in the mathematics context was adapted to suit the students' level. In particular, derivatives were conceptualized as the slope of the tangent line, while integrals were conceptualized as the area under the curve. Following test administration, a learning path was organized employing online modules similar to the ones described above, together with other strategies designed for the specific student population. The test was then administered again as a post-test to evaluate the efficacy of the learning path. The pilot version of the project was conducted in 2019/20 with a group of self-selected students and was reiterated in 2020/21 with a more representative group. Figure 5 shows the scores distribution for the pre- and post-tests. The results are promising and support the effectiveness of the learning modules in fostering students' mathematization skills.

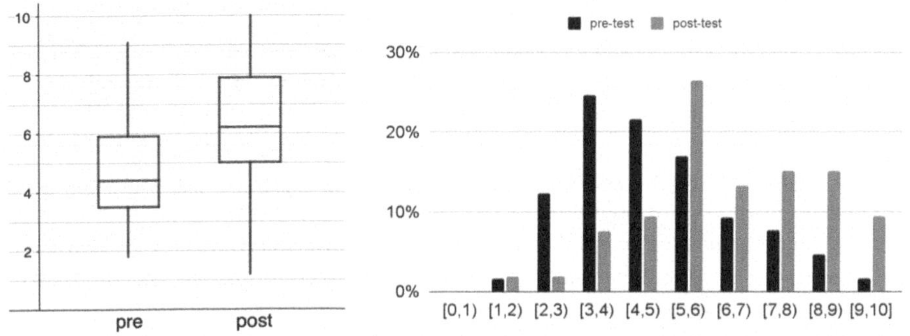

Fig. 5 Comparison between pre- and post-test results for the high-school version of the TCV-MP

The test and the learning materials were used again in 2021 with 174 students from 7 classes (article in preparation). In 2022, a project was launched to extend the experimentation beyond the pilot school. Currently, five schools (including the original one) are experimenting the test and the learning path with different classes.

4.5 Conclusions

The project described here is part of our effort to improve students' mathematization skills in physics. We believe that the TCV-MP can be a valuable resource for both students and instructors. Combining the test with the online learning modules can further help students develop the mathematization skills they need to successfully engage in physics learning.

5 University Students' Reasoning for Understanding of Moment of Inertia

5.1 Introduction

Undergraduate physics courses typically begin with mechanics, which is a review of high school physics for most of the students and one may expect that students will not have significant difficulties with it. However, as it has been shown, understanding the kinematics and dynamics of circular motion present difficulties for high school and university students (Klammer 1998). Therefore, students' difficulties grow up when a particle is substituted by a rigid body.

The research on the students' understanding of rigid body rotation at the University level has shown many difficulties. Frequently, students mix the concept of torque with the concept of force. In the same way, they do not establish any relation between the torque and the angular acceleration, they think that a constant torque results on a constant angular velocity and they show difficulties when taking into account the line of action of a force besides its point of application (Duman et al. 2015).

There is little research on rotational dynamics of a rigid body and its moment of inertia. This investigation shows that students often have difficulties in correctly relating the magnitude moment of inertia to other quantities involved in rotational dynamics such as mass or angular acceleration (Rimoldini and Singh 2005). However, we have found no research that deals with the understanding of the qualitative meaning of the moment of inertia in relation to the rotational motion of a body, nor with the relationship between the use of the analytical definition of the moment of inertia and the meaning with which it is used. Therefore, in the presented investigation we analyse the understanding of students in introductory physics courses of the qualitative meaning of the moment of inertia as the "resistance" of the body in changing its state of rotation and its relation to the use of the operational definition in familiar situations in academic contexts.

5.2 *Experimental Design and Methodology*

We conducted a study with students from first course of engineering of the University of Basque Country (UPV/EHU). All students take two physics courses involving topics on mechanics electromagnetism during postcompulsory education (16–18 years old). The students were randomly distributed among the first-year engineering groups. We gave students an open-ended questionnaire (7 questions) after they had studied the rigid body rotation, where rotation was considered to be around a fixed axis in space.

In order to validate the questionnaire, once it was prepared, three professors concluded that the objectives of each question were the ones we aimed for and were clear. Moreover, we carried out a draft test with 25 first-year course students, which confirmed that students had no problem understanding how the questions were formulated. Finally, the questions were included in the first-year students' test in the form of a post-test for first-year students who had already completed the semesters of Mechanic. Students' responses were analysed and classified in different explanatory categories that emerge on the analysis. The categorization process was validated by three researchers, and the classification of students' answers into categories was supported by Cohen's kappa ($\kappa = 0.92$). The questionnaire was given to 97 first-year university students.

The questions on the questionnaire are familiar to students in the academic context and are usually mentioned in textbooks as examples of rigid body rotation. In Fig. 6 we present one of questions related with the qualitative comprehension of the moment of inertia as the resistance to change the rotational motion of a body. This question will serve as an example to show how we carried out our investigation on students' understanding of the moment of inertia.

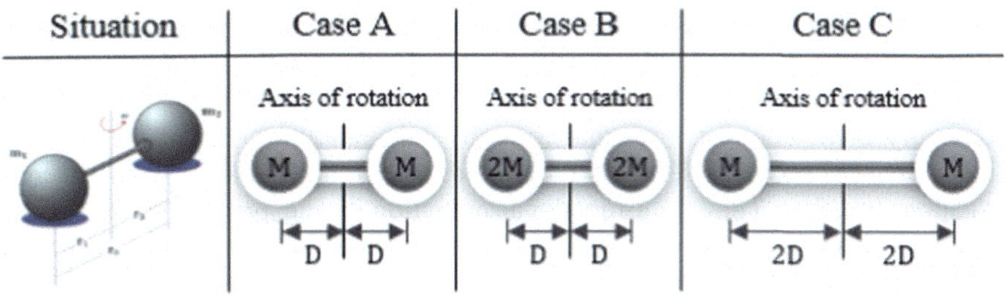

Fig. 6 To inquire about moment of inertia, a rotational motion of two balls joined by a rod is proposed to students. They need to analyse the moment of inertia of the system in three different cases and sort those cases according to easiness to rotate with respect to the axis. The concept moment of inertia is not mentioned in the question

Question Q1. Let us suppose a situation where two balls with the same mass and the same radius joined by a rod of negligible mass can rotate with respect to an axis located in the middle of the rod. The following three cases are presented:

In which case will it be easier to rotate the system? Sort the cases from greatest to least according to how easy it is to rotate the system. EASIEST ____ ____ ____ HARDEST. Argue your answer.

Question Q1 aims to investigate whether students understand the meaning of the moment of inertia in a rotating particle system. According to the theoretical framework of classical physics, the moment of inertia measures the resistance that a body that rotates around an axis opposes to the change of rotational motion and is defined as: $I = \sum m_i r_i^2$. The moment of inertia about an axis has an analogous role to that of mass in translational motion.

The three different cases in the question Q1 differ the distance from the balls to the axis of rotation and the balls' mass, so that the students can make an analysis when justifying their choice of ordering the cases according to the analysis of variables that influence on the change of rotation, values of moment of inertia and its meaning.

5.3 *Results*

We first present the five categories that emerged from the analysis of the answers to question Q1 regarding the concept of moment of inertia (see Table 1). Then, we show a summary of the results, highlighting the main findings and discuss their implications.

Category A includes explanations that imply an understanding of moment of inertia as a resistance of a body to change its rotation state. We separated this category into two subcategories A1 and A2. Most of the answers in category A take into account both variables that influence the moment of inertia explicitly, the mass and the distance to the rotation axis, in a proper way according to the definition of moment of inertia $I = \sum m_i r_i^2$ (subcategory A1). The other responses in category A make an explicit qualitative analysis of the variables without putting any equation in, they only indicate the calculation of the equation to justify their answer or they do not get the calculation right due to some small mistake (subcategory A2).

Category B include responses that correctly calculate the value of the moment of inertia but do not analyse the two variables, mass and distance from the axis of rotation, together. They do a functional reduction giving priority to one of the variables, without justifying why is one of them prioritized over the other one. This could be related with linear causal explanations that establish a chain of simple causes in which any change is the result of a previous change, and in turn, is the cause of the subsequent effect, For example:

"The axis of rotation will be on the centre and the balls on the edges, the closer the balls from the axis the easier will be to turn the system." (Student No. 3).

Around 25% of the answers (see Table 1) were explanations that focus their analysis on one of the two variables (mass and distance to the axis of rotation) that influence the moment of inertia and were classified as category C. The reductionist analysis of the moment of inertia, in this case, leads to incorrect answers. These incorrect answers reflect a lack of understanding of the concept of moment of inertia.

"The closer the turning objet from the axis of rotation, the harder it will be for the system to turn because of its smaller moment of inertia. If the distance between the axis of rotation and the object is the same, the smaller the mass the greater its moment of inertia, then it is easier to turn the system." (Student No. 5).

5.4 Discussion and Conclusions

Our study includes individual student's questionnaire using open-ended questions for inquiring students' learning on momentum of inertia. Here we have presented one of the questions on the questionnaire, but the student difficulties we have described are common to other questions. We have described some common mistakes students make when answering questions about moment of inertia in a context of rotation of a particle system (Q1). A third of the students answer question Q1 correctly (category A1), and we understand that about half of the students show a correct understanding of the moment of inertia and its meaning (categories A1 and A2). However, we note that many others have difficulties in analyzing the influence that moment of inertia has on a body's rotation. Many students do not identify the moment of inertia as a quantity that measures the resistance of the system to the change of rotation, and then it is likely that they have difficulties in establishing relationships between moment of inertia of a body and its angular acceleration of rotation. It occurred for around a third of students in Q1 (category C). Furthermore, some students, although a minority of cases, correctly indicate the value of the moment of inertia, the lack understanding of its meaning leads to erroneous reasoning patterns. We find valuable the idea of moment of inertia as the resistance of the system to the change of rotation, because the learning difficulties with the idea of "inertia" also extend to similar ones with linear motion in high school students (Lehavi & Galili 2009).

The study also shows that there can be a learning progression about calculating and understanding the moment of inertia in comparison with high school students (24% for Q1) (Olazabal et al. 2021). Beginning with students who are not able to calculate the moment of inertia, nor understand its meaning (categories C), going through other students who know how to calculate it from the analysis of the mass and distance to the axis of rotation or either qualitatively understand the meaning of moment of inertia (category B). Finally, those who understand the meaning of the moment of inertia and are able to calculate it and understand it properly as a resistance to change body's state of rotation (category A).

In traditional instruction, it is first taught to calculate the moment of inertia separately as a mathematical integration problem and later, the moment of inertia is analyzed in the context of Newton's second law for rotation. We show that a significant part of students does not relate the moment of inertia with the analysis of change in rotational motion of a body. This suggest that instruction could be more effective if the moment of inertia is taught from the beginning in relation to its meaning of opposing the change of rotation of the system and, being aware the problems that the students will have in the study of the second Newton's law of rotation.

References

ACHERET (2005) ACHERET Center. https://www.acheret.org.il/

Adams WK, Perkins KK, Podolefsky NS, Dubson M, Finkelstein ND, Wieman CE (2006) New instrument for measuring student beliefs about physics and learning physics: the colorado learning attitudes about science survey. Phys Rev Spec Topics Phys Educ Res 2(1):010101

Berland LK, Steingut R (2016) Explaining variation in student efforts towards using math and science knowledge in engineering contexts. Int J Sci Educ 38(18):2742–2761

Carli M, Lippiello S, Pantano O, Perona M, Tormen G (2020) Testing students ability to use derivatives, integrals, and vectors in a purely mathematical context and in a physical context. Phy Rev Phys Educ Res 16:010111

Duman I, Demirci N, Sekercioglu A (2015) University students' difficulties and misconceptions on rolling, rotational motion and torque concepts. Intern J New Trends Educ Implic 6(1):46–54

Guisasola J (2019) Based alternatives to traditional physics teaching at university and college. Upgrading physics education to meet the needs of society. Springer, Cham, pp 127–131

Holland DC, Lachicotte W, Skinner D, Cain C (1998) Identity and agency in cultural worlds. Harvard University Press

Ibrahim B, Rebello NS (2012) Representational task formats and problem solving strategies in kinematics and work. Phys Rev ST Phys Educ Res 8:010126

Ibrahim N, Damio SM, Zulkipli ZA, Dalim SF, Yusof MMM (2022) Epistemological beliefs and attitudes towards physics and physics learning among Malaysian STEM students in Selangor. Asian J Univ Educ 18(4):919–932

Olazabal L, Sarriugarte P, Zuza K, Guisasola G (2021) Analyzing the upper secondary school students' difficulties in the rotation of rigid body. J Phys Conf Ser 1929:012008

Suárez Á, Baccino D, Monteiro M, Marti AC (2022) Differences in the attitudes and beliefs about science of students in the physics-mathematics and life sciences areas and their impact on teaching. arXiv preprint arXiv:2202.11680

Suárez Á, Baccino D, Monteiro M, Martí AC (2021) What are the attitudes and beliefs about Science of the Physics teachers and future Physics teachers in Uruguay? arXiv preprint arXiv: 2103.13775.

Suwonjandee N, Mahachok T, Asavapibhop B (2018) Evaluation of Thai students and teacher's attitudes in physics using Colorado learning attitudes about science survey (CLASS). J Phys Conf Series 1144(1):012124

Uhden O, Karam R, Pietrocola M, Pospiech G (2012) Modelling mathematical reasoning in physics education. Sci Educ 21:485–506

The Human-Centered Design Framework as a Means of Bridging the Research-Practice Gap

F. Bernstein, S. Schmeling, and T. Wilhelm

Abstract Design-based research is a genre of research in which the development of theoretical understanding and the design of practical solutions to real-world problems are pursued at the same time. With this approach comes the aspiration to contribute to everyday educational practice in a meaningful way. Therefore, questions of adoption, implementation and scaling have been on the design-based research agenda since the beginning. However, there is little specific guidance within the Physics Education Research (PER) community available on how to maximize the probability of adoption by practitioners systematically. In contrast, Human-centered design (HCD), rooted in industry practices, offers a promising avenue for bridging this gap, owing to its inherent emphasis on adoption rates and scalability within the commercial sphere. To what extend insights and optimization strategies derived from Human-centered design can enhance the adoption of educational innovations is currently being explored in a research project on the usability of modern physics experiments at the Goethe University Frankfurt, where two Michelson interferometers a subjected to comparative usability tests.

1 Design-Based Research and the Implementation Challenge

Design-based research (DBR) has become a well-established and even mainstream methodology in educational research (Tinoca et al. 2022), aiming to solve practical problems while enhancing theoretical understanding of educational settings and challenges. While it can be intricate to exactly define what design-based research entails or what the methodology comprises (McKenney and Reeves 2019; Philippakos et al.

F. Bernstein (✉) · T. Wilhelm
Goethe-Universität Frankfurt am Main, Institut für Didaktik der Physik, Max-von-Laue-Str. 1, 60438 Frankfurt, Germany
e-mail: bernstein@physik.uni-frankfurt.de

S. Schmeling
CERN, 1211 Geneva, Switzerland

S. Faletič and J. Pavlin (eds.), *Teaching and Learning Physics Effectively in Challenging Times*, Challenges in Physics Education, https://doi.org/10.1007/978-3-031-72541-8_5

2021), there seems to be a consensus that, generally speaking, three core processes are at the heart of design-based research: (a) Analysis, (b) design and (c) evaluation of designs and proposed solutions (Bakker 2019; McKenney and Reeves 2019) (see Fig. 1a). However, there is a fourth aspect that is equally important, which has to do with the fact why design-based research was introduced to the field of education in the first place: the so-called research-practice gap or disillusionment with too much educational research, often descriptive in nature, which does not contribute to improving educational practice in the short or long term (Hirschkorn and Geelan 2008; McKenney and Reeves 2019; The Design-Based Research Collective 2003). As Arthur Bakker puts it: "Most educational research describes or evaluates education as it currently *is*. [...] Design research, however, is about education as it *could be* or even as it *should be*." (Bakker 2019) In a similar fashion, McKenney and Reeves identify the "discrepancy between the existing and desired situations" as a major driver of design-based research (McKenney and Reeves 2019). This is why implementation and scaling of proposed solutions are also key considerations in design-based research and are usually addressed within the research process.

However, when it comes to methodology, there seems to be little advice or guidance, beyond very general recommendations (McKenney and Reeves 2019; Rogers 2003), on how to achieve adoption or implementation in practice. While this can partly be attributed to the wide variety of research questions, subjects, and educational contexts addressed within the design-based research community, which, in turn, call for different implementation strategies, this still does pose a challenge to the researcher trying to ensure adoption and scaling of designs and research products.

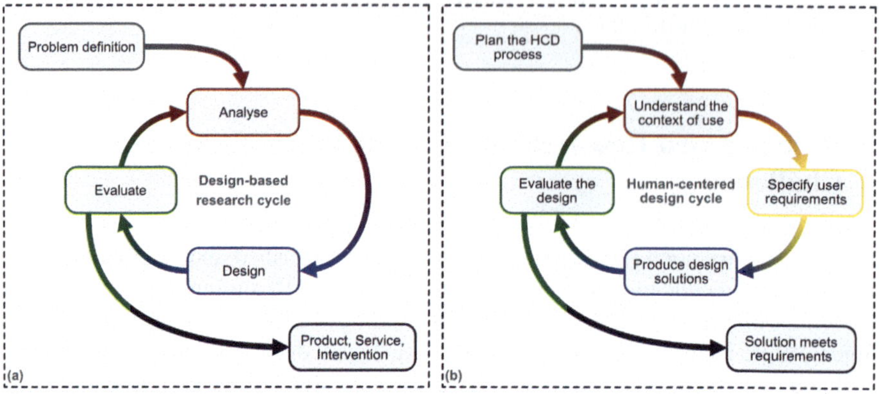

Fig. 1 **a** Design-based research cycle. **b** Human-centered design cycle

2 Human-Centered Design as an Implementation Strategy in DBR Projects

With this state of affairs, one way forward could be to draw on methods and approaches from other disciplines where similar problems have been successfully tackled. One promising candidate is the field of human–machine interaction and, more precisely, Usability Engineering and Human-centered design (HCD). Human-centered design is an industry framework in which the user's needs and preferences are at the core of the design process, with the goal of creating products and services that provide better customer experience, thus leading to higher adoption rates. Notably, akin to the less standardized methodology known as "Design Thinking" (Brown 2008, 2019; Plattner et al. 2011), a profound comprehension of the challenges and obstacles users encounter when striving to accomplish specific objectives stands as a central tenet of this approach. To our understanding, this is of particular importance in the educational field, given the institutional and spatial separation that often exists between the origin of innovations (e.g. universities, research institutes) and their place of deployment (e.g. schools). From a theoretical and practical perspective, HCD and Usability Engineering are particularly promising since they are dealing with similar challenges, have been shown to be effective, and are, furthermore, specified in several ISO standards (ISO 2018, 2006, 2019), which provide a rich body of definitions, concepts, and methods that can be readily applied in practice.

2.1 Fundamentals of the Human-Centered Design Process

The basic framework of Human-centered design is depicted in Fig. 1b. Similarly to design-based research, HCD proceeds in an iterative manner, moving from analysis to the derivation of user needs and requirements, design of possible solutions to the evaluation of the designs against the user needs. The starting point of a Human-centered design process is, therefore, a thorough understanding of the user (or users): His goals and tasks, the context in which he operates, and the resources available to achieve the desired goals. In Human-centered design, these aspects are subsumed under the term "context of use" (see Fig. 2) (ISO 2019). Only profound insight into what drives the user allows for the derivation of user needs and requirements, which form the basis of design solutions that take these considerations into account. Ideally, multiple design solutions are produced and evaluated against the user's needs. This, in turn, leads to a better understanding of the context of use and thus forms the basis of iterative product development.

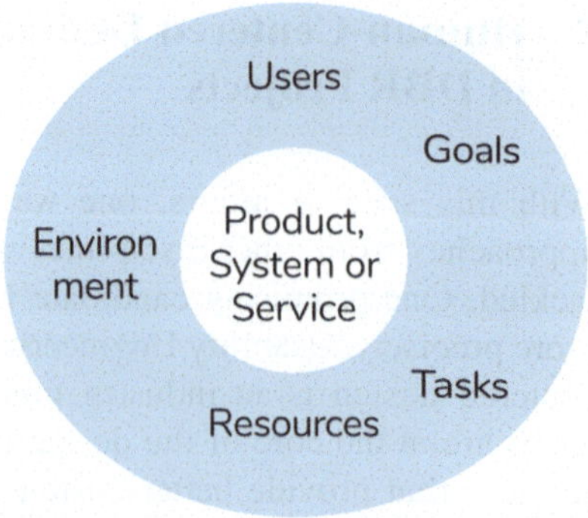

Fig. 2 The "context of use" in HCD

2.2 *"Usability" and "Usability Testing" in the Human-Centered Design Framework*

In the framework of Human-centered design, the notion of "usability" plays a decisive role. According to ISO 9241-11:2018, usability refers to the extent to which a product or service can be used by its intended users to achieve specific goals effectively, efficiently, and with satisfaction in a given context of use (ISO 2018). Effectiveness, efficiency, and satisfaction are explicitly defined in the standard and can be quantified using metrics such as "Task Success" (percentage of users who successfully completed a given task) and "Time on task" (time required by the user to complete a task). Various questionnaires have been developed to measure satisfaction, e.g., the System Usability Scale (Lewis 2018) or the User Experience Questionnaire (Laugwitz et al. 2008), which allow for A-B tests as well as benchmark comparisons (Schrepp et al. 2017).

While there are different techniques for evaluating designs against user needs, such as surveys, field reports, expert evaluations, and usability tests, the latter is of particular interest since it allows for a systematic evaluation of the usability of a product based on observational data, both qualitatively and quantitatively. A usability test, therefore, is *not* about surveying users about their behavior, product features, assessments, etc., a fact that is often neglected or misunderstood. Instead, a usability test is an evaluation in which representative users perform representative tasks with an interactive system in a controlled setting, which serves to identify usability problems or to measure effectiveness, efficiency, and user satisfaction (Rubin et al. 2008).

Two main types of usability tests can be distinguished: formative usability tests and summative usability tests. Formative usability tests are used to identify and classify usability problems during the product development process; their primary purpose is to contribute relevant information for product iteration. Summative usability tests, on the other hand, are used to compare the usability of a product with a benchmark or another product (ISO/TS 20282-2 2013). Therefore, summative usability tests rely on quantitative data, while in formative usability tests, qualitative data is often more important.

3 Why Human-Centered Design in Design-Based Research?

From a methodological point of view, there are marked similarities between the design-based research cycle and the Human-centered design cycle. Both consist of analysis, design, and evaluation in an iterative manner. However, there are also decisive differences: the primary focus in many design-based research projects has been, in one way or another, on improving student learning or on other educational or didactical goals (Tinoca et al. 2022), *not* on increasing the usability of products or solutions generated in the design-based research process. This is where Human-centered design steps in. Human-centered design in education could offer a systematic way to increase the probability of adoption by target groups by means of optimizing the usability of the research product while it is still in the design phase.

For example, HCD could offer the insight that an improvement in student learning resulting from the use of a certain intervention or research product might not necessarily be a sufficiently large incentive, from the perspective of the practitioner, to actually adopt, implement or sustain this intervention or product in a given educational environment. This is to say that, e. g. student learning—or any other legitimate objective of a design-based research project—might not be the only consideration of those in charge of deciding if an intervention or product will be (repeatedly) used in the classroom. In those cases where the primary concepts of usability—i.e., efficiency, effectiveness, and satisfaction—play a decisive role in the decision-making process of the relevant stakeholders, Human-centered design could offer a clear methodological path on how to maximize the implementation probability.

Human-centered design could therefore complement the traditional design-based research cycle by providing an additional layer of analysis, aiming at systematically optimizing the probability of widespread adoption during the design-based research cycle while simultaneously pursuing the goals set in the design-based research project. This way, the probability of implementation would be an optimization objective during the design-based research cycle and not left as an afterthought once the DBR cycle, driven by a different set of objectives, has already been completed.

4 Case Study: HCD for the Design of Modern Physics Experiments

4.1 Motivation and Research Question

In the case of the use of experiments in the physics classroom, there are, indeed, clear indications that student learning alone, be it conceptual or methodological, is not the only, or even the prime, concern of teachers deciding whether or not to deploy a particular experiment in their classroom. From the 1963 Kerr study on "practical science in school" (Kerr et al. 1963) to the Beatty and Woolnough (1982) and Thompson (1975) follow-up studies to the more recent research of Donnelly (1998) and Jonas-Ahrend (2003), it has become somewhat evident that practical considerations play a decisive role in the decision-making process of physics teachers.

This has been confirmed by a 2018 reasoned action approach study with British, Italian, and German teachers (N = 127) conducted at CERN (Bernstein et al. 2020b), which demonstrated that time for preparation and setup, classroom time spent, reliability of the experiment and the availability of documentation and student worksheets all are of significant importance from the perspective of teachers when deciding whether or not to do experiments in class. Since these practical issues can be directly framed in terms of efficiency, effectiveness, and satisfaction, HCD has been chosen as a framework to explore how experiments on modern physics can be designed in such a way that the probability of adoption by high school teachers is maximized.

4.2 Methodology

In the course of the Human-centered design research project carried out at CERN in conjunction with the Goethe University Frankfurt, multiple experiments on modern physics, such as a LINAC model (Bernstein et al. 2020a), a 3D-printed experiment of the photoelectric effect (Bernstein et al. 2022) or a 3D-printed interferometry set were developed. One of these experiments, a 3D-printed Michelson interferometer (Bernstein et al. 2020) (see Fig. 3a), was subsequently subjected to a usability test to evaluate where teachers would struggle with the usage of the experiment and how the usability of the interferometer could be improved. The interferometer was selected for usability testing since these types of interferometers tend to be extremely sensitive and, therefore, somewhat difficult to set up and prone to usability issues. For comparison purposes, a standard lab interferometer of a major German physics equipment supplier for schools (see Fig. 3b) was subjected to the same test.

The test sample of teachers was recruited from the physics student laboratory of the Goethe University Frankfurt (N = 11). Teachers coming to the lab with their

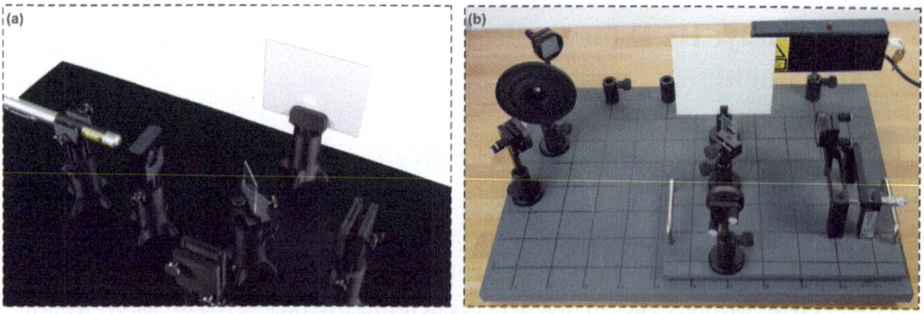

Fig. 3 **a** 3D-printed Michelson interferometer. **b** Michelson interferometer for schools

Table 1 Metrics and Instruments of the usability test

	Usability test plan	Metrics and instruments			
(1)	PreTest-interview	Demographics, reasoned-action-approach questionnaire			
		Effectiveness	**Efficiency**	**Satisfaction**	**Usability problems**
(2)	Test task 1	Task success	Time on task	Single Ease Question	Usability testlog
(3)	Test task 2	Task success	Time on task	Single Ease Question	Usability testlog
(4)	PostTest-interview	System Usability Scale and User Experience Questionnaire			

classes were asked to participate in a two-hour usability test while their students were supervised by the university staff. The usability test took place in a separate room, where the teacher was put in front of one of the interferometers and asked to complete specific tasks that might as well occur in a school setting. These tasks entailed setting up the Michelson interferometer to generate an interference pattern on a screen. Documentation of the experiment was provided, but no additional assistance was offered during the usability test. To better understand where and why participants struggled or failed, usability problems were logged during the process. Additionally, video and audio of the test sessions were recorded to allow for systematic analysis once the test session was completed. The teachers were asked to think aloud to better understand their thought processes while setting up the interferometer. Once the experiment was completed, satisfaction was evaluated by means of the System Usability Scale and the User Experience Questionnaire. Time on task was taken as a metric for efficiency and task success as a measure of effectiveness. The usability test plan and the metrics used are depicted in Table 1. Once the usability test with the first interferometer was completed, teachers were asked to repeat the same tasks with the second interferometer. The order in which the two different interferometers were presented was randomized between participants.

4.3 Preliminary Results of the Study

The preliminary qualitative results indicate recurring problems that may discourage teachers from using the interferometers. For example, the different lengths of the mirror mounts of the commercial interferometer are a severe source of error and frustration for most teachers. The four mirrors come in three different lengths without any explanation in the documentation as to which mirror goes to which position. This leads to a trial-and-error approach, with teachers eventually discovering that the setup they chose does not work since the laser beam cannot be adequately aligned. A similar type of problem occurs with respect to the lens holder. The lens is a critical piece of equipment in the setup in that it must be very carefully aligned to generate an interference pattern. However, the lens mount is very short and limited in the height it can be set to. Therefore, many times the teachers set up all the other components only to find out in the end that their setup was either too high or too low—the lens could not be inserted into the setup at the right height. As a result, the whole setup needed to be readjusted, which is a very time-consuming process and a frustrating experience for the teachers. Similarly, the usage of the 3D-printed laser mount for the 3D-printed interferometer is apparently non-intuitive for most teachers. It requires a screw to permanently hold down the pushbutton of the laser pointer, a concept that is seemingly not obvious to the teachers. Since the laser pointer mount is unique in its design, it offers no affordances in how to use it.

4.4 Conclusion

In conclusion, we argue that integrating a Human-Centered Design (HCD) approach holds substantial potential for enhancing design-based research projects. By including metrics of effectiveness, efficiency, and user satisfaction in the research process, it should be possible to raise the adoption rates of educational innovations by practitioners. In the case of teachers' use of experiments in the classroom, existing research consistently underscores the pivotal role of practical considerations, such as preparation time and reliability, in teachers' decision-making processes regarding the incorporation of experiments into their teaching. HCD offers a tried and tested methodological framework for addressing these and similar concerns. Therefore, we advocate for a proactive approach, suggesting that questions of adoption and scalability should not be relegated to an afterthought in the design-based research process; instead, they should be tackled concurrently by applying Human-Centered Design principles.

To explore this notion, an ongoing study at Goethe University Frankfurt is currently underway, investigating the usability of two Michelson interferometers by means of usability testing. Thus far, our qualitative findings reveal substantial variability in teachers' experimental skills while permitting us to systematically pinpoint the origins of errors and frustrations, many of which can be traced back to design decisions. Consequently, this opens up a promising pathway for substantially enriching the overall user experience for teachers as they engage with these experiments. Furthermore, our quantitative data also points in this direction and will be reported upon the completion of all tests ($N_{final} = 20$). While we anticipate that improved usability will translate into higher adoption rates of these experiments, further research is necessary to validate this hypothesis.

References

Bakker A (2019) Design research in education: a practical guide for early career researchers. Routledge Taylor & Francis Group, London, New York

Beatty JW, Woolnough BE (1982) Practical work in 11–13 science: the context, type and aims of current practice. Brit Educ Res J 8(1):23–30

Bernstein F, Schmeling S, Wilhelm T (2020) 3D-gedruckte Experimente zur modernen Physik. Vom Michelson-Interferometer zum Quantenradierer. Plus Lucis 2020(4):33–39

Bernstein F, Keller O, Schmeling S, Wilhelm T, Woithe J (2020a) Ein LINAC zum Selberbauen. Funktionales Modell eines elektrostatischen Linearbeschleunigers. Unterricht Physik 180:33–35

Bernstein F, Schmeling S, Wilhelm T, Woithe J (2020b) Saliente Überzeugungen von Physiklehrkräften zum Experimentieren. In: Proc Ges f. Didaktik d. Chemie u. Physik Annual Conference (Vienna) vol 40. Wien, University Duisburg-Essen, Duisburg-Essen, pp 86–89, 9–12 Sept 2019

Bernstein F, Keller O, Schmeling S, Wilhelm T (2022) Der äußere Fotoeffekt in neuem Licht. Ein 3D-gedrucktes, Mikrocontroller-gestütztes Schülerexperiment neuer Art. MNU-J 75(5):402–405

Brown T (2008) Design thinking. Harv Bus Rev 86(6):84

Brown T (2019) Change by design: how design thinking transforms organizations and inspires innovation (revised and updated). Harper Collins Publishers, New York

Donnelly JF (1998) The place of the laboratory in secondary science teaching. Int J Sci Educ 20(5):585–596

Hirschkorn M, Geelan D (2008) Bridging the research-practice gap: research translation and/or research transformation. Alberta J Educ 54(1). https://doi.org/10.11575/ajer.v54i1.55207

ISO (2006) ISO 9241-110:2006 ergonomics of human-system interaction—part 110: dialogue principles (ISO)

ISO (2013) ISO/TS 20282-2:2013 usability of consumer products and products for public use—part 2: summative test method

ISO (2019) ISO 9241-210:2019 ergonomics of human-system interaction—part 210: human-centred design for interactive systems (ISO)

ISO (2018) ISO 9241-11:2018 Ergonomics of human-system interaction—part 11: usability: definitions and concepts (ISO)

Jonas-Ahrend G (2003) Physiklehrervorstellungen zum Experiment im Physikunterricht. Studien zum Physiklernen, vol 34

Kerr JF, Boulind HF, Scott DW, Rolls MJ, Stafford E (1963) Practical work in school science: an account of an inquiry sponsored by the Gulbenkian foundation into the nature and purpose of practical work in school science teaching in England and Wales. Leicester University Press, Leicester

Laugwitz B, Schrepp M, Held T (2008) Construction and evaluation of a user experience questionnaire. In: Holzinger A (ed) HCI and usability for education and work, vol 5298. USAB 2008. Springer, Berlin, Heidelberg, pp 63–76

Lewis JR (2018) The system usability scale: past, present, and future. Int J Human-Comput Interact 34(7):577–590

McKenney S, Reeves TC (2019) Conducting educational design research, 2nd edn. Routledge, London, New York

Philippakos ZA, Howell E, Pellegrino A (eds) (2021) Design-based research in education: theory and applications. The Guilford Press, New York, London

Plattner H, Meinel C, Leifer L (eds) (2011) Design thinking: understand—improve—apply. Springer, Berlin, Heidelberg

Rogers EM (2003) Diffusion of innovations, 5th edn. Free Press, New York

Rubin J, Chisnell D, Spool J (2008) Handbook of usability testing: how to plan, design, and conduct effective tests, 1st edn. John Wiley & Sons, Indianapolis

Schrepp M, Hinderks A, Thomaschewski J (2017) Construction of a benchmark for the user experience questionnaire (UEQ). Int J Interact Multimedia Artif Intell 4(4):40–44

The Design-Based Research Collective (2003) Design-based research: an emerging paradigm for educational inquiry. Educ Res 32(1):5–8

Thompson JJ (1975) Practical work in sixthform science: an enquiry into the aims and methods of the teaching of practical work in Sixthform Science as perceived by teachers of physics, chemistry and biology. Science Centre, Oxford

Tinoca L, Piedade J, Santos S, Pedro A, Gomes S (2022) Design-based research in the educational field: a systematic literature review. Educ Sci 12(6):410

Improving the Effectiveness of Physics Teaching

Valentina Bologna, Simon Peter Leban, Francesco Longo, Maria Peressi, and Paolo Sorzio

Abstract Physics teachers need to understand subject matter knowledge tailored specifically to teaching. It can be reached by extending the teachers' background with in-service training activities and professional development. We report the results of a recent experience involving some Physics teachers engaged in a deep revision of their Pedagogical Content Knowledge (PCK). This allowed investigation of the standard features in their instructional to identify their specialised Content Knowledge for Teaching (CKT). Two representative stories are reported here, selecting a young teacher and an experienced one. They revised their specific way of Physics teaching, adopting the ISLE (Investigative Science Learning Environment) approach and encouraging the students' development of reasoning skills.

1 Introduction and Theoretical Framework

Content Knowledge for Teaching (CKT) is an operational extension (Etkina 2018; Ball 2000) of the well-known and widely used Pedagogical Content Knowledge (PCK) concept by Shulman's work (2011) and others for Physics teaching (Keller et al. 2017; Magnusson et al. 1999; Zeidler 2002; Etkina 2010).

V. Bologna (✉) · F. Longo · M. Peressi
Physics Department, University of Trieste, via Alfonso Valerio 2, 34128 Trieste, Italy
e-mail: valentina.bologna@units.it

F. Longo
e-mail: francesco.longo@infn.ts.it

M. Peressi
e-mail: peressi@units.it

S. P. Leban
ISIS Paolino D'Aquileia, via Dell'Istituto Tecnico Agrario 42, 33043 Cividale Del Friuli, Udine, Italy
e-mail: simonleban@paolinodaquileia.edu.it

P. Sorzio
Department for Humanities, University of Trieste, via Lazzaretto Vecchio 8, 34124 Trieste, Italy
e-mail: psorzio@units.it

S. Faletič and J. Pavlin (eds.), *Teaching and Learning Physics Effectively in Challenging Times*, Challenges in Physics Education, https://doi.org/10.1007/978-3-031-72541-8_6

"CKT is broader than PCK because it includes disciplinary knowledge and pedagogical applications of that knowledge" (Seeley et al. 2018). It focuses empirically on the specialised disciplinary knowledge teachers act on in their classrooms (Seeley et al. 2018). Recently, some efforts have been made to define CKT in Physics (Etkina 2018; Seeley et al. 2018). It is still developing and needs time to categorises and assess its domains.

CKT allows us to individuate the peculiar features of Physics teaching better. We can recognise if a teacher shows foundational content knowledge or an elaborative type, as discussed by Seeley et al. (2018). The difference between the two types resides in the knowledge of scientific practice (Seeley et al. 2018). If teachers are not prepared enough or skilled in these practices, they may have difficulties building their CK for Physics Teaching (Etkina 2018; Seeley et al. 2018).

Focusing on Physics teaching practices and how teachers could change in practicing towards an elaborative CK is the aim of our study. Therefore, our research question is:

> *How to guide teachers in the revision of their Physics teaching practices developing a deeper and aware Content Knowledge for Teaching?*

Our brief contribution intends to sustain the education research discussion about how "it is critical to prioritize knowledge and strategies that will best support teachers in responding productively to student thinking" (Seeley et al. 2018).

To answer the research question, we present the professional in-service training program we developed to promote teachers' revision of their practices. The goal of the revision was to coach teachers in adopting the ISLE approach (Etkina et al. 2019). We report a brief extract of their stories to witness the effect of the improvement of their CKT (Seeley et al. 2018) as a consequence of the ISLE's adoption (Etkina et al. 2017). This approach requires teachers to change many practices, outlined as tasks of teaching (Etkina 2018). "Tasks of teaching describe the key activities through which teachers enact practices that promote and support student learning" (Seeley et al. 2018).

2 Method and Activity Development

The methodology used for this study is based on phenomenological case study research (Vagle 2014; Creswell and Clark 2017) to "develop a composite description of the essence of the experience for all of the individuals" (Creswell and Clark 2017) and to validate the training program we used to. We collected the data of each case study presented by monitoring and reporting teachers' artifacts, activities and extracts of conversations. In the analysis of the case studies, we tried to highlight how teachers enacted a change in their practices, that means in their tasks of teaching (Etkina 2018), as a consequence of participating in the program they were involved in. We analysed the case studies following the teachers' training program phases.

Table 1 Physics teachers' program phases

a.y.	Phase/goal	Method
2019/20	**Phase I:** Building teachers' awareness about PCK with particular attention to the one regarding the Phys-Math interplay	Observing classroom activities, discussing with teachers, and collecting information through teachers' interviews
2020/21	**Phase II:** Training teachers from PCK's towards awareness of exploring a new CKT, based on content knowledge integration and use of Multiple Representations	Clinical classroom practices: planning, performing activities, assessments and revision. Video-recording teachers' discourses
2021/22	**Phase III:** Training teachers towards the adoption of content knowledge management based on the ISLE approach—inquiry student-centered	Clinical classroom practices, with planning, revisions, and discussions

Our program of Physics teachers' professional training consists of three phases, as reported in Table 1.

As a theoretical reference for our program, we chose to adopt the one used for pre-service professional preparation, called DHAC (Developing Habits through Apprenticeship in a Community) (Etkina et al. 2017). We did not go into detailing and motivating our choice, as this is not the aim of this contribution. Here, we strictly focus on the effect of the program adopted under the stories of two teachers, one expert (later referred to as Teacher #1) and one novice (later referred to as Teacher #2).

2.1 Phase I

We monitored a group of Italian Physics teachers to outline the feature of prevalent PCK concerning their Phys/Maths interplay patterns (Bologna 2022; Lehavi 2017). Teachers #1 and #2 pinpointed their own PCK features leading to re-think the way to enact the integration between the two disciplines.

We could refer to this phase as a *teacher's diagnostic tool*: it helped teachers to become aware of their PCK, mainly referring to the one involved in the Phys/Math interplay (Bologna 2022).

2.2 Phase II

The second phase focused on two main actions. The first one concerned investigating a different way of learning sequence building to overcome the constraints of the individuated diagnostic phase. In this *transitory* phase, teachers moved the first steps towards the ISLE's adoption, at the beginning by the introduction of the use of Multiple Representations in learning sequence activities, then promoting the ISLE's process through Observational, Testing and Application Experiment (Etkina et al. 2019).

The second action engaged teachers in revising how they conducted discourses in classroom activities, making them aware of the need to change from teacher-centred lessons to student-centred ones.

By analysing the video-recorded lessons, teachers reinforced their motivation for changing practices.

2.3 Phase III

In the third phase, teachers really enacted the change, planning many clinical classroom activities based on the ISLE approach, preparing materials, and revising the *old* practices in the "light" of the *new* ones. This phase could be referred to as the *implementation phase*, guiding teachers towards the acquisition of the specific CKT based on the ISLE approach (Etkina et al. 2019), which means improving well-defined tasks of teaching (Etkina et al. 2017).

3 Case Studies' Analysis

3.1 Teacher #1

Teacher #1 has been a permanent Maths/Phys teacher in a secondary school for scientific studies for almost six years. This position was gained after eight years of temporary tutoring at college instruction and nine years in other local secondary schools.

After the Master's degree and PhD in Maths studies, this teacher attended professional pre-service training for becoming a teacher, the so-called SSIS (this acronym stands for Scuola di Specializzazione per l'Insegnamento Secondario—Specialised School for Secondary Teaching; (Dal Passo et al. 1848)). This biennial instructional program for teachers' certification was activated from 1999 to 2009.

The school context of Teacher #1 is highly competitive, facing high performance and preparation levels from students' and teachers' points of view. The Physics curriculum has been recently revised, and all Physics school teachers must implement it to guarantee school homogeneity. This is a formal constraint for teaching innovation and changes. The constraint is substantiated by disciplinary recovery courses activated during schooling periods. If a student has difficulty in Physics, he/she could attend these extra-curricular courses to overcome difficulties. The course is not prepared by the same classroom teacher but by another teacher in that school. This means that the teacher's work must be aligned with the others in the content and the "regular types" of problems used for assessing and consolidating student content knowledge. Teachers could decide on lab activities, but the school has only one Physics lab (for fifty classes), so it isn't easy to plan experiments regularly. Other school projects for curriculum improvement are not mandatory; every teacher chooses them according to their teaching needs.

If teachers wish to change, they could sometimes feel isolated and have difficulty moving out of the homogeneity constraints. Table 2 reports the main features of Teacher #1's engagement in the training program.

The phase I helped this teacher to reflect upon PCK. Teacher #1 mostly recognised three features about herself:

1. The prevalent application pattern of Phys-Math interplay in her instructional practices.
2. Students' difficulties developing reasoning skills and deep understanding or simply describing physical phenomena.
3. Assessing knowledge through Maths requests by applying formulas in problem-solving exercises.

Teacher #1 acknowledged that it is very difficult to pinpoint students' p-prims (diSessa 1993) and use them as resources (Hammer 1996) for conceptual change (diSessa 2014) in the process of building knowledge.

> "In a "transmissive" process of building knowledge, nothing addresses this task. You think what you have to say, not what students think about or how to make visible their thinking. What is dominant is your thinking and process, not the students'".

Table 2 Teacher #1's engagement in the program

Type of activity	Academic year		
	2019/20	2020/21	2021/22
	Phase I	Phase II	Phase III
Classroom Observations	16h		
Teaching Activities		20h	80h
Discussions		50h	70h
Audio/Video recorded		8h	

Reading the transcriptions of the video-recorded lessons, this teacher became aware of how teacher-centred the process of building knowledge was.

> "I even tried to include students in my discourse. I was unaware that I included them in my reasoning, not their own. Of course, my reasoning was thoroughly formal and clear and correct. But it was not their reasoning. Thus, I noticed they could not replicate it as I asked them to."

Likewise, the kinds of questions posed during the classroom discourses were teacher-centred.

> "I asked students to respond to my thinking process. I understood that they often did not respond to anything because they did not follow the reasoning that it was not theirs".

This would mean changing the way of planning and designing lessons. It meant finding the time when students' ideas could unfold, emerge and be part of the process of building knowledge.

In the second phase, Teacher #1 started to re-design specific topics reflecting on Content Knowledge. In detail, Teacher #1:

- explored the knowledge building of Geometrical Optics starting from Observational Phenomena;
- devoted more time to students' engagements in phenomenological descriptions by natural language (discourses);
- completed the content knowledge building through phenomenological relations in maths representations, using DESMOS[1] platform for online activity;
- assessed the students' knowledge by problem-solving mainly based on maths formulas.

The main difficulties Teacher #1 encountered in this process were lesson-time management and classroom setting. The teacher made many efforts to overcome them but met two constraints.

First, students were not usually to be engaged in active learning practices. Also, even if they got involved, they sometimes behave as if active participation means less *formal/effective* lessons (as already seen in a higher level of instruction, (Deslauriers 2019)).

Second, the school management of Teacher #1 does not allow using working groups in classroom activities to prevent COVID-19 infection. This restriction is still mandatory and limits the activation of sharing knowledge building in peers-works.

In the teacher's words, we recognised an unavoidable claim of the need to change:

[1] https://teacher.desmos.org.

"I need to expand our classroom discourses to activate students' participation! I have to stimulate discourses more than I was used to doing before; it's become a teaching necessity."

Phase III featured the change in building learning sequences. Teacher #1 changed the way of designing and sequencing learning experiences and activities. Before the training program, the teacher presented the specific content topic describing phenomenological situations. Based on collected evidence, the description was portrayed without any experimental activities. The focus was on a mathematical model of the phenomenon, as the teacher declared:

"I ever preferred to emphasise maths representations of physical phenomena more. I always thought that was the most important thing I had to do in my lessons. Now, I'm confident many other things make students aware of Physics knowledge."

The teacher revised content knowledge by changing lesson preparation and learning sequences.

"I changed my content knowledge. I know Physics in a conceptual way that I did not really understand before. The first conceptual change was mine. Then in my students!"

The changes mainly regarded the adoption of Observational and Testing experiments, multiple modes of representations and multiple strategies to build conceptual knowledge. Teacher #1 became skilled in using experimental activities during learning sequences.

"Demonstrative lab activities are useful because they show how to take, elaborate and use experimental data. These experiments show the phenomena and that these phenomena are real. However, they do not help students build knowledge and never test hypotheses or design experiments. Observational, Testing and Application Experiments are completely different. When I use them, they work on students' building concepts. I did not always need to go to a laboratory room. My classroom is a good setting for using online video experiments."

One effect of Teacher #1 changes was the structure of the written tests for assessing students' learning. Introducing multiple representations would mean sustaining students' learning and deep conceptual understanding. The teacher implemented written tests and changing quizzes (with respect to the previous ones used) to assess learning goals. The previously written tests consisted of quizzes/questions to investigate the so-called "theory-based" knowledge and quizzes/exercises to investigate "application or problem-based" knowledge. The first aimed to measure knowledge acquired, whereas the second measured the procedural application of the maths formula describing a physical law. The new types of written tests are completely different. There is no distinction between the theory and application parts. Each exercise is based on real-life everyday phenomena students could have experienced, seen, or recalled to something known.

"Changing the way of assessments, asking more "why", "represent", and "reason" than "what" or "how many", my students were engaged in a deep learning process".

This teacher recognised how assessing differently is relevant to engaging students in meaningful learning.

Therefore, despite school constraints, despite the large amount of work for producing *new* materials and resources for lessons and activities, Teacher #1 continues to improve *new* practices sustaining the change in her tasks of teaching (Etkina 2018).

3.2 Teacher #2

Teacher #2 is now a permanent Physics and Optics teacher in a secondary school for professional and technological studies, mainly in an Optics School. During the research project, he had a temporary position as a Mathematics and Physics teacher in a high secondary school for ancient classic studies for Slovenians who live in Italy. He was a novel teacher with only 2 years of teaching experience. He started his studies in Physics Education during his apprenticeship and the Bachelor's Thesis Degree in Physics, implementing an ancient galvanometer for educational purposes (Leban 2020). Since his master's degree in Physics Education and History of Physics, he has been involved in the training program for in-service Physics teachers (Table 3).

So, he was a novel teacher with a theoretical background in Physics Education, which sustained him to move his first steps in teaching practices.

In his former education, he was trained towards learning about Physics teacher PCK (Magnusson et al. 1999; Etkina 2010), how students build Physics knowledge and what is conceptual change (Hammer 1996; diSessa 1993; Levrini and diSessa 2008) and how epistemological beliefs influence this knowledge (Elby 2010). This highlights his ability to build and reflect on his teaching practices under these frameworks.

Table 3 Teacher #2's engagement in the program

Type of activity	Academic year		
	2019/20	2020/21	2021/22
		Phase I–II	Phase III
Teaching activities		24h	122h
Discussions		33h	48h
Audio/video recorded		6h	

In the first phase, Teacher #2 laid the foundation of his PCK by making a lot of effort to cover all the patterns in the Phys-Math interplay in his instructional practices, promoting classrooms' discourses for reasoning skills development and assessing knowledge also with no-traditional testing exercises (Bologna and Leban 2022). He dedicated many efforts to defining the features of his teaching under two intentionalities:

- he would not teach as he learnt Physics;
- he would teach, encouraging students' centrality in the learning process.

He soon realised that there was a consistent difference between studying and teaching Physics.

> "I learnt Physics in a traditional way, with frontal lessons, no interactivity, no more labs activities, and more Maths, everywhere Maths. I feel this was a limit to overcome for my professional development."

Focusing on the mathematisation of Physics contents was a limit he wanted to overcome to build his teaching practice. In the secondary school where he taught, students attended this school to avoid scientific subject matters.

This context constraint supported, even more, the need to come over the mathematisation process in the conceptual building as a unique way of presenting content.

> "I needed to open my mind and try to find something different regarding conceptual Physics knowledge building."

The first problem encountered was the lack of teaching tools. In order to find something different in terms of conceptual Physics knowledge building, the urge to create and prepare new materials emerged because there were no available textbooks, exercises, or anything else that could help and support learning for his students.

Under these reflections and needs, the second phase and the third meant pinpointing the characteristics of Content Knowledge for Teaching in ISLE-based learning.

Teacher #2 engaged students in Observational activities with the online DESMOS platform both for Circular Motion and Energy Conservation, planned more lecture time for discussions with less talking about formulas and assessed knowledge by inserting open questions in order to develop students' reasoning skills.

What is remarkable is how he adopted the ISLE approach slowly during the first year and the second school year, fully in the classes of 11th grade, whereas in the ones of 12th grade and the last ones only for some activities.

There are two main aspects characterising the practice of this teacher:

1. the learning sequence employed in the conceptual building based on Observational and Testing Experiments;
2. the use of Multiple Representations to enact students' reasoning skills.

> "I understood that all these instructional practices enact a way of building knowledge deep and durable. It is not the temporary knowledge acquired by memorising formulas."

The difficult part was the assessment. It isn't easy to assess students in written tests, creating exercises which do not assess memorising procedures but involve representations and reasoning about them.

> "We have to change completely the way of thinking about how to assess students' knowledge. This is still something I have to improve."

At the same time, Teacher #2 transcribed the video-recorded of his lessons. He made the transcriptions himself because the video-recorded lessons were in the Slovenian language. By listening again to classroom discourses, he was encouraged to improve his questioning (which kind of questions are used to guide students) during the scaffolding of students' concept-building process. He noticed he needed to let students talk in classroom discourses more than he realised to do during his lessons' hours.

Lastly, he was not sustained by his school colleagues but by the community of teachers involved in the program. Sharing ideas and difficulties and exchanging materials and experiences supported his incoming towards becoming a Physics teacher.

4 Results and Conclusions

These two teaching experiences have disclosed the potential gain of professional development facing specialised CKT (Etkina 2018; Seeley et al. 2018). Analysing the two case studies, we found some interesting points:

- Changing teachers' instructional practices requires a long time, meaningful goals, targeting objectives, and useful and replicable teaching/learning paths.
- Teachers need to acquire awareness about their PCK as a starting point for all possible changes to promote and improve. Their self-consciousness is the key feature for increasing motivation in changing and overcoming difficulties (curricular design, institutional targets... colleagues, and much more).
- The change becomes effective when they explore Content Knowledge for Teaching, which clearly shows how to conceptualise the knowledge from qualitative to quantitative acquisition. That's a crucial point: they left maths representations (that is quantitative) for the qualitative description through experimental observation of physical phenomena.
- The most significant change is about assessment. When a teacher changes his/her kind of tests and exercises he/she administers to his/her students, this means he/she has changed the way of teaching. This is a sort of "cartina tornasole" (litmus paper). A teacher asks what a teacher teaches: improving the skill of reasoning means asking to reason (from explanation to examination) through different languages (multiple representations) integration.

These findings focus on the results of the effect of this training program, answering our research question. We aimed to validate this program for improving Physics teaching practices. Changing teachers' practices mean developing *new* habits (Etkina et al. 2017). These case studies are examples of how to promote teachers' improvements in the effectiveness of Physics Teaching. Of course, they encourage us to continue adopting this kind of program for in-service teachers; the teachers' stories are case studies we could investigate in more research on teachers' development of habits.

Competing Interests The authors have no conflicts of interest to declare that are relevant to the content of this chapter.

Acknowledgements We want to acknowledge the teachers who participated in this research project and their schools.

References

Ball D (2000) Bridging practices: intertwining content and pedagogy in teaching and learning to teach. J Teach Educ 51(3):241–247

Bologna V et al (2022) Monitoring PCK physics teachers' strategies for math and physics languages integration: the teacher footprint. J Phys: Conf Ser 2297:012034

Bologna V, Leban SP (2022) The puzzle of problems: multiple representations tools to support teachers' instructional practice. Gior Fis 63(s01):99–108
Creswell JW, Clark VLP (2017) Designing and conducting mixed methods research. Sage Publications, New York
Dal Passo F, Laurenti A (2017) La scuola italiana. Le riforme del sistema scolastico dal (1848) ad oggi. Novalogos, Roma
Deslauriers L et al (2019) Measuring actual learning versus feeling of learning in response to being actively engaged in the classroom. PNAS 116(39):19251–19257
diSessa AA (1993) Toward an epistemology of physics. Cogn Instr 10:105–225
diSessa AA (2014) A history of conceptual change research: threads and fault lines. In: Sawyer RK (ed) The Cambridge handbook of the learning sciences. Cambridge University Press, pp 88–108
Elby A (2010) Helping physics students learn how to learn. Am J Phys 69:S54–S64
Etkina E (2010) Pedagogical content knowledge and preparation of high school physics teachers. Phys Rev Spec Topics Phys Educ Res 6:020110
Etkina E et al (2018) Design of an assessment to probe teachers' content knowledge for teaching: an example from energy in high school physics. Phys Rev Phys Educ Res 14:010127
Etkina E, Gregorcic B, Vokos S (2017) Organizing physics teacher professional education around productive habit development: a way to meet reform challenges. Phys Rev Spec Topic Phys Educ Res 13:010107
Etkina E, Brookes DT, Planinsic G (2019) Investigative science learning environment. Morgan and Claypool Publishers, San Rafael
Hammer D (1996) More than misconceptions: multiple perspectives on student knowledge and reasoning, and an appropriate role for education research. Am J Phys 64:1316–1326
Keller MM, Neumann K, Fischer HE (2017) The impact of physics teachers' pedagogical content knowledge and motivation on students' achievement and interest. J Res Sci Teach 54:586–614
Leban SP et al (2020) Il riscatto del galvanometro: storia di uno strumento antico utilizzato per una didattica innovativa. Quaderni CIRD 20:64–81
Lehavi Y et al (2017) Classroom evidence of teachers' PCK of the interplay of physics and mathematics. In: Greczylo T, Debowska E (eds) Key competences in physics teaching and learning. Springer Proceedings in Physics, vol 190. Springer International Publishing, Switzerland, pp 95–103
Levrini O, diSessa AA (2008) How students learn from multiple contexts and definitions: proper time as a coordination class. Phys Rev Spec Top Phys Educ Res 4:010107
Magnusson S, Krajcik J, Borko H (1999) Nature, sources, and development of pedagogical content knowledge for science teaching. In: Gess-Newsome J, Lederman NG (eds) Examining pedagogical content knowledge: the construct and its implications for science education. Kluwer Academic Publishers, Dordrecht, pp 95–132
Seeley L, Etkina E, Vokos S (2018) The intertwined roles of teacher content knowledge and knowledge of scientific practices in support of a science learning community. In: Uzzo S, Graves S, Shay E, Harford M, Thompson R (eds) Pedagogical content knowledge in STEM. Springer International Publishing, Switzerland, Advances in STEM Education, pp 17–48
Shulman L (2011) Knowledge and teaching foundation of the new reform. Harv Educ Rev 57:1–21
Vagle DM (2014) Crafting phenomenological research. Left Coast Press, Walnut Creek
Zeidler DL (2002) Dancing with Maggots and Saints: Visions for subject matter knowledge, pedagogical knowledge, and pedagogical content knowledge in science teacher education reform. J Sci Teach Educ 13:27–42

Exploring the Relationships Between Undergraduate and High School Students' Self-Efficacy, Engagement, and Attitudes Towards Physics

Danilo Catena and Italo Testa

Abstract The aim of this study was to investigate the relationships between attitudes, engagement and self-efficacy in physics. The analysis was based on three Likert-scale surveys designed to measure the constructs addressed and administered online to N = 1,971 Italian undergraduate and high school students. After exploring several possible models, we validated a structural model in which attitudes play the role of the independent variable, self-efficacy represents the dependent variable, and engagement acts as the overall mediator. The results have implications for teaching in terms of understanding the mechanism underlying the promotion of students' self-efficacy in physics.

1 Introduction

Over the last twenty years, there has been growing concern about the declining interest and retention of students in physics worldwide (Eurydice 2012; NSB 2014; NSF 2017; OECD 2008; Oon and Subramaniam 2011; Stokking 2000). As a consequence, society is experiencing a shortage of highly qualified research personnel (EU 2010; UN 2012; UNESCO 2010) and a lack of general knowledge about the physical principles underlying the functioning of technological devices (Cajas 2001; Osborne and Dillon 2008). A large body of research suggests that a relevant construct useful for framing the above issues is that of attitudes towards physics (Adams et al. 2006; Angell et al. 2004; Bøe and Henriksen 2013; Irving and Sayre 2015; Regan and DeWitt 2015; Reid and Skryabina 2002). In this study, we conceptualize this construct as a set of feelings or evaluative judgments formed by an individual about physics that are capable of influencing choices and decisions (Ajzen 2001; Barmby et al. 2008; Crano and Prislin 2006; Kind et al. 2007; Ramsden 1998). Previous

D. Catena
Department of Mathematics, Computer Science and Physics, University of Udine, Udine, Italy

I. Testa (✉)
Department of Physics "E. Pancini", University of Naples "Federico II", Naples, Italy
e-mail: italo.testa@unina.it

S. Faletič and J. Pavlin (eds.), *Teaching and Learning Physics Effectively in Challenging Times*, Challenges in Physics Education, https://doi.org/10.1007/978-3-031-72541-8_7

work shows that physics is perceived as a difficult school subject (Williams et al. 2003), that hard work is necessary to succeed (Mujtaba and Reiss 2013), and that expectations of good grades are lower (Barmby and Defty 2006). Finally, physics is perceived as abstract and of little use (Kessels et al. 2006). Such attitudes can range from a sense of inadequacy (Eccles 2007) to a research-driven motivation for physics (Bøe and Henriksen 2013).

Therefore, in order to better describe issues related to students' declining interest and retention in physics, we included two other psychological constructs in this study, self-efficacy and engagement. Self-efficacy in physics can be defined as an individual's perception of their ability to perform certain physics-related activities (Bandura 2001). In particular, we consider two dimensions specific to physics (Sawtelle et al. 2012): perceived ability to solve problems and construct demonstrations, and perceived ability to perform experiments. Engagement in physics can be defined as the participation, behaviour, action, emotion, investment and motivation in a given activity to sustain commitment and self-regulated learning processes (Ben-Eliyahu et al. 2018). For physics, engagement measures the intensity of students' productive involvement in physics-related activities. In this study, we will focus on only two dimensions of the Fredricks et al. (2004) model (Fredricks et al. 2004)—cognitive engagement (self-regulation in the study of physics) and behavioural engagement (participation in activities)—because the dimension of emotional engagement, operationalised as interest in physics, is included in the attitude construct (Testa et al. 2021).

While the above constructs have been widely studied in science education literature, their interrelationships have not been clearly identified and have rarely been studied in physics. Therefore, the overarching aim of this study was threefold: (1) to analyse patterns of attitudes towards physics and to test whether the findings reported in (Testa et al. 2021) are confirmed; (2) to analyse the dimensions of engagement and self-efficacy in physics and to test whether the models reported in the theoretical framework are confirmed; (3) to explore structural relationships between students' attitudes, engagement and self-efficacy. The specific research questions that guided the study were:

- (RQ1) What are the dimensions underlying the attitudes towards physics identifiable for the sample involved in the research?
- (RQ2) What are the dimensions of engagement and self-efficacy identifiable for the sample involved in the research?
- (RQ3) What relationships exist between the dimensions underlying the attitudes towards physics and the dimensions of engagement and self-efficacy?

2 Theoretical Framework

2.1 The Semiotic Cultural Psychological Theory

To answer our research questions, we will analyze attitudes towards physics using the framework of Semiotic Cultural Psychological Theory (SCPT). SCPT conceptualizes psychological processes as sensemaking processes of interpretation, which in turn shape experience (Salvatore et al. 2018). In this way, people feel, think and behave by performing sensemaking in terms of non-rationally justified assumptions about what the world is and how it works. These assumptions are based on the individual's environment, feelings, thoughts and social agency (Testa et al. 2021). To identify such assumptions, the SCPT models the sensemaking process as a dynamic of sign transitions within a semiotic space constituted by signs (e.g., verbal statements, emotional states, or nonverbal expressions) that activate a particular instance of meaning (Salvatore et al. 2018). The sensemaking process is then modelled as a trajectory, i.e. how one sign follows another in semiotic space, whose dimensions correspond to the latent dimensions of meaning represented by oppositional lines. In this way, the SCPT framework allows the set of attitudes towards physics to be conceptualized as a dynamic entity, which has an advantage over the other static descriptions of attitudes mentioned above (see Sect. 1).

This framework was used in (Testa et al. 2021) to identify the main dimensions that define the semiotic space describing attitudes towards physics. The three main dimensions were interpreted according to the following oppositions between the negative (−) and positive (+) polarity of the factorial axis:

1. Relevance (−)/Passivity (+): an interesting and not abstract characterization of physics (−), is opposed to a not negative, but neither too positive, connotation of physicists and their role in today's society (+).
2. Refusal (−)/Acceptance (+): physics is characterized as useless for daily life and for the future (−), in opposition to a society's view where physics plays an important role for solving the world demands (+).
3. Wish to be (−)/Out of touch (+): the desire of being more engaged in physics and the request for greater recognition of its relevance from the society (−), is opposed to a characterization of physics as too demanding and abstract (+).

Following the SCPT framework, these dimensions are supposed to be stable as the research sample varies.

2.2 Engagement in Physics

As briefly mentioned in the introductory section, engagement is a multi-dimensional construct that measures the intensity of students' productive involvement in any activity (in our case related to physics). Unlike other definitions in the literature, this one does not include aspects related to motivation, which we include in attitudes towards physics.

There are measures of engagement that use some self-regulated learning strategies as indicators of a cognitive component of this construct, on the assumption that such strategies imply a certain level of engagement. In fact, engagement represents an iterative cyclical process of monitoring and control that leads to the calibration of goals and learning strategies (Ben-Eliyahu et al. 2018).

Following the model reported in Ben-Eliyahu et al. (2018), it is possible to split this construct into a cognitive and a behavioral dimension, although some studies show that in some processes the two dimensions act as a single factor. In particular, the cognitive component represents the extent to which the individual reflects on the learning activity or is focused on a particular task. This differs from the nature of the cognitive processes involved (e.g. verification, elaboration or conceptual reorganization), as these vary in relative intensity depending on the demands. Instead, the behavioral dimension focuses on what students involved in an activity do (or do not do) in terms of some typical attitudes and their frequency (e.g. asking for clarification during a lesson or discussing a topic with classmates).

2.3 Self-Efficacy in Physics

As briefly mentioned above, self-efficacy is not a single characteristic of the learner, but a dynamic set of beliefs. In line with Bandura's social cognitive theory (Bandura 2001), the level of self-efficacy directly influences the amount of energy expended in performing a task and the duration of the effort. It is therefore an important predictor of academic performance and persistence. It is therefore possible to hypothesize a positive correlation between engagement and self-efficacy, although there is no general agreement on a possible causal relationship between the two constructs.

Finally, this measure refers to assigned activities or tasks (e.g. solving kinematics problems) but not to specific demands (e.g. solving a particular kinematics problem). In physics, it is useful to explore self-efficacy in relation to experiments, as they tend to represent an area where the skills required are significantly different from those traditionally required to solve exercises.

3 Methods

3.1 Participants

For this study, we involved a convenience sample constituted by 1,971 Italian students (864 female students), divided into four groups:

- *Subsample 1:* 708 high school students (female students = 50.6%) who voluntarily participated in a public engagement program at the university of the first author, focused on physics-related activities.
- *Subsample 2:* 449 high school students (female students = 45.9%) who voluntarily participated in a public engagement program at the university of the first author, focused on non-physics activities. Approximatively each student of the first two subsamples participated to a total of about 20 h of extra-curricular activities.
- *Subsample 3:* 427 freshman engineering students (female students = 37.5%) who were attending different degree programs: electronics, telecommunication, informatics and biotechnology.
- *Subsample 4:* 387 undergraduate physics students (female students = 36.2%).

The female under-representation of the last two groups reflects that of STEM university courses.

3.2 Instruments

The collected data consist of the answers to three self-report surveys, submitted online:

- *Attitudes towards physics* (36 items): the first part concerns the perceived value of physics in today's society (4 items), the recognition of the difficulty of physics (6 items), the intrinsic interest in physics (4 items), the perceived personal relevance of dealing with physics in the future (3 items), the perceived relevance of learning physics at school (4 items); the second part concerns the recognition of the role of physicists in society (9 items) and the perceived value of the career of physicists (6 items). The same instrument was used in (Testa et al. 2021).
- *Engagement in physics* (20 items), which measures aspects concerning self-regulation (5 items), information processing (10 items) and participation in lessons (5 items).
- *Self-efficacy in physics* (7 items), which measures self-beliefs about designing experiments (4 items), problems solving and presenting contents orally (3 items).

The students were asked to state their degree of agreement on the survey's items, by using a 5-point Likert scale (1 = not at all; 2 = not very; 3 = fairly; 4 = mostly; 5 = completely).

3.3 Data Analysis

To answer RQ1, we conducted a Multiple Correspondence Analysis (MCA) of the students' responses regarding their attitudes towards physics. MCA is a technique used in psychological research to deal with categorical scales (Salvatore et al. 2018).

All extracted factors are orthogonal to each other, and each student is assigned a score for each axis, so that the average of the scores for the entire sample is zero. As usual, only the first three extracted dimensions with the highest percentage of explained variance are reported: these represent the lines of force that shape the semiotic space of the cultural milieu of physics.

The interpretation of these dimensions is based on the modalities with the highest *t-value* (in absolute value): in fact, these values measure the distance of a modality from the origin, i.e. its contribution to the factorial axis. Consequently, each component marks a latent dimension of meaning, interpreted in terms of a generalized opposition between a positive and a negative polarity. The process of interpretation consists in coherently linking the main modalities that emerged from the analysis by means of logical connectives, in order to obtain the synthesis of a semantic opposition that identifies the main characteristic (in this case, attitudes towards physics) represented by the factorial axis in question.

To answer RQ2, two confirmatory factor analyses (CFAs) were carried out to validate the multi-dimensional models of both engagement and self-efficacy in physics. The dimensions were discussed above. To assess the models' fit the main CFA indexes were calculated and compared with standard thresholds [34]: Root Mean Square Error of Approximation (RMSEA < 0.08), Normed Fit Index (NFI > 0.9), Tucker-Lewis Index (TLI > 0.9), Incremental Fit Index (IFI > 0.95) and Comparative Fit Index (CFI > 0.9). Finally, to establish the convergent and divergent validity, the Average Variance Extracted (AVE > 0.5) and the Composite Reliability (CR > 0.7) were calculated for each factor. As usual, to get the best model, we used modification indexes to define new parcel variables, taking the average scores of the pairs of variables with high modification indexes (Schreiber et al. 2006).

Finally, to answer RQ3, different structural equation models were tested to identify the cause-effect relationships between the dimensions of the constructs investigated. In all models, the dimensions of the construct of attitudes towards physics were used as independent variables, while the dimensions of engagement and self-efficacy were subsequently used as mediators and dependent variables, respectively. A covariance constraint was imposed between the mediating components, while a single dependent variable was used. The theoretical justification for these models lies in the fact that attitudes towards physics include qualitative aspects of the discipline, including extracurricular aspects, which are a priori present and are hypothesized to influence levels of engagement and self-efficacy. On the other hand, the cause-effect relationship between these last two variables is not sufficiently established in the literature, especially in the context of science learning (Lin 2021), so we tested the different models to explore the correct model (if the results provide one and only one solution).

For the structural equation modelling (SEM), we have reported the same indices as for the CFA, and for the standardized regression coefficients we have also considered the significance of the effects, distinguishing between direct and indirect effects: the latter are those of the independent variable on the dependent, and are equal to the product of the direct effects. The total effect is the sum of all direct and indirect effects of an independent variable on a dependent variable. When the model fit was satisfactory, a path analysis was finally performed.

Multiple correspondence analysis was performed using the SPAD v.5.5 package, while confirmatory factor analysis, structural equation modelling and other statistical analyses were performed using IBM AMOS v.26.

4 Results

4.1 Multiple Correspondence Analysis

The multiple correspondence analysis extracted 133 factorial dimensions. After applying the Benzécri/Greenacre simplified formula of re-evaluation, these factorial dimensions account for the 85.66% of the total inertia (respectively, 51.29%, 27.37%, 7.00%). The description and interpretation of the three factorial axes confirm the results reported in Sect. 2.1, so we conceptualized the semiotic space describing attitudes towards physics in terms of the following oppositions: *Relevance/Passivity, Refusal/Acceptance,* and *Wish to be/Out of touch.* Thus, the hypothesis regarding the stability of the generalized oppositions, when changing the sample, is confirmed.

4.2 Confirmatory Factor Analysis

The confirmatory factor analysis of the self-efficacy and engagement items required the removal of some items to achieve a good fit, but the hypothesized dimensions were confirmed. Figures 1 and 2 show the standardized estimates of the regression coefficients for self-efficacy and engagement, respectively.

For the engagement construct, we had to carry out some items' parceling. Three new variables were defined, as reported in Table 1, taking the average scores of the items reported in the table.

4.3 Structural Equation Modeling

After having explored all the possible models, only the one shown in Fig. 3 showed an acceptable fit ($\chi^2[6] = 7.299, p < 10^{-3}$; RMSEA = 0.057; CFI = 0.992, TLI = 0.979). All direct effects are significant, except for those associated with the dimension "Wish to be → Out of touch". Indirect effects are also present, but those associated with the third factor extracted from the analysis of the questionnaire on attitudes towards physics are not significant. The remaining four possible paths are significant. The results show that the mediation of the cognitive dimension of engagement is stronger than that of the behavioral component.

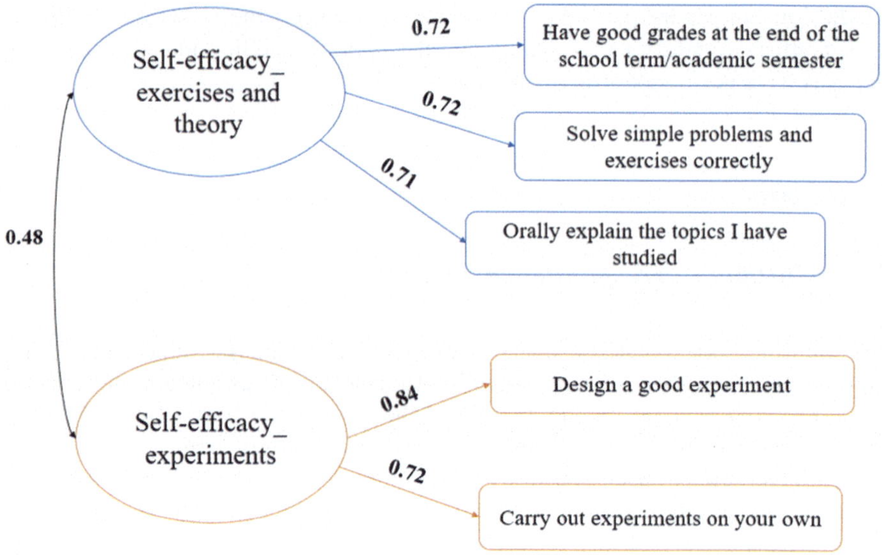

Fig. 1 Standardized regression weights of the confirmatory factor analysis for self-efficacy. The CFA indexes have the following values: χ^2 [11] = 92.984, p < 10^{-3}; RMSEA = 0.062, CFI = 0.979, TLI = 0.961, NFI = 0.977, IFI = 0.980, AVE > 0.514, CR > 0.759

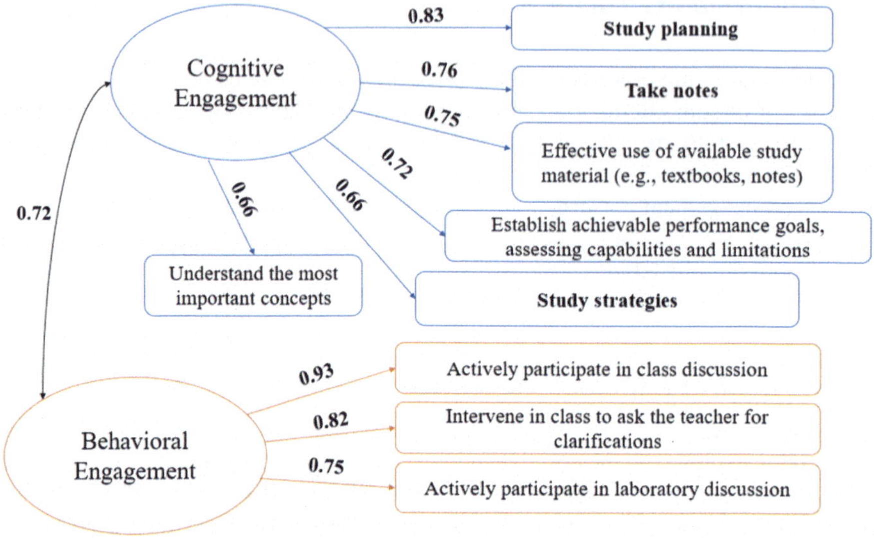

Fig. 2 Standardized regression weights of the confirmatory factor analysis for engagement. The new variables "parcels" are reported in bold. The CFA indexes have the following values: χ^2 [26] = 323.443, p < 10^{-3}; RMSEA = 0.076, CFI = 0.969, TLI = 0.957, NFI = 0.966, IFI = 0.969, AVE > 0.536, CR > 0.871

Table 1 Items' parceling following the CFA on engagement

New variables ("parcels")	Items corresponding to the parcels
Study planning	Always respect the established study plan Maintain a constant pace of study throughout the school/academic year Organize your home study time effectively Plan your study according to the difficulty of the topics
Study strategies	Adequately elaborate in written form the topics studied Make use of appropriate and effective learning strategies (e.g., diagrams, maps)
Take notes	Take clear notes when the teacher explains Grasp and re-elaborate the essential contents of a lesson Grasp and re-elaborate the essential contents of an experiment

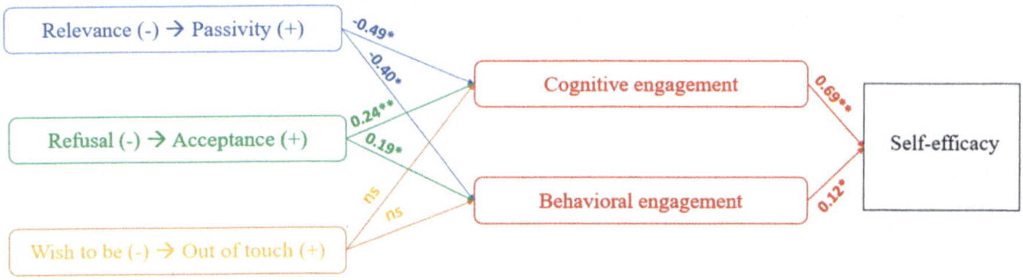

Fig. 3 Validated model of cause-and-effect relationships between attitudes, engagement, and self-efficacy in physics. The standardized regression weights related to direct effects are reported, with their statistical significance follows this classification: *ns* → $p > 0.05$, * → $0.01 \leq p \leq 0.05$, ** → $p < 0.01$

5 Discussion

5.1 *What Are the Dimensions Underlying the Attitudes Towards Physics Identifiable for the Sample Involved in the Research?*

The results reported in Testa et al. (2021) about the interpretation of the three main generalized opposition of the semiotic space regarding attitudes towards physics are confirmed. Here we deepen the interpretation of these dimensions, to subsequently attribute a more precise meaning to the relationships we found through the structural equation modeling.

The first dimension extracted from the responses to the attitudes scale "*Relevance → Passivity*" indicates that the main axis of the semiotic space is represented by the perception of the level of abstraction in physics, namely how much it can represent real objects and describe real phenomena. This factor has been interpreted in terms of the perceived relevance of the discipline, and the interpretation follows the assumption that any language, regardless of its degree of complexity and completeness, loses its importance if it is not useful for solving practical problems.

The school (and university) experience regarding physics seems to play a more decisive role in determining the position along the dimension "*Refusal → Acceptance*". This type of experience determines, according to the interpretation provided, the level according to which one is willing to interface with the discipline. This level has to be interpreted in a quantitative way (i.e., how much time one is willing to spend studying physics), especially in comparison to other disciplines. Furthermore, as mentioned above, the two extreme polarities (total refusal and total acceptance) are not equidistant from the origin of the factorial axis: our sample is on average polarized towards the negative pole of this dimension, as well as for the first factor.

Finally, the semantic line "*Desire to be → Out of touch*" represents the degree of perceived repulsion that physics exerts towards students. This repulsion is determined not only by the perceived degree of commitment required in studying physics, but mostly by the consequent social recognition, also in practical terms of job opportunities. Therefore, the perception of the social recognition of physicists' role is linked with the frustration due to the repulsion exercised by the discipline. Vice versa, the satisfaction due to the good relationship with the discipline coexists with, and probably causes, a request for more opportunities for those who undertake a career closely related to physics.

5.2 What Are the Dimensions of Engagement and Self-Efficacy Identifiable for the Sample Involved in the Research?

The hypothesized structure of the engagement and self-efficacy constructs was confirmed. In the first case, this result shows that the self-regulation and processing aspects form a single dimension. It was therefore possible to adopt the subdivision into cognitive and behavioral factors, as interpreted in the theoretical framework. The former can be further subdivided into sub-dimensions, as shown by the operation of item parceling. It is interesting to note that the self-regulation component is highly coherent and has the highest regression coefficient, while the aspects related to processing seem to be more independent of each other. Furthermore, for the behavioral component, participation in laboratory activities seems to have a lower weight in determining the factorial scores. This does not mean that the sample has a low perceived ability to carry out these activities, but rather that they are not sufficiently explored in the teaching practice and are therefore perceived by the students in a—qualitatively—different way than the other activities. On the other hand, experimental activities in physics are often absent both at school and at university level.

Regarding self-efficacy, the bi-dimensional model is confirmed, as the CFA suggests that the skills related to experimental activities belong to a different component than the skills related to problem solving and oral exposure. The correlation coefficient between the two dimensions is not high (<0.5), confirming the different way in which laboratory activities are perceived by students in their school experience.

5.3 What Relationships Exist Between Attitudes Towards Physics and the Dimensions of Engagement and Self-Efficacy?

Only one structural model shows an acceptable fit, reinforcing the importance of the solution. Engagement acts as a full mediator through its two components: this suggests that perceived levels of self-regulation, information processing and participation skills influence perceived ability to perform a standard task. Furthermore, our data do not fit the model when the dependent variable is represented by experimental self-efficacy, showing that this scheme cannot be applied to activities that are perceived as qualitatively different from non-laboratory activities.

Since the effects (both direct and indirect) associated with the dimension "Desire to be → Out of touch" are not significant, it can be concluded that, in determining the role of this dimension in the structural equation, the aspect related to the recognition of the role of physics and physicists in society is predominant, so that this factor does not influence the level of self-efficacy. Regarding the first two components of attitudes towards physics, we found that the shifts towards the polarities "Relevance" and "Acceptance" are related to a higher level of engagement, which in turn implies a higher level of self-efficacy. However, the cognitive component has a more significant effect than the behavioral one. This is not obvious, as passivity and refusal could lead to lower engagement. Conversely, the results show that these attitudes mostly imply low perceived processing and self-regulation skills, probably because they require activation at a more fundamental cognitive level.

6 Conclusions

From our results, it can be concluded that the hypothesized relationships between the constructs of attitudes, engagement and self-efficacy in physics have been substantially confirmed. Furthermore, the construct of attitudes towards physics, analyzed in the framework of the SCPT, has allowed us to validate a model in which qualitative views of the discipline (in particular those interpreted in terms of the generalized oppositions "relevance → passivity" and "rejection → acceptance") indirectly influence self-efficacy (at least when the part related to experiments is excluded), with the cognitive and behavioral dimensions of engagement acting as total mediators.

Further efforts are needed to test the validity of the hypothesis that the whole sample can be treated as a single cultural entity, which is a necessary condition for adopting the SCPT framework. Furthermore, the stability of the model should be tested by varying the cultural milieu in which the research sample developed their attitudes towards physics, as this may strongly influence the results. Finally, the role of other possible moderating variables of the identified relationships needs to be investigated, as, for instance, gender, the nature of prior experience with physics, and the performance in physics.

References

Adams WK, Perkins KK, Podolefsky NS, Dubson M, Finkelstein ND, Wieman CE (2006) New instrument for measuring student beliefs about physics and learning physics: the colorado learning attitudes about science survey. Phys Rev Spec Topics Phys Educ Res 2(1):010101
Ajzen I (2001) Nature and operation of attitudes. Annu Rev Psychol 52(1):27–58
Angell C, Guttersrud Ø, Henriksen EK, Isnes A (2004) Physics: frightful, but fun. pupils' and teachers' views of physics and physics teaching. Sci Educ 5(88):683–706
Bandura A (2001) Social cognitive theory: an agentic perspective. Annu Rev Psychol 52:1–26
Barmby P, Defty N (2006) Secondary school pupils' perceptions of physics. Res Sci Technol Educ 24(2):199–215
Barmby P, Kind PM, Jones K (2008) Examining changing attitudes in secondary school science. Int J Sci Educ 30(8):1075–1093
Ben-Eliyahu A, Moore D, Dorph R, Schunn CD (2018) Investigating the multidimensionality of engagement: affective, behavioral, and cognitive engagement across science activities and contexts. Contemp Educ Psychol 53:87–105
Bøe VM, Henriksen EK (2013) Love it or leave it: Norwegian students' motivations and expectations for postcompulsory physics. Sci Educ 97(4):550–573
Cajas F (2001) The science/technology interaction: implications for science literacy. J Res Sci Teach 38(7):715–729
Crano WD, Prislin R (2006) Attitudes and persuasion. Annu Rev Psychol 57:345–374
Eccles J (2007) Where are all the women? In: Ceci SJ, Williams WM (eds) Why aren't more women in science? American Psychological Association, pp 199–209
EU (2010) Europe 2020. A European strategy for smart, sustainable and inclusive growth. European Commission
Eurydice (2012) Key data on education in Europe 2012. Eurydice
Fredricks JA, Blumenfeld PC, Paris AH (2004) School engagement: potential of the concept, state of the evidence. Rev Educ Res 74:59–109
Irving PW, Sayre EC (2015) Becoming a physicist: the roles of research, mindsets, and milestones in upper-division student perceptions. Phys Rev Spec Topic Phys Educ Res 11(2):020120
Kessels U, Rau M, Hannover B (2006) What goes well with physics? Measuring and altering the image of science. Br J Educ Psychol 76(4):761–780
Kind PM, Jones K, Barmby P (2007) Developing attitudes towards science measures. Int J Sci Educ 27(7):871–893
Lin T (2021) Multi-dimensional explorations into the relationships between high school students' science learning self-efficacy and engagement. Int J Sci Educ. https://doi.org/10.1080/09500693.2021.1904523
Mujtaba T, Reiss MJ (2013) Inequality in experiences of physics education: secondary school girls' and boys' perceptions of their physics education and intentions to continue with physics after the age of 16. Int J Sci Educ 35(11):1824–1845

National Science Board (2014) What is the S&E retention rate in U.S. 4-year institutions? https://www.nsf.gov/nsb/sei/edTool/data/college-10.html

National Science Foundation (2017) Improving undergraduate STEM education: education and human resources (IUSE: EHR). https://www.nsf.gov/pubs/2017/nsf17590/nsf17590.htm

OECD (2008) Encouraging student interest in science and technology studies. Global Science Forum

Oon PT, Subramaniam R (2011) On the declining interest in physics among students—from the perspective of teachers. Int J Sci Educ 33(5):727–746

Osborne J, Dillon J (2008) Science education in Europe: critical reflections. A report to the Nuffield Foundation

Ramsden JM (1998) Mission impossible? Can anything be done about attitudes to science? Int J Sci Educ 20(2):125–137

Regan E, DeWitt J (2015) Attitudes, interest and factors influencing STEM enrolment behaviour: an overview of relevant literature. In: Henriksen E, Dillon J, Ryder J (eds) Understanding student participation and choice in science and technology education. Springer

Reid N, Skryabina E (2002) Attitudes towards physics. Res Sci Technol Educ 20(1):67–81

Salvatore S, Fini V, Mannarini T, Veltri GA, Avdi E, Battaglia F et al (2018) Symbolic universes between present and future of Europe: first results of the map of European societies' cultural milieu. Plos One 13(1):e0189885

Sawtelle V, Brewe E, Kramer LH (2012) Exploring the relationship between self-efficacy and retention in introductory physics. J Res Sci Teach 49(9):1096–1121

Schreiber JB, Nora A, Stage FK, Barlow EA, King J (2006) Reporting structural equation modeling and confirmatory factor analysis results: a review. J Educ Res 99(6):323–338. https://doi.org/10.3200/JOER.99.6.323-338

Stokking KM (2000) Predicting the choice of physics in secondary education. Int J Sci Educ 22(12):1261–1283

Testa I, De Luca Picione R, Scotti di Uccio U (2021) Patterns of Italian high school and university students' attitudes towards physics: an analysis based on semiotic-cultural perspective. Eur J Psychol Educ

UNESCO (2010) Engineering: issues, challenges and opportunities for development. United Nations Educational, Scientific and Cultural Organization

United Nations (2012) Millennium development goals. www.un.org/millenniumgoals/

Williams C, Stanisstreet M, Spall K, Boyes E, Dickson D (2003) Why aren't secondary students interested in physics. Phys Educ 38(4):324–329

The Role of Domain-Specific Growth Mindset (Implicit Theories) at the Beginning of STEM University Studies: Comparing New and Old Measures

Malte Diederich and Verena Spatz

Abstract Mindsets according to Dweck can have a great impact on learning and play a particularly important role when facing challenges such as transition from school to university. This study compares classic domain-general (intelligence) mindset measures to a subject-specific measure for STEM university studies in Germany. We conducted a questionnaire study with first year students at two survey time-points (survey 1 at the beginning and survey 2 two month into the first semester). With the data taken at the first time-point, we performed an exploratory factor analysis on 17 mindset items from different scales and formed a short subject-specific scale. We checked construct validity by comparing it to four scales from the "mindset meaning system". For the domain-general scale, we collected further evidence that students' interpretation of the word "intelligence" impacts scale values. In the second survey, we applied the new measure and found significant correlations between a subject-specific growth mindset and lower intentions to change mayor or drop out, while the domain-general scale did not show significant correlations. Finally, we observed an interaction between academic self-concept and mindset with no correlations in survey 1 and significant correlations in survey 2 for the subject-specific scale. Based on this results we recommend using a subject-specific mindset scale in this context.

1 Theoretical Background

Starting university studies for a degree in STEM poses many new challenges for students compared to school. With typically only one big exam in every lecture at the end of semester, the German university system expects a high degree of self-motivation and self-organization from its students. Not all of them master these

M. Diederich
Technical University of Darmstadt, Hochschulstrasse 12, 64289 Darmstadt, Germany
e-mail: malte.diederich@physik.tu-darmstadt.de

V. Spatz (✉)
Technical University of Darmstadt, Hochschulstrasse 12, 64289 Darmstadt, Germany
e-mail: verena.spatz@physik.tu-darmstadt.de

S. Faletič and J. Pavlin (eds.), *Teaching and Learning Physics Effectively in Challenging Times*, Challenges in Physics Education, https://doi.org/10.1007/978-3-031-72541-8_8

challenges successfully and the dropout rate in these subjects has been above average for years (Heublein et al. 2022). This is problematic given the continued high demand for graduates. A possible level of support here is on the individual-motivational level. Dresel and Grassinger (2013) observed a dramatic drop in motivation for a significant proportion of students within the first semester, calling for support.

1.1 Mindset Research

Mindset theory according to Dweck (1999) might offer one possible way to support students in this critical phase. Mindset is the belief that personal characteristics such as intellectual abilities can be developed ("growth mindset" or incremental theory) or are largely predetermined and unchangeable ("fixed mindset" or entity theory, ebd.). These beliefs were originally referred to as implicit theories or lay theories (ebd.). Similar to a scientific theory, they form a framework for interpreting the meaning of everyday events and making predictions about possible effects. In contrast to a scientific theory, they are highly subjective and rarely made explicit (Yeager and Dweck 2012).

Years of research have shown that a fixed mindset is associated with unfavorable learning reactions and attitudes (f.e. Burnette et al. (2013)). Four constructs found connected to a fixed mindset are higher negative effort beliefs (i.e. associating higher effort with lower talent), higher performance-avoidance goals (i.e. focusing on avoiding to look dumb), lower learning goals (i.e. focusing less on learning new things and more on getting good grades) and helpless responses (i.e. connecting failure to lack of talent), which have been summarized as the "mindset meaning system" (Yeager and Dweck 2020).

Large studies (e.g. U.S. National Study of Learning Mindsets (8th grade), N = 12,490 (Yeager et al. 2019); PISA 2018, N > 600,000 (Gouëdard 2021)) have also found relations between mindset and academic performance. Other studies have found null results (e.g. Czeck university applicants, N = 5653 (Bahník and Vranka 2017), German students (11th grade), N = 463 (Dinger et al. 2013)). One possible explanation lies in heterogeneity between cultures and within groups (Yeager and Dweck 2020). In the 2018 PISA study, correlations between mindset and test results in Germany were among the lowest together with Czeck Republic while in most countries moderate correlations were found (Gouëdard 2021), which stresses the need for further analysis. Another explanation might lie in how mindset items are translated and interpreted in different languages.

1.2 Defining and Measuring Mindset

The most common way to measure mindset in an academic setting is with students' agreement to several statements about the malleability of intelligence, measured on a 6-point Likert scale (e.g. "Your intelligence is something about you that you can't change very much", single fixed mindset item used in the PISA study (Gouëdard 2021)). While some studies differentiate between growth and fixed mindset formulations (f.e. Limeri et al. (2020)), creating two separated scales, mindset has been largely considered as a one-dimensional spectrum (f.e. Dweck 1999; Hong et al. 1999). Other studies differentiate between "My intelligence" and "Your intelligence" (de Castella and Byrne 2015), but all items are still very similar in wording and correlations are very high (0.80, ebd.). Recent studies discuss the effect of individual definitions of intelligence on the scale values. Depending on their definition, students might understand items differently, threatening overall construct validity. Students who define knowledge as part of intelligence lean more towards a growth mindset on the scale (Limeri et al. 2020; Santos et al. 2021), which should be investigated further empirically.

Similar to other motivational components like academic self-concept (Dresel and Lämmle 2017), mindset believes might be different depending on the specific subject. Several studies try to operationalize a subject-specific mindset by replacing the word "intelligence" by a subject-specific term like "chemical intelligence" (Limeri et al. 2020) or "talent for physics" (Broome 1998). However, problems with interpretation of these terms might persist. In the last years, our research group has developed a physics specific scale. We started by interviewing students (Spatz and Goldhorn 2021; Diederich and Spatz 2021) and teachers (Spatz and Lippmann 2018) to identify their mindset. Rather than asking about the malleability of intelligence or talent directly, we discussed the consequences of talent and intelligence for acquiring physics skills: *Could you get a good understanding of physics, even if you are not talented? Will some people never be good in a subject, even if they try really hard and study a lot?*

From these interviews, we developed two sets of items about the role of talent and effort for success, which have been tested with physics university students (Rehberg et al. 2019). In recent time, there have been similar approaches (f.e. physics Malespina et al. (2022), STEM Limeri et al. (2023)), but to our knowledge there has not been a comparison to existing measures. One can therefore come to the conclusion that although some scales have been proposed for measuring mindsets, these are constructed differently and also provide heterogeneous findings. What is missing is a systematic, empirical comparison of the scales, which also includes a consideration of the correlation of individual scales with other constructs.

2 Objectives and General Approach

In this study we want to apply our physics specific scale to the STEM context and compare it to several other approaches to measure mindset. Correlations with the mindset meaning system and academic self-concept are used check for construct validity. By examining different definitions of intelligence, we want to further investigate potential problems of the domain-general scale. The first goal is to form a short subject-specific mindset measure and to improve the way of measuring mindset in a STEM university context. The second goal is to apply this measure to investigate the predictive power of mindsets for intentions to changing ones' major or even dropping out of university education. To reach these goals, a questionnaire study with two survey time-points (survey 1 at the beginning and survey 2 two month into the first semester) has been conducted. The data taken will allow to answer the following research questions:

- (Q1a) How do domain-general and subject-specific mindset items relate to each other?
- (Q1b) How do different mindset scales correlate to the mindset meaning system and academic self-concept at the start of university studies?
- (Q1c) How do the definitions of intelligence influence the score in different mindset scales?
- (Q2) How are domain-general and subject-specific mindset scales correlated to academic self-concept and intentions to change major or drop out two month into the first semester?

3 Survey 1—Method

We conducted an online survey in introductory STEM courses at our university in the fall 2021 semester (local ethics committee approved) during the first week of the semester. Students participated voluntarily during lecture time. Of 826 students, 21 students were excluded for answering items in an average time of less than 1,5 s, leading to 805 participants (28 % of all first semester students in STEM subjects at TU Darmstadt, 31 % female, 75 % in first semester). Participants come from in total 24 different STEM fields (largest groups: mechanical industrial engineering (16 %), informatics (10 %), electrical engineering (8 %), mechanical engineering, medical technology, physics, mechatronics, mathematics (each 6 %)).

3.1 Measuring Instruments

The survey consisted of 17 mindset items, 20 items from four scales of the mindset meaning system and 3 items measuring subject-specific academic self-concept. All items were measured on a 6-point Likert scale from 1 (completly disagree) to 6 (completly agree). These mindset items can be found in Table 1 and will be explained in more detail here:

3.1.1 Domain-General Mindset

We included two fixed mindset items ("My intelligence") from a German scale (Spinath and Stiensmeier-Pelster 2001), a growth mindset item (''My intelligence")

Table 1 Descriptive statistics

Variable	Mean	SD	% Agree	Curtosis	Skew
Malleability of intelligence (domain-general)					
I1: I can learn new things, but I can't really change by intelligence	3.02	1.16	33.4 %	−0.57	0.33
I2: Your intelligence is something about you that you can't change very much	2.85	1.34	28.8 %	−0.47	0.51
I3: I have a certain amount of intelligence, that I cannot change much	3.11	1.20	37,0 %	−0.41	0.21
I4: With enough time and effort I think I could significantly improve my intelligence level	4.34	1.20	79.6 %	0.21	−0.70
Malleability of talent (subject-specific)					
M1: I can learn new things, but I can't really change, how talented I am in my field of studies.	3.28	1.22	44.7 %	−0.70	0.01
M2: My talent for my field of studies is a part of me that I can't change much.	3.21	1.12	41.5 %	0.21	−0.70
M3: I have a certain amount of talent for my field of studies, and I can't do much to change it.	2.97	1.06	31.2 %	−0.42	0.04
Math item					
Y1: Being a "math person" or not is something that you really can't change. Some people are good at math and other people aren't	3.22	1.29	44.1 %	−0.70	0.02
Role of effort and talent (subject-specific)					
T1: You will never be good in my field of study, if you did not inherit a certain talent for it	2.77	1.22	27.8 %	−0.45	0.36
T2: If you are not talented in your field of study, there is not much you can do to improve significantly	2.28	0.96	10.7 %	0.19	0.59
T3: You need to inherit a specific talent for my field of study	2.47	1.17	20.4 %	−0.48	0.46
T4: You either understand my field of studies or you don't	2.54	1.22	20.2 %	−0.12	0.62
T5: I believe that some people are good in my field of studies and others will never really learn it, no matter how much they try	2.94	1.21	30.9 %	−0.34	0.34
E1: If they really want it and make an effort, any of my fellow students can successfully study my field of studies	4.66	1.12	85.6 %	0.25	−0.77
E2: Even if they invest a lot of time: Not every one of my fellow students could successfully study my field of study	3.05	1.28	38.5 %	−0.70	0.16
E3: Each of my fellow students can understand my subject in depth. You just have to do enough for it	4.45	1.11	81.7 %	−0.04	−0.56
E4: Even if they try hard in my field of study, some of my fellow students could never understand the content in depth	2.89	1.19	29.7 %	−0.30	0.33

All items are Likert scaled from 1 (completly disagree) to 6 (completly agree). N = 805, survey 1 the at start of the semester. Italic items are incremental items (growth mindset), five bolded items are used in short subject-specific scale. Note that the original items are in German. Translating some terms is difficult and might change the meaning slightly (f.e. "Begabung"—talent, "Anstrengung"—effort) translated from [16] and the item used in the PISA survey ("Your intelligence") [10], (I1 to I4, see Table 4, $\alpha = 0.87$). Additionally, we included one math item (Y1), which has been recommended as a short three item scale together with two fixed "Your intelligence" items in a recent paper discussing mindset controversies by leading mindset researchers Dweck and Yeager (see appendix of Yeager and Dweck (2020)).

3.1.2 Malleability of Talent (Subject-Specific)

We included three physics items from [20], which we adapted in wording by changing "physics talent" to "talent for my field of study". The items are very similar in wording to the intelligence scale items ((M1 to M3, $\alpha = 0.68$).

3.1.3 Role of Talent and Effort (Subject-Specific)

We included nine physics items developed by our research group (Rehberg et al. 2019), which we also adapted in wording by changing "physics" to "my field of studies". Five items are about the role of talent (T1 to T5) and four items are about the role of effort (E1 to E4) for succeeding in one's field of study ($\alpha = 0.84$).

3.1.4 Mindset Meaning System

We included four scales found related to mindset in other studies. All scales were translated to German. For negative effort beliefs, we combined two items from Dweck (1999) with a third item developed from interview material (e.g. "If you have to work very hard for your studies, you are not so gifted for it.", $\alpha = 0.71$). For performance avoidance goals, we used a four item scale from Midgley et al. (2000) (e.g. "One of my goals in class is to avoid looking like I am having trouble doing the work.", $\alpha = 0.81$). For learning goals, we included three items from Dweck (1999). α was not acceptable (0.57), with one item loading on a the helpless response scale in EFA. Droping this item lead to an aceptable score with only two items (e.g. "It's much more important for me to learn something in class than it is to get the best grades.", $\alpha = 0.71$). For helpless response, we included 9 items from Robins et al. (2002). The original items were separated into two subscales (helpless response and mastery response). In our survey, both scales could not be separated into two factors and two items loaded scattered. Additionally, two items had small factor loadings and were excluded (-0.38), while the item from the learning goal scale clearly loaded on the factor. We combined these six items into one helpless response scale by inverting the mastery response items (e.g. "When I fail to understand something, I become discouraged to the point of wanting to give up.", $\alpha = 0.75$).

3.1.5 Subject-Specific Academic Self-Concept

We used a three-item scale (confidence in one's intelligence-scale) by Dweck (1999) adapted to the university context by Dresel et al. (2013). As an exception, the items consisted of a pair of statements about low and high confidence on a 6-point answer scale (e.g. "I feel pretty confident-" versus "I'm not very confident about my intellectual abilities, which I need in my field of studies" $\alpha = 0.76$).

3.1.6 Definition of Intelligence

After the 40 items, we asked students how they define the term intelligence by selecting up to five given statements. The statements (see Table 5) were generated by comparing definitions stated by chemistry students (Santos et al. 2021) to the definitions given by 13 German students in an interview study (Diederich and Spatz 2021).

4 Survey 1 Results

In the following, the results for the first three research questions (Q1a, Q1b, Q1c) are discussed separately,

4.1 Q1a: How Do Domain-General and Subject-Specific Mindset Items Relate to Each Other?

To answer the question, we conducted a exploratory factor analysis (EFA). Table 2 shows the results of MAP test and eBIC for models with different numbers of factors and the assigned eigenvalues. The values were calculated with the psych package in r (fa(data, nfactors = 2 or 3, SMC = TRUE, fm ="pa", rotate = "promax")). For all 17 mindset items, Kaiser-Guttman criterion suggests three factors, while MAP test suggests only two factors. The factor loadings of both models can be seen in Table 3 (consult Table 1 for full item names). Factor 1 includes all subject-specific items, while factor 2 can be identified as the domain-general items (I1 to I4). In the three factor model, the incremental items (E1 and E3) build the third factor. E2 and E4 are inverse formulations of E1 and E3 and load mixed on F1 and F3. This makes it difficult to interpret the third factor. The intelligence items all clearly load on one factor, independent from the incremental formulation (I4) or the difference between "Your intelligence" (I2) and "My intelligence" (I1 and I3).

When excluding the three incremental items, the eigenvalue of factor three drops to 0.91, suggesting only two factors (Table 2). Looking again at the factor loadings

Table 2 Statistics summary for different factor numbers in EFA

# of factors	17-item version			14-item version			8-item version		
	EV	MAP	cum. var.	EV	MAP	cum. var.	EV	MAP	cum. var.
1	6.37	0.0024	37%	5.68	0.0023	41%	3.78	0.010	47%
2	1.95	**0.0006**	49%	**1.55**	**0.0010**	52%	**1.38**	**0.009**	65%
3	**1.16**	0.0009	56%	0.91	0.0030	58%	0.68	0.045	73%
4	0.86	0.0033	61%	0.84	0.0051	64%	0.55	0.110	80%
5	0.85	0.0053	66%	0.74	0.0153	69%	0.52	0.224	86%

EV, Eigenvalues; MAP, Velicer's Minimum Average Partial; cum. var., cumulative variance explained

in table 2, the malleability of talent scale (M1, M2, M3) loads mixed on both scales. The math item Y1 from Yeager et al. (2020) loads on the subject-specific factor F1 and not on the domain-general factor F2. In the STEM university context, being a "math person" or "being good in my field of studies" seems to be highly related.

To create clear factors, we excluded the malleability of talent scale and reduced the subject specific items down to five by dropping items with lower factor loadings and still keeping the different content facets of the scale. The resulting eight items lead to two clearly separable factors (F1 subject-specific mindset and F2 domain-general mindset) with a correlation of 0.55 (see Tables 2 and 3).

4.2 *Q1b: How Do Different Mindset Scales Correlate to the Mindset Meaning System and Academic Self-Concept at the Start of University Studies?*

Table 4 shows descriptive statistics and the correlations of the domain-general and the subject specific mindset scale. Correlations to the mindset meaning system (negative effort believes, learning goals, performance avoidance goals, helpless response) are significant in the expected direction for both scales, while there are no correlations with subject-specific academic self concept.

4.3 *Q1c: How Do the Definitions of Intelligence Influence the Score in Different Mindset Scales?*

Table 5 shows the mean values for both mindset measures depending on whether students agree or disagree with a statement about intelligence. Students could select several definitions. The group of students including knowledge and experience in

Table 3 Results of EFA for different numbers of mindset items

	I1	I2	I3	**I4**	Br1	Br2	Br3	Y1	B1	B2	B3	B4	AB5	**A1**	A2	**A3**	A4	correl F	
17 items																		F2	F3
F1					0.34	0.37		0.59	0.53	0.66	0.66	0.56	0.74	−0.49	0.66	−0.59	0.70	0.59	
F2	0.89	0.79	0.84	−0.81	.37														
Com.	0.76	0.60	0.73	0.49	0.40	0.32	0.24	0.37	0.35	0.44	0.46	0.25	0.50	0.26	0.40	0.31	0.43		
F1			0.30	0.32	0.49	0.67	0.49	0.59	0.38	0.42	0.48	0.54	0.34				0.33	0.49	0.65
F2	0.78	0.70	0.71	−0.85															0.33
F3										0.32			0.46	−0.69	0.46	−0.70	0.42		
Com.	0.75	0.59	0.73	0.61	0.43	0.42	0.29	0.40	0.35	0.44	0.46	0.29	0.50	0.42	0.41	.45	.42		
14 Items (all in fixed mindset direction)																			
F1					0.29	0.32		0.60	0.52	0.65	0.62	0.59	0.73		0.62		0.70	0.65	
F2	0.92	0.80	0.94		0.42	0.33	0.34												
Com.	0.72	0.56	0.76		0.42	0.34	0.26	0.40	0.35	0.44	0.45	0.28	0.48		0.38		0.42		
8 Items																			
F1										0.62	0.55		0.71		0.68		0.73	0.55	
F2	0.88	0.77	.84																
Com.	0.76	0.59	0.74							0.43	0.39		0.49		0.45		0.47		

Bolded items are incremental items (growth mindset). Com.—Communalities of the items in different models. correl F—Correlation between factors

Table 4 Means, standard deviations, and correlations with confidence intervals

Variable	M	SD	1	2	3	4	5	6
1. Domain-general mindset (Intelligence)	2.99	1.14						
2. Subject-specific mindset	2.73	0.87	0.46** [0.40, 0.51]					
3. Negative effort believes	2.64	0.98	.36** [0.29, 0.40]	0.40** [0.34, 0.45]				
4. Learning goals	3.72	1.09	−0.15** [−0.21, −0.08]	−0.18** [−0.25, −0.11]	−0.16** [−0.23, −0.09]			
5. Performance avoidance goals	3.50	1.08	0.14** [0.07, 0.21]	0.21** [0.14, 0.27]	0.16** [0.09, 0.22]	−0.21** [−0.27, −0.14]		
6. Helpless Response	2.60	0.77	0.20** [0.13, 0.26]	0.22** [0.15, 0.28]	0.23** [0.16, 0.29]	−0.23** [−0.30, −0.17]	0.28** [0.21, 0.34]	
7. Subject-specific academic self-concept	3.84	1.03	−0.05 [−0.12, 0.01]	−0.01 [−0.07, 0.06]	−0.02 [−0.09, 0.05]	0.07* [0.00, 0.14]	−0.15** [−0.22, −0.08]	−0.32** [−0.38, −0.26]

N = 805; survey 1 at the start of the semester. M and SD are used to represent mean and standard deviation respectively. Values in square brackets indicate the 95 % confidence interval. * $p < 0.05$; ** $p < 0.01$. For mindset, higher values correspond to a fixed mindset

Table 5 Mean values for subject-specific and domain-general mindset in dependence of intelligence definition

Definition	% Agree	M SuSp	p SuSp	M DoGe	p DoGe
Total		2.73		2.98	
Intelligence is the ability to think abstractly and reasonably	67.9%	2.75	0.300	3.05	0.018
Intelligence is different in separated areas (f.e. math intelligence and speach intelligence)	68.7%	2.70	0.140	2.94	0.096
Intelligence includes knowledge and experience	41.8%	2.65	0.026	2.56*	<0.0001
The word intelligence includes sozial and emotional intelligence	54.6%	2.62*	<0.0001	2.78*	<0.0001
Intelligence is what is measured in an IQ-test	16.0%	2.89	0.027	3.66*	<0.0001

N = 779, as some students did not fill out this part of the survey. M SuSp—Mean subject-specific mindset; M DoGe—Mean domain-general mindset; p value for two-sided t-tests, * $p < 0.001$ significant with Bonferroni-Holm adjustment for ten measurements. Higher values correspond to a fixed mindset their definition of intelligence score significantly towards a Growth Mindset in the domain-general scale, while the group of students that define intelligence as what is measured in an IQ-test score towards a fixed mindset. Students that include social and emotional intelligence in their definition score more towards a Growth Mindset on both scales. The latter is the smallest effect on the domain-general scale.

5 Survey 2 Method

The second online survey was conducted in the same STEM classes two month into the first semester, again with voluntary participation during the lecture. 335 students (34 % female, 90 % in first semester, largest groups: mechanical engineering 19 %, physics, biology, industrial engineering, mathematics, informatics (7–9 %)) participated in the survey.

5.1 Measuring Instruments

The survey consisted of the *domain-general* (3 items, $\alpha = 0.89$) and the *subject-specific* (5 items, $\alpha = 0.82$) mindset scale described in Sect. 4.1 and the *subject-specific academic self-concept* scale (3 items, $\alpha = 0.89$) from survey 1. In addition, we included two new scales, *Intentions to change major* and *Intentions to drop out* (3 items each, e.g. "At the moment I don't feel suitable for my mayor/for studying"., $\alpha = 0.79$ and 0.86, both on Likert scale 1–6) adopted from the differentiated process model of student dropout by Bäulke et al. (2022).

6 Survey 2 Results

6.1 Confirmatory Factor Analysis of Mindset Scales

To validate the factor structure found in study 1, we performed a confirmatory factor analysis of the 8 mindset items using the lavaan package in r. A one-factor model showed poor fit (df = 20, $\chi^2 = 278.3$, rmsea = 0.196, srmr = 0.111, cfi = 0.791), while the proposed two-factor model showed good fit parameters (df = 19, $\chi^2 = 37.7$, rmsea = 0.054, srmr = 0.035, cfi = 0.985).

Table 6 Means, standard deviations, and correlations with confidence intervals; N = 335; survey 2 two month into the first semester

Variable	M	SD	1	2	3	4
1. Subject-specific mindset	2.94	0.93				
2. Domain-general mindset (intelligence)	3.11	1.15	0.54** [0.46, 0.62]			
3. Subject-specific academic self-concept	3.86	1.19	−0.16** [−0.26, −0.05]	−0.09 [−0.19, 0.02]		
4. Intentions to change major	2.43	0.94	0.26** [0.16, 0.36]	0.11* [0.00, 0.21]	−0.58** [−0.65, −0.50]	
5. Intentions to drop out	2.09	0.97	0.12* [0.02, 0.23]	0.04 [−0.07, 0.15]	−0.40** [−0.48, −0.30]	0.48** [0.39, 0.56]

M and SD are used to represent mean and standard deviation, respectively. Values in square brackets indicate the 95 % confidence interval. *$p < 0.05$.; ** $p < 0.01$. For mindset, higher values correspond to a fixed mindset

6.2 *Q2: How Are Domain-General and Subject-Specific Mindset Scales Correlated to Academic Self-Concept and Intentions to Change Major or Drop Out Two Month Into the First Semester?*

Pearson correlations can be found in Table 6. The subject-specific fixed mindset scale correlates significantly with higher intentions to change major or drop out. The domain-general scale also correlates with higher intentions, but correlations are smaller. Also, there is a significant correlation between a subject-specific fixed mindset and a lower academic self-concept.

7 Discussion and Limitations

In this study, we compared several ways to measure mindset. While small differences between growth mindset and fixed mindset formulations or the difference between "my intelligence" and "your intelligence" are measurable in other studies, we found the difference between a domain-general mindset (intelligence) and a subject-specific mindset to be more important. The malleability of talent approach by changing the word "intelligence" to "talent for my field of study" loaded mixed between both scales.

We propose a short subject-specific mindset scale (5 items) for the university STEM context. In order to reduce social desirability effects, we included only entity (fixed mindset) items. The scale is similar to the math item proposed by Yeager et al. (2020). In contrast to the short scale proposed by Yeager et al., where intelligence items and the math item are summarized in a single factor, the new scale is separated into a distinct factor in survey 1 (EFA) and survey 2 (CFA). The new scale shows the expected correlations to the mindset meaning system, which supports construct validity of the scale. In the data of survey 2, the subject-specific fixed mindset scale correlated significantly with higher intentions to change major (0.26) and drop out (0.14), while the domain-general scale correlated weaker. While at the start of the semester, academic self-concept and mindset was independent, there was a significant correlation in survey 2 two month into the semester. There seems to be an interaction between mindset and academic self-concept over the course of the semester. The direction of this interaction can only be speculated from this study and needs further research.

Our study confirms that the definition of intelligence has an impact on mindset scale values. E.g., including experience in your definition leads to a shift to a "growth mindset" in the domain-general scale, while defining it as an IQ value leads more toward a "fixed mindset". Both effects do not show as strongly in the subject-specific scale.

While this threatens scale validity, it has to be noted that the domain-general scale has been used successfully to show e.g. the relation to the mindset meaning system in many studies. In general, mindset is an often implicit construct that we are trying to measure with an explicit scale - a critic that also applies to the subject-specific approach. Dweck et al. (1995) describe fixed and growth mindset as two separate knowledge structures that can be triggered in certain situations. This helps explain why mindsets are both stable Robins and Pals (2002) and easily manipulable Hong et al. (1999) at the same time. We therefore should not use a single number from a mindset scale to put students into boxes of "fixed mindset" and "growth mindset". Instead, we should always keep the limitations of measurement in mind.

The main limitation of our study lies in a selection bias which should be particulary large in survey 2. We cannot generalize our findings without further studies. Furthermore, we could not differentiate between different STEM subjects. Effects for example in biology and mathematics might be different, as the level of difficulty in abstract thinking varies. Last, we only measured self reports. Correlations to test grades or the actual decision to change studies or drop out could give clearer insights.

This being said, findings in our study support the claim that mindset does matter in a university STEM context and a subject-specific scale can help to better describe the mindset of students and to detect further effects of a fixed mindset. Future research should focus on ways to support students in developing a growth mindset in such a challenging environment.

Competing Interests The authors have no conflicts of interest to declare that are relevant to the content of this chapter.

Ethics Approval Approval was granted by the Ethics Committee of Technical University of Darmstadt (Date 21.06.2021).

Acknowledgements The GroMM-project (Growing MINT Mindsets) was funded by the German Federal Ministry of Education and Research. The authors are responsible for the content of this publication.

Federal Ministry
of Education
and Research

References

Bahník Š, Vranka MA (2017) Growth mindset is not associated with scholastic aptitude in a large sample of university applicants. Pers Individ Differ 117:139–143. https://doi.org/10.1016/j.paid.2017.05.046

Bäulke L, Grunschel C, Dresel M (2022) Student dropout at university: a phase-orientated view on quitting studies and changing majors. Eur J Psychol Educ 37:853–876. https://doi.org/10.1007/s10212-021-00557-x

Broome P (1998) Implizite Begabungstheorien und erlernte Hilflosigkeit. Peter Lang, Berlin

Burnette JL, O'Boyle EH, VanEpps EM et al (2013) Mind-sets matter: a meta-analytic review of implicit theories and self-regulation. Psychol Bull 139(3):655–701. https://doi.org/10.1037/a0029531

de Castella K, Byrne D (2015) My intelligence may be more malleable than yours: the revised implicit theories of intelligence (self-theory) scale is a better predictor of achievement, motivation, and student disengagement. Eur J Psychol Educ 30(3):245–267. https://doi.org/10.1007/s10212-015-0244-y

Diederich M, Spatz V (2021) Validierung eines Mindset-Fragebogens für Physik- (Lehramts-) Studierende mittels Interviewstudie. In: PhyDid B—Didaktik Der Physik—Beiträge zur DPG-Frühjahrstagung. Available via DPG. https://ojs.dpg-physik.de/index.php/phydid-b/article/view/1135. Accessed 16 Jul 2024

Dinger FC, Dickhäuser O, Spinath B et al (2013) Antecedents and consequences of students' achievement goals: a mediation analysis. Learn Individ Differ 28:90–101. https://doi.org/10.1016/j.lindif.2013.09.005

Dresel M, Lämmle L (2017) Motivation. In: Götz T (ed) Emotion, Motivation und selbstreguliertes Lernen. Ferdinand Schöningh, Paderborn

Dresel M, Grassinger R (2013) Changes in achievement motivation among university freshmen. J Educ Train Stud 1(2):159–173. https://doi.org/10.11114/jets.v1i2.147

Dweck CS (1999) Self-theories: their role in motivation, personality, and development. In: Essays in social psychology. Psychology Press, New York

Dweck CS, Chiu C, Hong Y (1995) Implicit theories and their role in judgments and reactions: a word from two perspectives. Psychol Inq 6(4):267–285. https://doi.org/10.1207/s15327965pli0604_1

Gouëdard P (2021) Can a growth mindset help disadvantaged students close the gap? In: PISA in focus. OECD Publishing. Available via OECD. https://doi.org/10.1787/20922f0d-en. Accessed 22 Jul 2024

Heublein U, Hutzsch C, Schmelzer R (2022) Die Entwicklung der Studienabbruchquoten in Deutschland. DZHW-Brief 5. https://doi.org/10.34878/2022.05.dzhw_brief

Hong Y, Chiu C, Dweck CS et al (1999) Implicit theories, attributions, and coping: a meaning system approach. J Pers Soc Psychol 77(3):588–599. https://doi.org/10.1037/0022-3514.77.3.588

Limeri LB, Carter NT, Lyra F et al (2023) Undergraduate lay theories of abilities: mindset, universality, and brilliance beliefs uniquely predict undergraduate educational outcomes. CBE Life Sci Educ 22:ar40. https://doi.org/10.1187/cbe.22-12-0250

Limeri LB, Choe J, Harper HG et al (2020) Knowledge or abilities? How undergraduates define intelligence. CBE Life Sci Educ 19:ar5. https://doi.org/10.1187/cbe.19-09-0169
Limeri LB, Carter NT, Choe J et al (2020) Growing a growth mindset: characterizing how and why undergraduate students' mindsets change. Int J STEM Educ 7(1):35. https://doi.org/10.1186/s40594-020-00227-2
Malespina A, Schunn CD, Singh C (2022) Whose ability and growth matter? Gender, mindset and performance in physics. Int J STEM Educ 9(1):28. https://doi.org/10.1186/s40594-022-00342-2
Midgley C, Maehr ML, Hruda LZ et al (2000) Manual for the patterns of adaptive learning scales. University of Michigan. https://public.websites.umich.edu/~pals/PALS%202000_V13Word97.pdf. Accessed 22 Jul 2024
Rehberg J, Wilhelm T, Spatz V et al (2020) Pilotierung eines Mindsetfragebogens mit Physik-(Lehramts-)Studierenden. In: Habig S (ed) Naturwissenschaftliche Kompetenzen in der Gesellschaft von morgen. Beiträge zur Jahrestagung der Gesellschaft für Didaktik der Chemie und Physik, Wien
Robins RW, Pals JL (2002) Implicit self-theories in the academic domain: implications for goal orientation, attributions, affect, and self-esteem change. Self Identity 1(4):313–336. https://doi.org/10.1080/15298860290106805
Santos DL, Gallo H, Barbera J et al (2021) Student perspectives on chemistry intelligence and their implications for measuring chemistry-specific mindset. Chem Edu Res Pract 22(4):905–922. https://doi.org/10.1039/D1RP00092F
Spatz V, Goldhorn L (2021) When it's more difficult, I just cram more! An exploratory interview study on students' mindsets in physics. Eur J Sci Math Educ 9(3):92–109. https://doi.org/10.30935/scimath/10948
Spatz V, Lippmann J (2018) Interviewstudie zu den Mindsets von Physiklehrkräften. In: PhyDid B—Didaktik Der Physik—Beiträge zur DPG-Frühjahrstagung. Available via DPG. https://ojs.dpg-physik.de/index.php/phydid-b/article/view/833. Accessed 16 Jul 2024
Spinath B, Stiensmeier-Pelster J (2001) Implicit theories about the malleability of intelligence and ability. Psychologische Beiträge 43(1):53–76
Yeager DS, Dweck CS (2012) Mindsets that promote resilience: when students believe that personal characteristics can be developed. Educ Psychol 47(4):302–314. https://doi.org/10.1080/00461520.2012.722805
Yeager DS, Dweck CS (2020) What can be learned from growth mindset controversies? Am Psychol 75(9):1269–1284. https://doi.org/10.1037/amp0000794
Yeager DS, Hanselman P, Walton GM et al (2019) A national experiment reveals where a growth mindset improves achievement. Nature 573(7774):364–369. https://doi.org/10.1038/s41586-019-1466-y

Experience-Based Teaching and Learning

Teaching Physics in a Swimming Pool

Marina Tamborini

Abstract The Physics high school teaching practice shows some serious shortcomings that cause either misconceptions and/or the fragmentation of students' knowledge. The students at all levels show severe difficulties in grasping the actual meaning of many physical phenomena. This paper presents a sequence of unusual and amusing experiments designed to be carried out by students themselves, under the teacher' supervision, while diving in a swimming pool. The most relevant of them regard the Newton's laws and the propagation of acoustic and electromagnetic waves. The core of the present teaching/learning proposal is to allow the students to observe how Physics laws act in a real-world contest and test if this is effective for them to learn how to derive from their experiences a unified and coherent description of how the nature works. This is done with the aim to increase also their environmental awareness. In the paper the experiences performed by the students are fully analyzed and the results of the final achievement test are discussed. This educational proposal received the 2022 GIREP Sara Barbieri Award.

1 Introduction

A significant fraction of students graduated from high school miss or give incorrect answers to relatively simple Physics questions on arguments that are part of all standard Italian high school curricula (Scientific Faculties Admission, 2017/2019). Table 1 shows a few examples of the results scored at the entrance exam by students applying for admission to the Science University Faculty in Italy. The test consists in a multiple-choice questionnaire.

The high number of unanswered questions is quite probably due to shortage of time. It evidences anyway that, when the student is asked to think logically instead

M. Tamborini (✉)
Liceo Scientifico Statale Piero Bottoni, Via Mac Mahon 96, 20155 Milano, Italy
e-mail: tamborini_marina@hotmail.com

Dipartimento Di Fisica, Università Degli Studi Di Milano, Milano, Italy

S. Faletič and J. Pavlin (eds.), *Teaching and Learning Physics Effectively in Challenging Times*, Challenges in Physics Education, https://doi.org/10.1007/978-3-031-72541-8_9

Table 1 Results scored at the entrance exam to university. Data was collected by the author (Audacity)

Topic	Sample size (year)	Right answers (%)	Wrong answers (%)	Missing answers (%)
Relationship force-pression-surface	749(2017)	27	21	52
Stevin's Principle	870(2017)	10	23	67
Buoyancy	797(2017)	22	27	51
Newton's third law of motion	806 (2017)	17	58	25
Relationship wave frequency-velocity	621 (2017)	13	51	36
Transmission of light from air to water	806 (2017)	8	42	50
Standing waves	749 (2017)	5	27	68

to apply mechanically his knowledge and the answer does not appear immediate, he chooses to dropout the question.

For younger students, attending their second/third year in a Scientific High School, serious misconceptions about Newton's laws and fluid laws are present. Before undertaking the swimming pool activities, the students' starting knowledge and misconceptions have been tested with a diagnostic test (Multiple-Choice questionnaire and explain your reasoning types). The distractors are based on widespread misconceptions. A few examples are reported in Tables 2, 3, 4 and 5. The data was collected by the author.

These discomforting results call for a deep revision of the teaching practice in the High School, which is still widely based on a formal approach. A largely diffuse

Table 2 Newton's laws. "The student is asked to identify the right statements in a set of four"

4 right answers	3 right answers	2 right answers	1 right answer	0 right answers
13%	25%	25%	20%	17%

Table 3 Archimedes' principle. "The student must explain why an untied balloon inflated with Helium moves upward. To be noticed in the answers the confusion between weight/mass/density and the misconception about an attractive force acting upward"

Air and Helium have different densities	Helium is lighter than air	Helium mass is less that air mass	Helium interacts with the air	A force attracts the Helium upward, does not exist a force toward the earth	Balloon internal pressure is less than external pressure
7%	55%	20%	4%	10%	4%

Table 4 Atmospheric pressure. "The student must explain how works a suction cup applied to a surface". The well-known misconception about the force of the vacuum scores 45% of the answers

The internal vacuum acts on the suction cup	The suction cup produces a pressure against the surface	Some kind of friction is acting	An external pressure is present	Some electric or magnetic attraction is present	The suction cup insulates the air inside and the air acts as glue
45%	5%	5%	9%	18%	18%

Table 5 Global warming. "Knowing that the Artic (North Pole) ice cap consists of fresh water ice floating on the sea (salt water) the student must explain how the sea level would change if the whole ice cup would melt"

Sea level would increase
91% students state that the sea level would increase. They know that the seas level would rise because the global warming, but are clearly unable to recognize the required physical context

practice in the school is to treat the main physics laws in a very simplified and ideal contest, often without even discussing the application limits. As examples of such a bad practice we may report: fluids treated as always homogenous, non-consideration of energy loss over long propagation distances, neglection of the frequency response of a given medium in waves' transmission and reflection. This lack of knowledge makes difficult, if not impossible at all, to treat correctly some very relevant topics as the atmospheric greenhouse effect, the underwater environment's preservation, or to understand some underwater life biological aspects, as the functioning of sight and hearing in marine animals.

In addition, the intuitive nature interpretation, a priori of any scholastic education, induces a spontaneous generalization of the physical rules regulating our everyday life contest to every kind of environment. It is very difficult for the students to use their school knowledge to predict that through a few water centimeters the visible radiation can be transmitted while at the same time the infrared radiation is absorbed or that the sound is almost not transmitted through the air/water interface.

Not a minor problem in teaching High School Physics is the relevant difficulty of analyzing very complex natural phenomena without making use of Mathematics knowledges too advanced for second year young students. The direct experimental observation is an effective starting point. The large amount of data collected during the activity in the swimming pool can afterward be resumed and deepened in the following school years, when the students will dispose of a wider and more complex Mathematics and Physics apparatus, suitable to learn how to model a physical phenomenon into equations.

The problems with a conceptual approach to Physics are well known to world-wide educational research and old dating. As reported in literature (Jones and Mooney 1981): "many students are unable to place the concepts in perspective. The result of this difficulty is that the students' knowledge and understanding of physics is frequently fragmented and compartmentalized and they never perceive a unity of the subject". As stated before, research's findings suggest that a good teaching practice should prefer a unified description of the physical world as respect to a detailed classification. Moreover, it is well stated that a direct experience and discover are much more effective on the students', especially young students', learning process. Therefore, the direct experience of unexpected effects precedes the theoretical explanation, as strongly suggested by educational research results (Scott et al. 2014).

The following educational project has been devised to encourage students to overcome a mnemonic knowledge, frequently leading to serious misconceptions, and to achieve a conceptual and sense-making learning. The research question is about the effectiveness of a teaching/learning path directed to very young students, lacking any advanced mathematical background, based on the direct students' experimentation in the real-world's contest of a swimming pool, often characterized by unexpected phenomena. The main goal is that the students learn how to derive a unified and coherent system of description of the real nature complexity from their experience, instead of being lectured on a mere theoretical ground. A swimming pool represents an extraordinary laboratory to experiment within many Physics topics included in the High School science curriculum, such as: optics, acoustics, heat, fluids and dynamics' laws. A final additional advantage is that the underwater laboratory takes place in an unusual, pleasant and relaxing atmosphere, most welcome by both teachers and students, and very favorable to capture students' attention and curiosity. This practice helps the students to transform the often-boring abstract book approach into a fascinating discovery of the functioning of one's own body and of the surrounding world.

2 The Didactic Project and the Research Methods

The leading idea refers to the student construction of knowledge methodology (Kapur 2019): "Instructional strategies should focus upon complex problems and not on problems with clear and correct answers".

Starting from the previous considerations we devised a teaching/learning sequence, coordinated between the school's departments of Natural Science and Sport Science Education with the help of the Physics Department of the Milano University UNIMI, that inquiries extensively the physics, chemistry, biology, medical and sports topics implied in the scuba diving activities. I had the role of Project coordinator.

Six classes of 15 years old students and ten teachers of a Scientific High School have been involved in the project. It requires 2 h for the diving instructors' introduction, 2 h for each class's experimentation and data taking in the swimming pool, about 8 h for classroom data analysis, discussion and conclusions and 2 h for the initial and final tests. As stated before, the experimental observations are more widely discussed and formalized in the following years' courses.

Most of the instrumentation has been found in the school laboratory and requires no sophisticated technology. The camera for underwater scenes has been lent to us by a friend and the thermal camera to study infrared images has been lent by the UNIMI Physics Department. As an alternative, a cheaper Sport Action Camera, with a waterproof case suitable for the underwater registration, can be got on the market. The camera of a water protected cell phone represents also a still useful second-best choice.

After a preliminary classroom discussion on the best way to perform experiments, 15-year-old students attending their second year in the high school, are allowed to experience in an underwater real environment, with the guidance of a pool of Milan City Police Diving Division trainers. They submerge, equipped with scuba tanks, in a 5 m deep swimming pool and perform, under the supervision of their teachers, a set of experiments regarding the laws of Mechanics, Fluids, Acoustics and Electromagnetic Waves. While being underwater, students can use their sensory organs to perceive the sharp changing in sight, earing, communications, equilibrium, weight, movement, body heat loss, as compared to their daily experience, in a world acted by gravity and surrounded by a gaseous atmosphere.

A simultaneous comparison between bodily sensations and simple instrumental measurements, followed by a classroom full analysis and discussion of the implications of the data collected, will help the students to fully organize their knowledge and comprehension of the involved Physics laws.

3 The Afforded Misconception and Didactical Issues

Below are listed the main didactical issues that were addressed in the activities.

3.1 Newtonian Mechanics Misconceptions

As it has been widely reported in literature since a long-time, the Newtonian Mechanics is one of the most misconstrued concepts in Physics. For extended discussion and other references see for example (Lark 2007). The colloquial expression “weightlessness” is the source of a widely spread confusion between the actions of to “delete” and to “equilibrate” a force with an opposite one. It is a known fact that most students think that on a ball thrown and moving upward is acting an upward force progressively eroded by friction against the air. This kind of answer shows an evident confusion between the force and velocity vectors and a misuse of the concept of vector sum. The third principle of dynamics is subject to originate another critical misconception. The students tend to think that the action and reaction forces are applied to the same body and produce a static equilibrium. In the swimming pool water surface, the human body float and move with little water friction and no ground friction at all, so the contest is ideal for fundamental Newtons’ laws testing.

3.2 Fluids Pressure Effects

The laws of fluids in static equilibrium are conceptually simple, but the values of fluid pressure attainable at school are very low. The availability of a few meters water depth in the swimming pool allows the students to experience the fluid statics' laws with some spectacular effects impossible to reproduce in the school laboratory.

3.3 Waves' Propagation

The dependence of the wave parameters from the properties of the propagation medium is not usually stressed in the school. This concerns in particular the different media frequency response, with the exception of the optical dispersion phenomenon, and their absorption properties. The role of the impedance of different media in wave transmission is treated only in specific electronics courses. The simultaneous existence of a signal reflected together with the transmitted one is not stressed. As a consequence, the critical question of the efficiency in a wave signal energy transmission/saving is not treated at all. The propagation of infrared electromagnetic radiation is not considered. While under water, students can perceive the sharp acoustic impedance mismatch that prevents sounds to propagate from air into water and vice versa and a relevant difference in frequency dependence of the electromagnetic absorption between air and water, giving rise to unexpected phenomena, such as respectively the long- range sound propagation in water and the water opacity to infrared electromagnetic radiation.

3.4 Unification of Waves' Description

Most textbooks explain the acoustic laws, the optical laws and the properties of electromagnetic waves in rigidly separated contexts, as if they were a different kind of phenomena instead of being the same phenomenon in different application contexts. The path presented here introduces the waves' phenomena in a unified framework aiming to highlight the many common features that identifies a wavelike phenomenon instead of disassembling the different fields of application. It overcomes the standard approach by early introducing the unifying concepts of impedance and absorption in order to explain the efficiency in energy transmission whatever kind of waves is considered, and affording only at a later stage a more extended waves' theoretical formalism.

3.5 Citizenship Competences

In the conclusive phase, the implications regarding environment preservation (oceanic noise pollution) and climate change (greenhouse effect) are stressed. A special opportunity is also given by the role of water rescuers acted by the Local Police Subs and by their reclamation and recovery of polluted rivers, canals and small lakes. In a special lecture section, the Subs explain their activities to the students, stressing how the Physics technological employments are exploited to contribute to society's welfare.

3.6 Universality of Physics Laws.

One of the strongest scientific requirements is that the Physics laws must be the same everywhere in the universe. To submerge in water is a disorienting experience, as would be to move on a different planet. Nothing seems to work as in every day's life: as like hearing, sight, and the body's equilibrium. That is a unique opportunity to discuss with the students that the apparently new phenomena are explained exactly by the same Physics laws acting in air, as only the medium's parameters are different, while the Physics laws apply universally.

4 The Experiments Performed in the Swimming Pool are as Follows

Below are listed the experiments performed in the swimming pool to address the didactical issues presented n the previous section.

4.1 Newton's Laws of Motion

As it is known the astronauts train to move as if they were weightless by floating in a swimming pool. This is also the first sensation experimented during the submerging and it offers the opportunity to discuss with the students that their weight has not "disappeared" but instead is equilibrated by the Archimedes' hydrostatic thrust that allows the buoyancy. A special feature is the opportunity to experiment the principle of action and reaction in a free body situation. Two students, better a boy and a girl of similar body structure, are asked to pull a short rope by the two ends while floating without swimming or reaching with their feet for the pool floor. No friction is possible and nobody wins the match, the students get closer to each other symmetrically with respect to a common center. The students recognize by direct experience that the action and reaction couple of forces cause always a motion and never an equilibrium state.

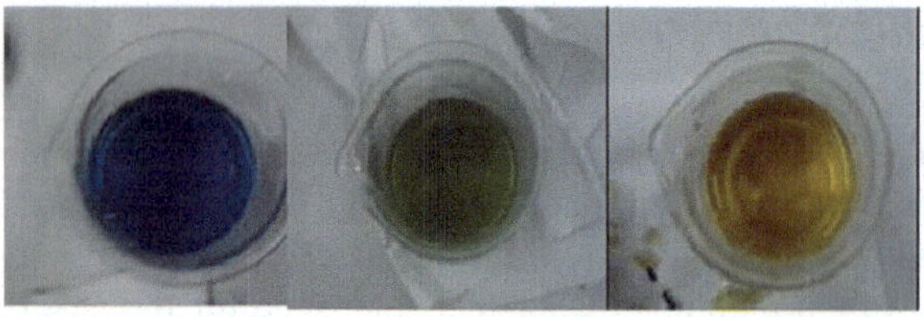

Fig. 1 Water acidity

4.2 The Henry's Law

During the preliminary conference, the diving instructors explain the functioning of the diving respiratory device, filled with pressurized air. They illustrate also the danger of decompression sickness, due to the forming in the blood of nitrogen bubbles by a sudden release of the pressure when the diver rises too quickly to the surface from an immersion at high depth. The phenomenon is described by the Henry's law that relates with a direct proportionality law the amount of a given gas that dissolves in a given type and volume of liquid to the partial pressure of that gas in equilibrium with that liquid. It is the well-known effect produced in a soda drink when the cap is open. The law has been previously investigated by mean of a few simple experiments realized in the school laboratory. The materials required are a bottle of sparkling mineral water, an acidity indicator like the Bromothymol Blue and the vacuum bell. Figure 1 shows three beakers filled with sparkling water at different H_2CO_3 concentration and a few drops of acidity indicator; lighter color indicates a higher acid concentration. On the right side: the sparkling water from the bottle rich in solute carbonic acid. In the middle: the water depressurized in the vacuum bell. On the left side: some still water from the tube where no carbonic acid is present.

4.3 The Laws of Fluid Dynamics and the State Equation of Ideal Gases

The availability of 5 m depth water in the swimming pool allows to investigate the fluids laws to an extent impossible in a school laboratory and with spectacular effects.

In Fig. 2a are visible the effects of the buoyancy force on a submerged balloon, held at one end by a sub, that assumes a drop like shape. When released from the bottom of the swimming pool the balloon resumes its original spherical shape and reaches the surface with high speed.

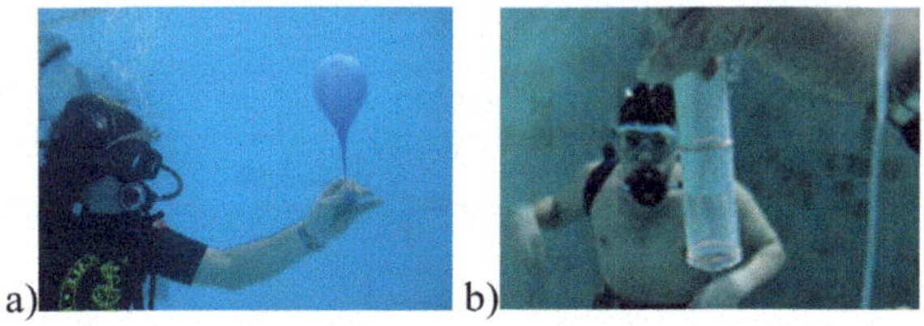

Fig. 2 **a** A balloon deformed by the buoyancy force. **b** The pocket of air in an inverted cylinder used to measure the pressure-depth relation

4.3.1 Measuring Hydrostatic Pressure

The pressure at the bottom of a 5 m deep pool is about 1.5 times the atmospheric pressure and acts on deformable objects. A small closed plastic bottle, filled with air, shrinks in a visible way; once brought back to the surface it resumes its original volume. To obtain a simple quantitative measure of the pressure as a function of the water's depth we devised to use a laboratory 1 mm graduated cylinder, shown in Fig. 2b and a rope held vertical and taut by means of a weight. On the rope are set a series of markers at a mutual distance of 1 m to measure the water depth. The pressure is measured indirectly by holding the cylinder inverted and measuring the volume of the air trapped inside. The volumes are converted in pressure units according to Boyle's law. Two elastic bands help the measurement. The bottom one marks the volume of the air at the surface; the upper one marks the volume at a given depth. A student sets the elastic bands while under water. Since an underwater camera was available, the measurement has been still simpler. A student holding the cylinder has been filmed while rising slowly from the pool floor to the surface along the rope. Then the frames have been scanned offline. The small temperature gap between the air and the warm swimming pool water is not relevant, so it is the fact that the air is not an ideal gas, since the measurement precision depends mainly from the inhomogeneity between the various camera frames positioning. The plot of the collected data is shown in Fig. 3.

4.4 Wave Propagation

4.4.1 Theoretical Background

There is no need of detailed theoretical demonstrations to make young High School students grasp the meaning of the following experimental sequence. Interested readers can find a more formal explanation in (Freegarde 2012). Instead of introducing only the optical refractive index as a function of the wave velocity in given medium, two main quantities have been defined: the characteristic impedance Z and the coefficient of absorption α. These describe any kind of waves' propagation in

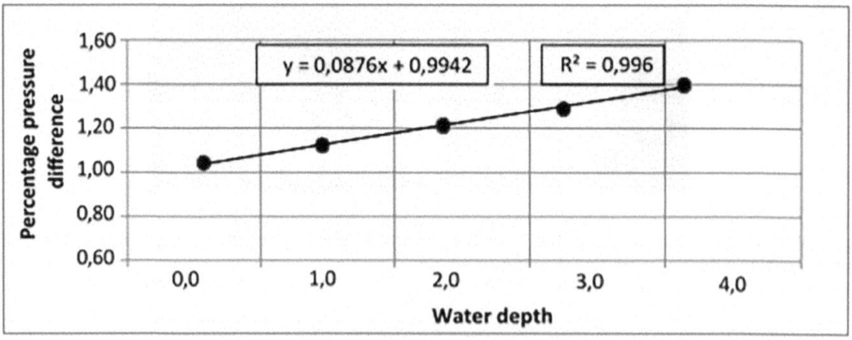

Fig. 3 Percentage pressure variation as a function of the water depth in meters

any medium, regardless of being a mechanical, acoustic or electromagnetic wave. Impedance and absorption coefficient depend on the specific medium properties and on the wave's frequency. For example for acoustic waves: $Z = \rho v$, where v stays for the wave speed and ρ is the medium density. For electromagnetic waves: Z = E/H, where E and H represents the intensities of the electric and magnetic fields. The transmission and reflection coefficients $C_{T/R}$ at a two media interface are defined as the ratio between transmitted (reflected) wave $A_{T/R}$ and the incident wave A_I amplitudes. They may be expressed as a function either of the propagation speeds v in the two media or of the impedances Z, regardless of the type of wave, as it is shown in the Eq. 1 where the indices 1 and 2 refer to the outcoming and incoming media. The sign of the reflection coefficient C_R expresses the shift in the reflected wave amplitude.

$$C_T = \frac{A_T}{A_I} = \frac{2Z_1}{Z_1 + Z_2} = \frac{2v_2}{v_1 + v_2}$$

$$C_R = \frac{A_R}{A_I} = \left(\frac{Z_1 - Z_2}{Z_1 + Z_2}\right) = \frac{v_2 - v_1}{v_1 + v_2} \tag{1}$$

In most applications, the interesting quantity is the ratios of transmitted and reflected energies ET/R per unit time at a two media interface. The two quantities are related to the media impedances as in Eq. 2:

$$\frac{E_T}{E_I} = \frac{4Z_1Z_2}{(Z_1 + Z_2)^2}$$

$$\frac{E_R}{E_I} = \left(\frac{Z_1 - Z_2}{Z_1 + Z_2}\right)^2 \tag{2}$$

the relevant parameters are the wave's intensity attenuation and the waveform distortion due to the dispersion relation. The signal of a point like source is attenuated in function of the distance by the well-known inverse square law. Additional causes contribute to the wave attenuation as, for a non- homogenous medium, the scattering against small discontinuities and obstacles of dimension similar to wave length, the medium viscosity and, for electromagnetic waves, molecular absorption. The related absorbance is usually described by the Lambert–Beer exponential law: $I(x) = I(0)e^{-\alpha x}$, where x is the thickness of the traversed medium and α is the absorption coefficient expressed in meters^{-1}. Both the impedance and attenuation coefficients depend on the physical properties of the medium and can be function of the wave frequency (response in frequency), producing a significant signal distortion. The water absorption coefficient is inversely proportional to the frequency of the sound and the density of the water, since high frequency sounds are absorbed and converted faster to heat.

4.4.2 Sound Propagation

The phenomenon of sound transmission between two different media is not stressed in school textbooks, despite the fact that it evidences the strict similarities between the acoustic waves' propagation and the optical ones usually fully treated. In the swimming pool the students are encouraged to listen to the sound of a rattle shaken by a sub either where source and listener are both underwater or where the source is in the air and the listener underwater or vice versa. Surprisingly, the sound transmitted from water to water is very clear while the sound emitted in the air is not perceived by an underwater listener. A sound quantitative measurement has been obtained by means of an amateur underwater camera. A student registered the sound and the recorded sound track has been subsequently analyzed in the classroom in order to explore in an effective way, without making use of a too complex formalism, the characteristics of waves transmission through different media. The sound track, elaborated in the classroom with the free software Audacity (Audacity), is shown in Fig. 4. On the plot the sound waveform is expressed either as intensity (Fig. 4a) or as sound level in decibels (Fig. 4b) as a function of the time. The data can now be easily discussed with the students.

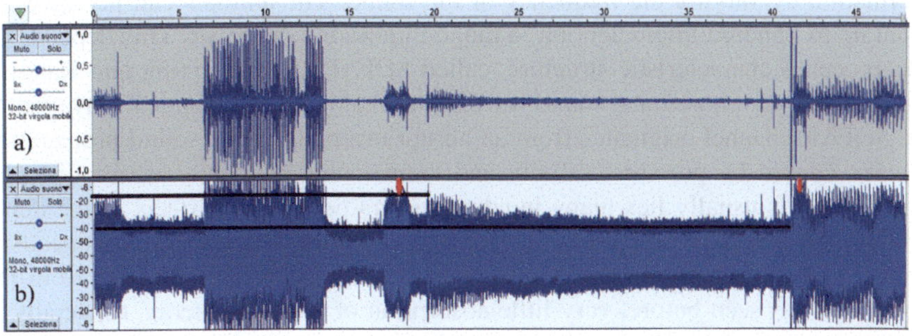

Fig. 4 The sound track of the rattle, **a** expressed in sound intensity and **b** expressed in decibels

At the start both the rattle and the camera are in the air. After about two seconds the camera is submerged and one can clearly perceive in the plot the sharp drop in the registered sound level. After about 6.5 s also the diver holding the rattle submerges and the sound abruptly rises to the original level. Then the diver gets close to the camera always keeping underwater and shaking the rattle. At about 18 s both the rattle and the camera get out in the air again. At 20 s, when the camera is taken away from the rattle till about 40 s, the plot shows the sound level's exponential decrease. At 41 s both camera and rattle are submerged again and the sound level is restored. A simple but clear quantitative explanation can be added, if expressed in terms of the acoustic impedance. The two respective sound impedances are: for air at 20 °C $Z_{air} = 415$ Pa s/m and for water: $Z_{water} = 1.48 \cdot 10^6$ Pa s/m. This gives for the reflection coefficient between air and water (formula 1) a value of $C_R =$ 0.9997 in both directions, explaining why the amount of acoustic energy transmitted through the air/water surface is negligible. Considering now the attenuation range: at the main rattle frequency of about 300 Hz, the water absorption coefficient is of the order of 104 km; for higher frequency, of the order of kHz, the coefficient in water is of the order of 1 km. This explains why the swimming pool water results "transparent" to the sound. The strong attenuation in air is instead mainly due to the spherical symmetry and not bounded sound propagation. The conclusions drawn are of general order and can be extended to other propagation media and any other type of waves.

4.4.3 Environment Preservation: The Whale and the Ship, Seas' Acoustic Pollution

The swimming pool experience can also be used for developing citizenship education. The sea water transparency to sound, better than to the light, makes hearing an important sensory mechanism for marine animals. During a lesson, a dive instructor witnessed of having perceived the so called "whale song", rising the students' curiosity. It is a directional sound sent by the cetaceans to communicate and locate objects that reflects from targets and is received as echoes. They can also gain more information by varying the frequency of the sound. The animals can perceive the signal up to hundred kilometers far, a range impossible in the air. This depends on a deep oceans characteristic structure, called SOFAR (Sound Fixing and Ranging Channel): a sort of horizontal sound waveguide, analogous to an optical wave guide. The SOFAR channel originates from an abrupt inversion in the sound propagation velocity, caused by pressure, salinity and temperature gradients, at two different depths. Its axis usually lies many hundred meters below the surface, at the depth of minimum sound speed (SOFAR). Just like an optical wave guide, this acoustic channel is able to trap the sounds by reflection and propagate them far from the source with, as seen before, very little absorption of acoustic energy, especially at low frequencies. This is very relevant for the preservation of the sea environment. It has been suggested that the ship noise could distress whales at the point of being responsible for their death by stranding (Nicolas 2019). The plot in Fig. 5 shows the

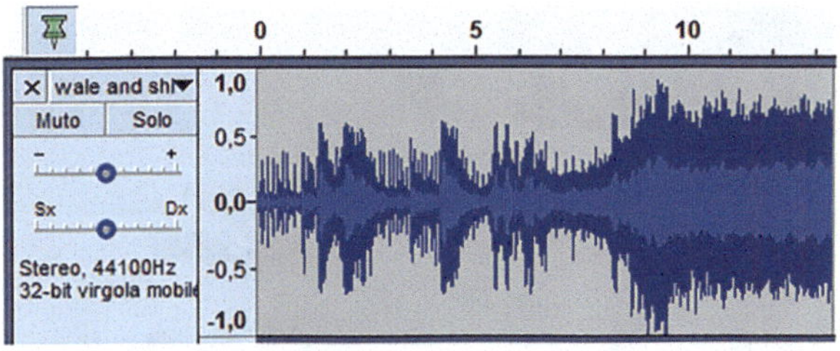

Fig. 5 The whale's and the ship's noise

track of the sounds recorded in a port. On the left side are visible the spikes due to a whale's signal; the big noise plateau on the right side, overcoming the animal's sounds, is due to an incoming ship (Ship noise).

On the converse, in some cases, the animal's sounds may cause a problem to scientific activities. The underwater KM3NeT (Kilo Meter Cube Neutrino Telescope) is a research infrastructure that is currently being deploying in the Mediterranean Sea. The experiment is aimed at the detection of cosmic neutrinos and a part of the detecting apparatus makes use of sonic signals that may be confused with the signals emitted by the cetaceans. The KM3NeT collaboration has started a program to involve the amateur scientists into discriminating ocean's noise from neutrino's signals. The required task is simple and can be used with the students of upper high school years as an unusual and amusing approach to particle's physics. For more information see (Deep Sea Explorers).

4.4.4 Light Propagation

Submerging in a swimming pool offers a unique opportunity to experiment how the light refraction/reflection work in a real contest. The human eye does not focalize very well underwater because the eye and water's refraction indexes are very similar. Analogously the fishes' eyes don't focalize well in the air. The diving mask interposes an air layer between the water and the eyes and allows the correct sight functioning, but the students perceive the objects as expanded and moved away, because of the refraction trough the mask glass. The phenomenon has been recorded with our camera lens half submerged, as can be seen in Fig. 6. A fish that would try at the same time to spot an insect prey outside the water surface and keep an eye on inside predators will face the same problem. (Fish Sensory System) describes a fish, the Anableps Anableps, that evolved special eyes to see simultaneously inside and outside water, each eye is split into two lobes with its own pupil and retina. Another well-known phenomenon that the students may experience is the total reflection on the water surface as can be seen and recorded from underwater.

Fig. 6 Image taken with the camera lens half submerged

4.5 *Infrared Absorption in Water*

The infrared radiation is in an electromagnetic frequency range fundamental for the life on the earth, but the difficulty to afford a full theoretical explanation discourages most teachers from explaining it. Again, the experimental observation constitutes a good simple alternative. The thermal camera we used in this study has been lent to us by UNIMI. It is a NEO THERMO TVS-700 NIPPON AVIONICS, with the following technical specifications:

- FOV: 26.0° × 19,6°
- Spatial resolution: 1.4 mrad
- Detector: Uncooled FPA, VOx microbolometer
- Temperature resolution: 0,08 °C at 30 °C (Blackbody)
- Wavelength: from 8 to 14 μm
- Measurement range: from −20 °C to + 500°C
- Effective pixels: 320(H) × 240(V) Pixel
- Frame time: 30 Hz

4.5.1 The Infrared Absorption

As a first step, the students observe on the thermal camera screen the "mysterious" disappearance of a classmate fellow progressively submerging in water. The screenshot taken during the swimming pool activity can be seen in Fig. 7. The image of the underwater student's body remains, obviously, well visible to a naked eye.

The students discover so that the water results opaque to thermal emission, whilst being well transparent in the optical range and to other kind of waves like, as seen, acoustic waves. The infrared transmission properties are not due to an impedance mismatch between water and air, like it was for sound, but instead to the water's strong infrared absorption properties. So, the conclusion is that a given medium's absorption properties depend both on the kind of waves and the spectral range considered. The impossible full theoretical calculation can be replaced by some simple considerations

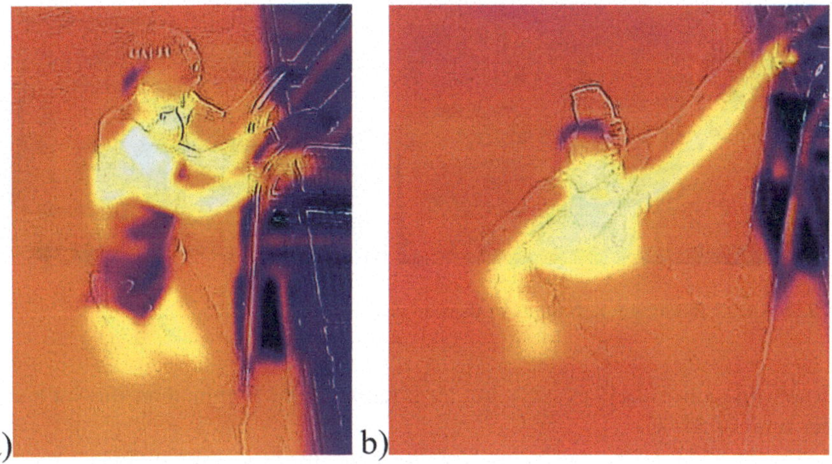

a) b)

Fig. 7 Thermogram of two semi-submerged students

on the structure of the water molecule. Some sketches showing its many degrees of freedom, are effective enough to give a hint about its strong infrared absorption properties, as a frequency function. Eventually, a look at the plot of water absorption coefficient in function of the frequency will confirm the net rise of the absorption coefficient moving from optical to infrared frequencies.

4.5.2 The Greenhouse Effect

The gained knowledge can now be used for a correct explanation of the spontaneous greenhouse effect's origin. In fact, this phenomenon, that ensures a warm habitable planet, is mainly due to the strong infrared water vapor absorption, since the atmosphere contains 13 million tons of water that account for about 70% of all atmospheric radiation absorption (Chaplin 2015).

4.5.3 Heat Transmission

As the subs explained, an important problem arising when diving takes place in very cold water is the effectiveness of the diving wet suit in protecting from excessive bodily thermal losses. We devise to use the thermal camera to investigate the issue. As a first step, to learn how to correlate the camera false colors images to the temperature's scale, the students observed the snapshot in Fig. 8a. It shows a cat's body at room temperature (I have taken previously the snapshot). The animal back appears the colder at about 20 °C (green color, it looks dark on a grayscale), since the fur is thick and protective against thermal losses. The naked ears and eyes look instead

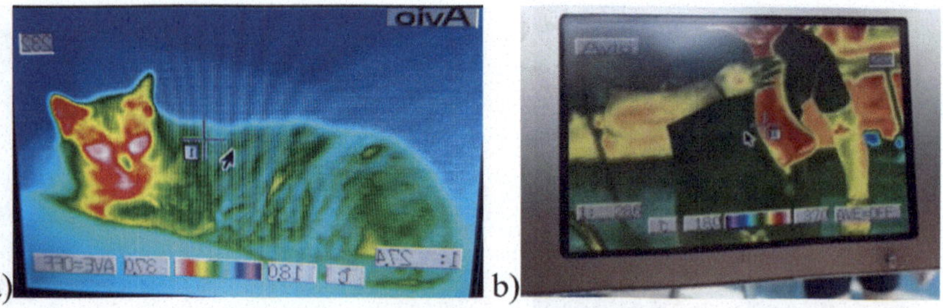

Fig. 8 **a** A cat's thermogram. **b** Study of thermal properties of the diving suit

yellow/white since they are very hot, at a temperature higher than the set camera upper limit of 37 °C.

In Fig. 8b is shown the thermogram of one of the divers accompanying us, recorded after about one-hour underwater activity. The temperature of his trunk skin, that has been protected by the wet suit, looks of about 30 °C (red color or medium grey in a grayscale) whilst the temperature of his arms and fingers is neatly lower, approximatively similar to that of the suit outer surface in direct contact with the water also visible in the background (about 20 °C, dark green color), so confirming the wet suit efficacity.

5 Results and Conclusions

After a full classroom analysis and discussion of the data collected in the swimming pool, the obtained achievements have been assessed with a Multiple-choice Questionnaire, asking the students to justify their choices by exposing in a clear way their motivations with an appropriate terminology and adequate language. The most significant examples are shown in Tables 6, 7 and 8.

A clear improvement in students' conceptual learning could be detected: 65% of students recognize the universality of Physics laws. As regarding the 20% of students that still confuse between "to cancel" and "to "equilibrate" the weight, the problem seems to be a language one, since it is especially related to students without an adequate knowledge of Italian. This must be taken into account when

Table 6 Universality of Physics laws. "The student is asked to identify the right statements in a set of four"

Objects float because buoyancy force cancels their weight	Objects on the ISS have no weight	Underwater Physics laws are different from earth surface	Underwater Physics laws are the same than in air, but physical constants differ	No answer
20%	2,5%	10%	65%	2,5%

Table 7 Newton's laws. "On the ISS a candle flame appears spherical. Explain why"

Atmospheric pressure is much lower than on the earth surface	The air density is much lower than on the earth surface	Due to reduced neat gravity conditions, it is meaningless to apply Archimede's principle	On the ISS physics laws are different from earth surface	No answer
30%	10%	57,5%	2,5%	0%

Table 8 Water infrared absorption. "Lakes and seas can absorb a large amount of solar heat. Explain why"

Water density is bigger than the air and disperse less solar heat	Water does not reflect light so solar radiation goes through without losses	The water molecule structure is adequate to absorb infrared solar radiation	The statement is not true, water does not absorb heat	No answer
20%	10%	67.5%	0%	2.5%

preparing lessons. Especially noticeable is the 67,5% of students that recognize that the properties of water to absorb infrared radiation is due to its molecular structure. The electromagnetic radiation is explained only in the last high school year and the argument was unknown to second year students. Some students motivate their answer by quoting explicitly the thermal camera observations in the swimming pool. The candle flame shape is a particularly subtle question and it is positively surprising that 57,5% of students are able to recognize that the application of Physics laws is not automatics but instead is related to environmental conditions. This development of students conceptual reasoning is a very significant achievement, proving the efficacity of a teaching approach through direct student involvement, in an informal and unusual real-world contest.

Acknowledgements Many thanks to the instructors of the Milan City Police Diving Division, for their expert and indefatigable help to our activities. We are deeply grateful to Professor Marco Giliberti of UNIMI for his constant support and explanations on teaching methodology and to Professor Nicola Ludwig of UNIMI for his precious help in understanding the infrared radiation physics.

References

Audacity. https://www.audacityteam.org. Accessed 30 July 2024

Chaplin M (2015) https://water.lsbu.ac.uk/water/water_vibrational_spectrum.html. Accessed 30 Jul 2024

Deep Sea Explorers. https://www.zooniverse.org/projects/reinforce/deep-sea-explorers/about/research. Accessed 30 July 2024

Fish Sensory System. http://www.ru.ac.za/media/rhodesuniversity/content/ichthyology/documents/Vision.pdf. Accessed 30 July 2024

Freegarde T (2012) Introduction to the physics of waves. Cambridge University Press, Cambridge

Jones H, Mooney R (1981) An approach to conceptual difficulties in physics. Phys Educ 16(6):356–359

Kapur R (2019) Constructivism in teaching-learning process. https://www.researchgate.net/publication/333507499_Constructivism_in_Teaching-Learning_Process. Accessed 30 July 2024

Lark AC (2007) Student misconceptions in newtonian mechanics (Master of science thesis)

Nicolas J (2019) Ocean uproar: saving marine life from a barrage of noise. Nature 7751:158–161

Scientific Faculties Admission test results (2017/2019) CISIA (Consorzio Interuniversitario Sistemi Integrati per l'Accesso) (Author's private communication)

Scott V, Sayre E, Clark J (2014) Traditionally taught students learn; actively engaged students remember. Am J Phys 82:798–801

Ship noise masking Whale Song: https://www.youtube.com/watch?v=nZHa2Wn5MzE. Accessed 30 July 2024

SOFAR. https://en.wikipedia.org/wiki/SOFAR_channel. Accessed 30 July 2024

Analysis and Evaluation of Inquiry-Based Tasks in Digital Learning Scenarios

Despoina Valsamouli and Ioannis Lefkos

Abstract This research aimed to analyze and evaluate inquiry-based tasks found in digital Inquiry Learning Scenarios on the Go-Lab platform, in terms of understanding the scientific concepts they offer, the scientific process skills involved, and their contribution to the understanding of the scientific inquiry. Inquiry-based tasks of 16 "Example" plus 10 "Top Rated" Inquiry Learning Scenarios were analyzed using the Inquiry-Based Tasks Analysis Inventory. The overall level of inquiry of the scenarios was investigated, as well as its possible dependence on various features of the scenarios. Our findings suggest that these ILS promote a high degree of understanding of scientific concepts and scientific inquiry processes, provide students opportunities to use all the inquiry process skills but not at the same frequency, and that the degree of authentic inquiry they offer, depends on internal characteristics of the ILS, e.g. using more interactive elements like apps and less static elements like images.

1 Introduction

Inquiry is the way scientists work, as well as the tasks through which students become familiar with both scientific concepts and scientific processes (Bybee 2006). Includes developing problem-solving skills (Pedaste and Sarapuu 2006) and aims to discover new relationships between concepts by creating hypotheses and evaluating them through experiments or observations (Pedaste et al 2012). Inquiry-based learning in science teaching is an increasingly popular model proposed and promoted in school curricula worldwide, international research, and teaching (Pedaste et al 2015). Technology plays an essential role in increasing the effectiveness of inquiry methods utilizing digital learning environments like online (de Jong et al. 2014; Molohidis et al. 2015) or stand-alone digital labs (Lefkos et al. 2011). When inquiry tasks are guided via the internet, they contribute to the improvement of different skills such as:

D. Valsamouli · I. Lefkos (✉)
University of Macedonia, Thessaloniki, Greece
e-mail: lefkos@uom.edu.gr

S. Faletič and J. Pavlin (eds.), *Teaching and Learning Physics Effectively in Challenging Times*, Challenges in Physics Education, https://doi.org/10.1007/978-3-031-72541-8_10

identifying problems, asking questions and formulating hypotheses, designing and carrying out experiments, collecting and interpreting data, etc. (Mäeots et al. 2008).

Inquiry is carried out in various phases. Pedaste et al. (2015), through their review, came up with a learning cycle that combines all the existing frameworks of inquiry approach, which includes five main phases (and sub-phases): (1) Orientation, (2) Conceptualization, which includes Questioning and Hypothesis Generation, (3) Investigation, which includes Exploration or Experimentation and Data Interpretation, (4) Conclusion, and (5) Discussion, which includes Reflection and Communication. Inquiry usually ends with the "conclusion" phase. The "discussion," as well as its sub-phases, may exist throughout the task and be linked to all other phases.

This inquiry learning cycle (Pedaste et al 2015) was materialized by the Go-Lab project. Go-Lab https://www.golabz.eu/ is an online platform that focuses on inquiry-based learning and includes online laboratories (Labs) and inquiry learning applications (Apps). According to de Jong (2015), the inquiry tasks of the Go-Lab platform compose Inquiry Learning Spaces (ILS), i.e., digital learning scenarios for inquiry. ILS created by teachers and include online labs and apps that are appropriate for their lesson plans (Dikke et al. 2014). There are seven types of scenarios available on the platform: Find The Mistake, Jigsaw, Learning By Critiquing!, Six Thinking Hats, Structured Controversy, The 5Es, and the Basic Scenario. The Basic Scenario consists of activities corresponding to the five inquiry phases of Pedaste et al. (2015). This way, educators using the Go-Lab platform can compose digital learning scenarios (ILS) following the proposed inquiry learning cycle (Pedaste et al. 2015).

Even though the Go-Lab platform allows all educators to create their ILS and submit it for publication, there are some particular ILS types in this repository. One of these types is the "Example ILS", which includes scenarios created by educators in the context of Go-Lab pilot programs and are thus recommended by the project. Another particular type of ILS is the "Top Rated", comprising scenarios that have been positively rated (voted) by the subscribed users of the Go-Lab platform.

According to Yang and Liu (2016), inquiry-based tasks should serve four essential functions: [a] to assist in the construction of understanding about scientific concepts, [b] to provide students opportunities to use inquiry process skills, [c] to contribute to the establishment of understandings about scientific inquiry and [d] to provide students with opportunities to develop higher-order thinking skills (Yang and Liu 2016). Considering these four functions, they developed an instrument for deploying content analysis of inquiry-based tasks, the Inquiry-Based Tasks Analysis Inventory (ITAI), which contains a scale comprising 22 items and detailed scoring rubrics for each item, that correspond to the functions [a], [b] and [c] (Tables 1,2 and 3). For function [d] (i.e., HOTS), they suggest a critical overview of the previous sections in the discussion section of a given research. Thus, their instrument covers the first three functions.

Very recently, ITAI was used by several researchers for the analysis of science textbooks in Primary or Secondary Education in Greece (Sideri and Skoumios 2021), Turkey (Dogan 2020), China (Yang et al. 2019), and Indonesia (Halawa et al. 2022). In most cases, they report quite similar findings like the following. Textbooks seem

Table 1 The first function of the Inquiry-based Tasks Analysis Inventory (ITAI) (Yang and Liu 2016)

Dimension 1: To assist in the construction of understandings about scientific concepts
ITAI 1. Scientific concepts involved in this task are consistent with the objectives of the lesson
ITAI 2. Understandings about the involved concepts contribute to learning core ideas

Table 2 The second function of the Inquiry-based Tasks Analysis Inventory (ITAI) (Yang and Liu 2016)

Dimension 2: In this task, students are expected to use the following skills
ITAI 3. Observing
ITAI 4. Inferring
ITAI 5. Measuring
ITAI 6. Communicating
ITAI 7. Classifying
ITAI 8. Predicting
ITAI 9. Controlling variables
ITAI 10. Defining operationally
ITAI 11. Formulating hypotheses
ITAI 12. Interpreting data
ITAI 13. Asking questions
ITAI 14. Formulating models

Table 3 The third function of the Inquiry-based Tasks Analysis Inventory (ITAI) (Yang and Liu 2016)

Dimension 3: The text of this task reflects the following understandings about scientific inquiry
ITAI 15. Scientific inquiry all begin with a question, but do not necessarily test a hypothesis
ITAI 16. There is no single set and sequence of steps or methods followed in all inquiries
ITAI 17. Inquiry procedures are guided by the question asked
ITAI 18. All scientists performing the same procedures may not get the same results
ITAI 19. Inquiry procedures can influence results
ITAI 20. Conclusions must be consistent with the data collected
ITAI 21. Scientific data are not the same as scientific evidence
ITAI 22. Explanations are developed from a combination of collected data and what is already known

to fail to provide students with satisfactory opportunities to develop their science process skills and facilitate a better understanding of scientific ideas and concepts (Sideri and Skoumios 2021), they seem to separate the inquiry process from scientific content (Yang et al. 2019), and they do not reflect a proper understanding about scientific inquiry (Dogan 2020; Yang et al. 2019; Halawa et al. 2022). Additionally, it was found that some of the inquiry skills are more emphasized, like observing and communicating (Halawa et al. 2022).

Even though there are a number of studies focusing on textbook inquiry-based tasks analysis, there is a lack of research aiming to analyze and evaluate digital inquiry-based tasks, like the ones included in digital learning scenarios. For example, Cellitti et al. (2018), used content analysis to examine the Science and Engineering Practices that were included in 40 web-based lesson units and found that (1) planning and carrying out investigations (2) developing and using models and (3) analyzing and interpreting data, were the ones most commonly used.

Our study aims to analyze and evaluate the inquiry tasks of Go-Lab's ILS using ITAI. The research questions were: Q1. What is the degree of support in understanding scientific concepts offered by Go-Lab's "Example" and "Top rated" ILS? Q2. Which scientific process skills are involved in Go-Lab's inquiry-based tasks, and to what extent? Q3.What is the degree of support of the Go-Lab's "Example" and "Top rated" ILS to understanding the scientific inquiry process? Q4. Are there specific content formats and applications of "Example" and "Top rated" ILS that result in a higher degree of authentic inquiry?

2 Research Methodology

The study was based on content analysis. The sample was the individual tasks included in 16 Greek "Example" and in 10 Greek "Top Rated" ILS available on the Go-Lab platform. The tool used for the analysis was the ITAI (Yang and Liu 2016). As mentioned earlier, ITAI contains 22 items, categorized into three dimensions: [a] Understanding about scientific concepts (items 1–2), [b] Use of scientific process skills (Items 3–14), and [c] Understanding about scientific inquiry (items 15–22). In this paper, we have examined all three dimensions of ITAI.

Each selected ILS was examined in terms of the involvement of each one of the 22 items of ITAI. Each ILS is considered a unit of analysis and receives a number code in the order of appearance in the list of available "examples" scenarios on the Go-Lab page or in the list of available "Top rated" scenarios and is analyzed with the first and third dimension of ITAI. Each phase of an ILS is considered an activity, encoded with a new number e.g., 1.1, where the first corresponds to the ILS´s number and the second to the phase number. Phase 3.2 for example, will correspond to the second phase/activity of the third ILS and is analyzed with the second dimension of ITAI. When an item was found in a task, it was rated with a "YES" response (1 point), while otherwise, it was rated with a "NO" response (0 points).

For each ILS there are three individual ITAI scores, one for each of the ITAI dimensions. The score for the first and third dimensions is calculated globally for each ILS. The score for the second dimension is calculated as the mean value of the sub-scores of the individual phases/activities of the ILS. Finally, an overall score is calculated as the sum of the above three dimensions, and this—according to the ITAI model—represents the level of authentic inquiry offered by the total ILS (see Fig. 1). According to ITAI, the higher the score in one ILS, the more elements of authentic inquiry are involved.

A random sample of 15 out of the 26 total ILS was reviewed by two additional inquiry-based learning experts/researchers for validity reasons, and any differences in their results were discussed and reached a consensus.

Additionally, each ILS is analyzed in terms of its various features such as Age Range, Big Ideas of Science, Subject Domains, Number of Phases, Apps, Virtual Laboratories, Images, Videos, and External Sources (see Fig. 2).

	DIMENSION 1			DIMENSION 2													DIMENSION 3									
ILS	ITAI 1	ITAI 2	ITAI 1 (max=2)	ITAI 3	ITAI 4	ITAI 5	ITAI 6	ITAI 7	ITAI 8	ITAI 9	ITAI 10	ITAI 11	ITAI 12	ITAI 13	ITAI 14	ITAI 2 (max=12)	ITAI 15	ITAI 16	ITAI 17	ITAI 18	ITAI 19	ITAI 20	ITAI 21	ITAI 22	ITAI 3 (max=8)	SCORE ITAI (max=22)
1	YES	YES	2													2,33	YES	YES	YES	YES	YES	YES	NO	YES	7	11,33
1.1				YES	NO	NO	NO	NO	YES	NO	NO	NO	NO	YES	NO	3										
1.2				NO	NO	NO	NO	NO	NO	NO	NO	YES	NO	YES	NO	2										
1.3				YES	NO	YES	NO	NO	NO	YES	NO	NO	NO	NO	NO	3										
1.4				NO	YES	NO	NO	NO	NO	NO	NO	YES	NO	YES	NO	3										
1.5				YES	NO	NO	NO	NO	NO	NO	NO	NO	NO	NO	NO	1										
1.6				NO	NO	NO	NO	NO	NO	NO	YES	NO	NO	YES	NO	2										

Fig. 1 ITAI analysis-scoring example

ILS	TITLE	Age range	Big Ideas Of Science	Subject Domains	Phases	Apps	Virtual Laboratories	Images	Videos	External Sources
1	Factors That Affect Photosynthesis	13-14 ,15-16, 16+	Energy Transformation, Organisms And Life Forms, Planet Earth	Biology, Botany, Ecology, Life Processes, Physics, Energy, Light	6	12	2	3	1	2
1.1	Let's think					2		2		
1.2	Let's hypothesize					4				
1.3	Let's experiment					2	2			2 V.Labs
1.4	Let's draw conclusions					1				
1.5	Let's reconsider					1		1	1	
1.6	Let's see what we learned					2				

Fig. 2 Analysis of characteristics of a sample ILS

Finally, a semi-quantitative analysis was performed comparing the two ILS with the highest overall ITAI scores, as well as the two ILS with the lowest overall ITAI scores in terms of their various design characteristics, i.e., the apps, virtual labs, videos, etc. included.

3 Results

3.1 ITAI Scores of "Example" Type ILS

Concerning the type "example ILS" our research results showed a very high degree (ITAI-1: 100% and ITAI-2: 87,50%) of support for the understanding of scientific concepts.

Additionally, we found that all process skills are involved in Go-Lab's inquiry-based tasks but to a different extent. Observing, Asking questions, and Defining operationally, are very frequently used. Measuring, Controlling variables, Communicating, and Formulating hypotheses were less frequently used and the skills of, Interpreting data, Predicting, Inferring, Classifying, and Formulating models were rarely involved.

Our research results showed a 78% degree of contribution of the Go-Lab's ILS to the understanding of the scientific inquiry process.

At the same time, it was observed that the scenarios with the highest ITAI score have a very high score on the third dimension of the ITAI. In other words, this dimension mainly contributes to the differentiation between the scores of the scenarios.

3.2 ITAI Scores of "Top Rated" Type ILS

On the other hand, on "Top rated" ILS, our research results showed a 95% degree of support for the understanding of scientific concepts.

Additionally, they provide students/users opportunities to use inquiry process skills but to a different extent. Observing is the most commonly presented inquiry process skill, with a degree of 82%. Asking questions and Interpreting data are very frequently used. Predicting, Controlling variables, Measuring and Communicating are less frequently used, and the inquiry process skills of Formulating hypotheses, Inferring, Classifying, and Formulating models are rarely involved.

On the third dimension of ITAI, the establishment of understanding about scientific inquiry our research results showed a 58, 75%°.

3.3 ITAI Scores of "Example" Versus "Top Rated" Type ILS

Comparing the results of the three dimensions of ITAI of the "Examples" ILS with those of the "Top rated" we found that their values are pretty close (see Figs. 3, 4, 5).

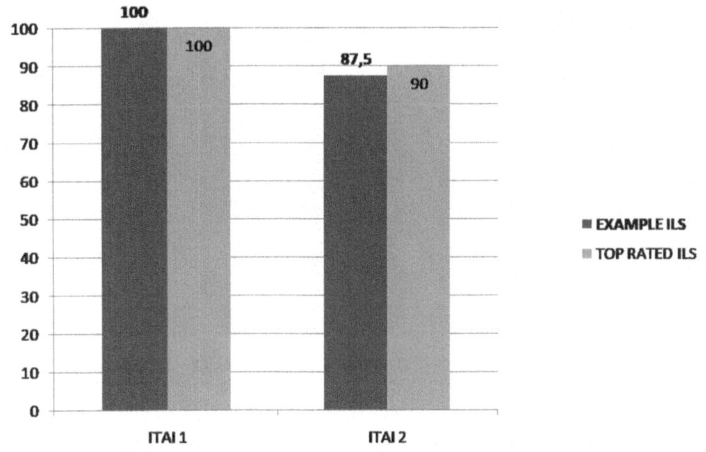

Fig. 3 Score (mean values) comparison of the **first dimension of ITAI** (understandings about scientific concepts), for the "Example ILS" versus "Top rated" ILS

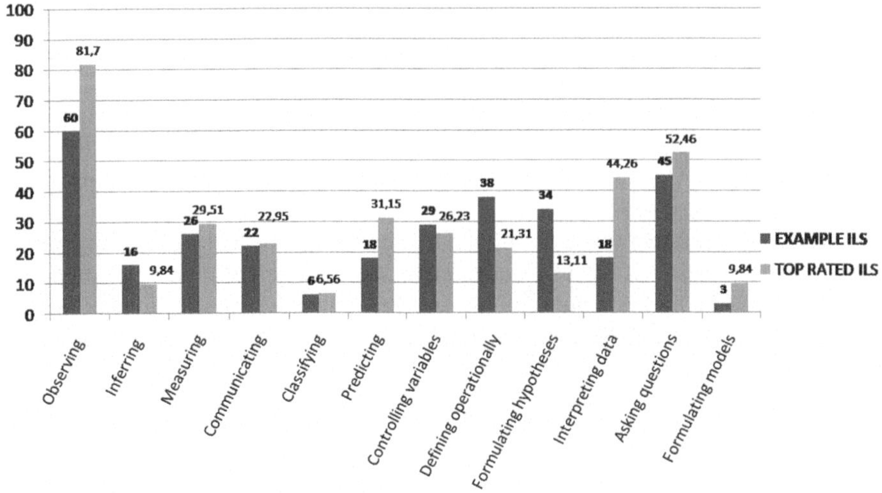

Fig. 4 Score (mean values) comparison of the **second dimension of ITAI** (inquiry process skills), for the "Example ILS" vs. "Top rated" ILS

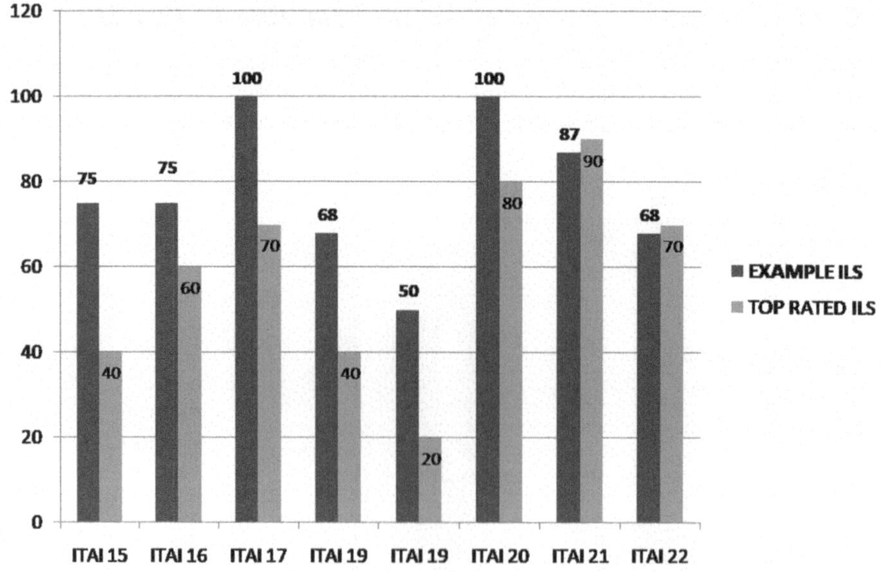

Fig. 5 Score (mean values) comparison of the **third dimension of ITAI** (understandings about scientific inquiry, for the "Example ILS" vs. "Top rated" ILS

3.4 ITAI Scores and Design Characteristics of ILS

Finally, regarding the dependence of the ITAI score on the design characteristics, such as the number of phases, apps, virtual labs, etc., from our analysis so far, no clear correlation has been found.

In this direction, we tried to get some generalized evidence using a semi-quantitative approach. Specifically, a comparison was made between the five ILS with the highest ITAI scores and the five ILS with the lowest scores in terms of the above characteristics (see Figs. 6, 7).

It was thus found that the ILS with the highest scores (a) have a very high score on the third dimension of ITAI (understanding about the scientific inquiry process), (b) they included a higher number of Go-Lab recommended applications per phase (e.g., hypothesis tool, experiment design tool) (de Jong 2015), and (c) include fewer images (see Fig. 8).

HIHG SCORED ILS	ITAI TOTAL SCORE (max 22)	ITAI 1 (understanding scientific concepts) (max2)	ITAI 2 (scientific process skills) (max 12)	ITAI 3 (understanding about the scientific Inquiry process) (max 8)	AGE	PHASES	RECOMMENDED APPS/ PHASE	V.LABS	PICTURES	VIDEOS
1	14,40	2	4,40	8	11-12, 13-14	5	YES	1	3	1
2	13,67	2	3,67	8	11-12	6	YES	1	14	2
3	13,57	2	3,70	8	13-14	7	YES	1	5	0
4	13,11	2	3,11	8	13-14, 15-16	9	YES	2	3	1
5	13,00	2	5,00	6	11-12, 13-14	5	YES	1	12	4

Fig. 6 Analysis of five ILS with the highest ITAI scores

4 Conclusions

Our findings suggest that the examined "Example" and "Top rated" ILS from the Go-Lab repository display similar characteristics: (a) they significantly promote the understanding of scientific concepts, (b) they highly contribute to the understanding of the scientific inquiry, and (c) they do not involve all the scientific process skills to the same degree.

These seem to be partly in contrast with findings reported from similar works in school textbooks. Most of the textbooks are mainly focused on the understanding of concepts, separating the inquiry process from the scientific content (Yang et al. 2019), thus failing to support a proper understanding of scientific inquiry (Dogan 2020; Yang et al. 2019; Halawa et al. 2022).

Concerning the science process skills, textbooks have also been found lacking in providing students with opportunities to develop them (Sideri and Skoumios 2021), or mainly promote just a few of them, like observing and communicating (Halawa et al. 2022). Quite similar results were reported from the analysis of web-based lesson units, where planning, modeling, and data handling were mainly promoted (Cellitti et al. 2018). Our findings resemble the fact that only some of the science process skills are mainly promoted (from 34%-60%), but in our case these skills

LOW SCORED ILS	ITAI TOTAL SCORE (max 22)	ITAI 1 (understanding scientific concepts) (max2)	ITAI 2 (scientific process skills)(max 12)	ITAI 3 (understanding about the scientific inquiry process)(max 8)	AGE	PHASES	RECOMMENDED APPS/ PHASE	V.LABS	PICTURES	VIDEOS
1	4,20	1	2,20	1	11-12, 13-14	5	NO	2	9	2
2	6,50	2	2,50	2	7-8, 9-10	6	NO	0	12	1
3	7,50	2	2,50	3	15-16, 16+	6	YES	2	21	0
4	8,29	1	3,29	4	9-10, 11-12	7	YES	3	8	4
5	9,67	2	2,67	5	11-12, 13-14, 15-16	9	NO	0	38	2

Fig. 7 Analysis of five ILS with the lowest ITAI scores

are: observing (similar to Halawa et al. 2022), asking questions, defining operationally, and hypothesizing. Controlling variables, measuring, and communicating are also fairly promoted (from 22%-29), thus partly coinciding with findings from the textbooks or web-based lessons (Halawa et al. 2022; Cellitti et al. 2018).

Moreover, our results suggest that the degree of the authentic inquiry-based learning promoted, depends on internal characteristics of the ILS, e.g. the number and kind of apps included. ILS with a larger number of Go-Lab recommended applications per phase (e.g., hypothesis tool, experiment design tool) and those that utilize a small number of images, seem to be more successful in fostering inquiry learning.

In our view, this means that using more interactive elements like the Go-Lab apps in learning scenarios seems to foster authentic inquiry. On the other hand, using a lot of static elements, like images in a scenario, does not seem to be successful in promoting authentic inquiry.

Web-based scenarios, like the ones hosted at the Go-Lab repository, are considered to be a powerful and flexible aid to the educators, since they can be used in a variety of educational contexts. Our findings can be useful to educators who would like to design or implement digital learning scenarios, providing a guideline (a) to choose

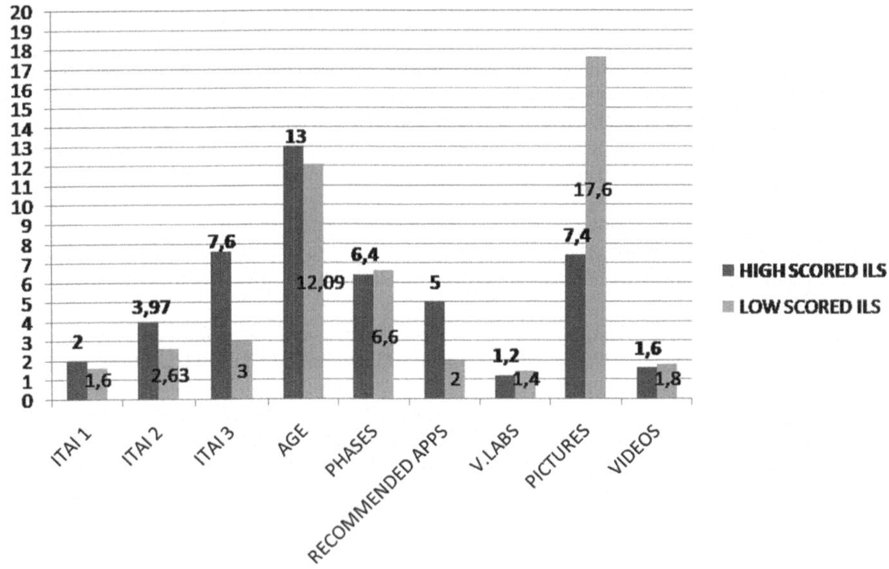

Fig. 8 Score (mean values) comparison between the five high scored ILS with and the five low scored ILS in terms of the above characteristics

the best possible ILS but also (b) to design new ILS depending on the process skills they wish to promote. It is thus recommended to use appropriately designed digital inquiry learning scenarios, both to support the understanding of the scientific inquiry process and to develop a variety of skills of scientific processes in students.

References

Bybee RW (2006) Scientific inquiry and science teaching. Sci Inq Nat Sci 25:1–14. https://doi.org/10.1007/978-1-4020-5814-1_1

Cellitti J, Likely R, Moy M, Wright C (2018) A content analysis of NGSS science and engineering practices in K-5 curricula. In: 2018 ASEE annual conference & exposition proceedings. https://doi.org/10.18260/1-2--29667

de Jong T, Sotiriou S, Gillet D (2014) Innovations in STEM education: the go-lab federation of online labs. Sci Inq Nat Sci. https://doi.org/10.1186/s40561-014-0003-6

de Jong T (2015) Go-lab deliverable D1.4 go-lab classroom scenarios handbook. In: hal. https://telearn.archives-ouvertes.fr/hal-01274922/document. Accessed 26 May 2024

Dikke D, Tsourlidaki E, Zervas P, Cao Y, Faltin N, Sotiriou S, Sampson DG (2014) Golabz: towards a federation of online labs for inquiry-based science education at school. In: Proceedings of the 6th international conference on education and new learning technologies EDULEARN 2014, Barcelona, 7–9 July, 2014, pp 3238–3248. https://library.iated.org/view/DIKKE2014GOL

Dogan OK (2020) Methodological? or dialectical? Reflections of scientific inquiry in biology textbooks. Int J Sci Math Educ 19:1563–1585. https://doi.org/10.1007/s10763-020-10120-7

Halawa S, Hsu Y-S, Zhang W-X (2022) Analysis of physics textbooks through the lens of inquiry practices. Asia Pac Educ Res 32:497–506. https://doi.org/10.1007/s40299-022-00671-4

Lefkos I, Psillos D, Hatzikraniotis E (2011) Designing experiments on thermal interactions by secondary-school students in a simulated laboratory environment. Res Sci Technol Educ 29:189–204. https://doi.org/10.1080/02635143.2010.533266

Mäeots M, Pedaste M, Sarapuu T (2008) Transforming students' inquiry skills with computer-based simulations. In: 2008 eighth IEEE international conference on advanced learning technologies. https://doi.org/10.1109/ICALT.2008.239

Molohidis A, Lefkos I, Taramopoulos A, et al (2015) Web-based virtual labs—a cosmos—evidence—ideas as a design framework leading to good practice. In: Proceedings of the 7th international conference on computer supported education. https://doi.org/10.5220/0005477204180423

Pedaste M, Sarapuu T (2006) Developing an effective support system for inquiry learning in a web-based environment. J Comput Assist Learn 22:47–62. https://doi.org/10.1111/j.1365-2729.2006.00159.x

Pedaste M, Mäeots M, Leijen Ä, Sarapuu S (2012) Improving students' inquiry skills through reflection and self-regulation scaffolds. Technol Instr Cogn Learn 9(1–2):81–95

Pedaste M, Mäeots M, Siiman LA et al (2015) Phases of inquiry-based learning: definitions and the inquiry cycle. Educ Res Rev 14:47–61. https://doi.org/10.1016/j.edurev.2015.02.003

Sideri A, Skoumios M (2021) Science process skills in the greek primary school science textbooks. Sci Educ Int 32:231–236. https://doi.org/10.33828/sei.v32.i3.6

Yang W, Liu E (2016) Development and validation of an instrument for evaluating inquiry-based tasks in science textbooks. Int J Sci Educ 38:2688–2711. https://doi.org/10.1080/09500693.2016.1258499

Yang W, Liu C, Liu E (2019) Content analysis of inquiry-based tasks in high school biology textbooks in mainland China. Int J Sci Educ 41:827–845. https://doi.org/10.1080/09500693.2019.1584418

Evaluation of the Effectiveness of an Introductory Mechanics Laboratory with Arduino and Smartphone

Eugenio Tufino and Giovanni Organtini

Abstract This article describes the reform of a physics undergraduate lab course by introducing new technologies such as Arduino microcontroller and smartphone, to help students experience a more authentic lab experience. This paper outlines the structure of the lab activities and presents findings from pre- and post-course questionnaires, including the E-CLASS survey, completed by participating students. Some interesting observations about teamwork and equity emerged from these questionnaires, particularly about female students. These results will guide us in designing more effective and equitable lab courses.

1 Introduction

In this paper we describe the reform of a mechanics laboratory course with the introduction of digital data acquisition devices. The idea of this reform dates back a few years but some resistance had been encountered from the faculty, so its implementation was facilitated by the emergence of the COVID-19 pandemic that forced Universities to find alternatives to in-person labs. A useful review of the wide range of approaches proposed by physics laboratory instructors in laboratory courses during the pandemic is available here (Fox et al. 2020). In general, in recent years, even before the pandemic, there has been a resurgence of interest in introductory physics lab courses also partly motivated by studies showing that students do not effectively learn physics concepts in traditional (confirmatory type), highly structured labs (Wilcox and Lewandowski 2016b; Wieman and Holmes 2015). We revised the overall approach of the course to introduce some active learning strategies, giving more space for students to make decisions and reducing the verification aspects in

E. Tufino (✉)
Department of Physics, University of Trento, Povo, Trento, Italy
e-mail: eugenio.tufino@unitn.it

G. Organtini
Sapienza Universitá di Roma and INFN-Sez. di Roma, Roma, Italy
e-mail: giovanni.organtini@uniroma1.it

S. Faletič and J. Pavlin (eds.), *Teaching and Learning Physics Effectively in Challenging Times*, Challenges in Physics Education, https://doi.org/10.1007/978-3-031-72541-8_11

lab activities. These in fact represent two of the most critical aspects of any lab course (Smith and Holmes 2021). The use of Arduino microprocessors represents a pedagogical support, students in fact were encouraged to build their own apparatus from scratch and decide on strategies for making measurements. The changes were implemented gradually. The 2021 edition, due to the pandemic, was held mostly remotely with students doing experiments from home in an active and collaborative approach with breakout rooms. The course lectures were given by a single lecturer, while for the laboratory part the students were divided into two groups, one (pilot lab) in which Arduino and smartphones were used for data acquisition, and Python language with Google Colab to do data analysis, while the other group used more traditional methods for data acquisition and data analysis. In the 2022 edition, the lectures and the lab activities were both run by one of the authors and held in person in the laboratory. To assess the lab transformation, we have applied the E-CLASS survey (Zwickl et al. 2014). The survey is intended to measure students' views on the nature of experimental physics, the role of experimentation in physics, and their confidence in their ability to conduct experiments. E-CLASS allows tracking of changes in students' attitudes and beliefs as a result of the course intervention, as well as to identify areas where students may need more support. To probe student perception of the laboratory classroom experience, an additional survey was administered at the end of the course. We also investigated group work in relation to gender. In this article, we report the details of the structure of the lab course and show the results on students' attitudes. Further details about this work can be found in (Organtini and Tufino 2022).

2 Course Structure

The introductory laboratory course held at University Sapienza of Roma focuses on selected topics of Mechanics. Apart from the laboratory activity, there are also some lessons on measurements and uncertainties, data analysis, probability theory and Bayesian reasoning. Seven experimental activities are planned during the semester course, two of which are conducted individually, while the others are carried out in groups. They are outlined in Table 1 and are common to all the channels of the course, both "traditional" ones (traditional in the sense that they used traditional and standard instruments for experiments) and the revised one. During the lab sessions in groups, students work in teams made of three or four students. At the end of the lab, they have to submit a lab report, which is reviewed with feedbacks and returned by the instructors before the next lab session.

2.1 Arduino and Smartphone

Arduino is a popular open-source platform and can be used very effectively in physics laboratories as a tool for data acquisition and data analysis. Considering the affordable

Table 1 Experiments proposed in the 2021/2022 course with different technologies: Arduino (A), smartphone (S), dedicated devices (D)

Experiments activities in 2021/22	Technology Used
Measuring the density of a body	D
Studying the motion of a pendulum	S
Studying the dynamics of a spring	A
Finding how the speed of sound depends on temperature	A
Radiation counters	D
Measuring the moment of inertia of a rolling cylinder	S
Studying the deformation of a slab	A

price, we asked students to purchase a simple kit consisting of an Arduino UNO board, a breadboard, a few connecting cables, an ultrasonic sensor, and a temperature sensor. This allows students to use Arduino at home for further experiments on their own. Since students have already taken a C programming course in the months prior to the lab course, they already know the syntax of Arduino programming and so a session of only two hours was sufficient to focus on specific statements for reading analog and digital pins, setting digital pins, etc. A program to read data from an ultrasonic sensor was used as an application example. In addition to Arduino, the smartphone is also used in the course to acquire data. Specifically, the Phyphox application (Phyphox 2023) is used, which is free, open-source and transforms smartphone sensors into measurement instruments. It can collect data from the accelerometer, gyroscope, magnetometer, video camera, microphone, light sensor and barometer.

To give some examples of the proposed activities, in the experiment on the dynamics of a spring, students used the ultrasonic sensor connected to Arduino (Fig. 1a) to determine the position of the end of a suspended spring and then, using this data, they estimated the period T of oscillation. They then plotted T^2 as a function of the suspended mass to determine the elastic constant with a linear fit. In studying the dynamic of pendulum (Fig. 1b), the smartphone's accelerometer was used to measure the period of the smartphone, which was used as the suspended mass. The gravitational acceleration was determined by fitting the distribution of T^2 values to the pendulum length.

Similarly, the study of the deformation of a slab can be done with either Arduino, measuring its deformation by means of the ultrasonic sensor (Fig. 2b), or with a smartphone, fixed to the slab and oscillated. The period of oscillation is proportional to the cube of the length of the slab, as well as its deformation.

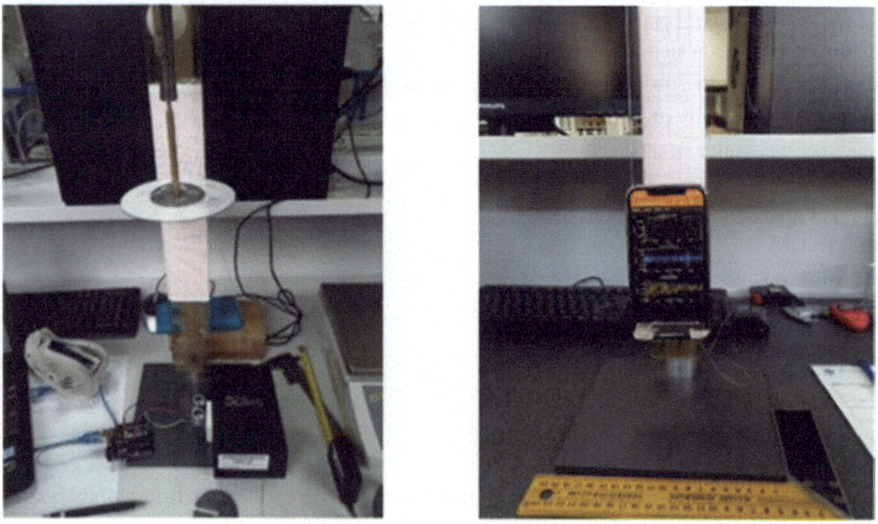

(a) (b)

Fig. 1 **a** An example of an experimental apparatus with Arduino and ultrasound sensor to study the dynamics of a spring. **b** Students' experimental setup for the pendulum experiment with smartphones and Phyphox

2.2 Data Analysis with Jupyter Notebook

For data analysis of experiments, students used Jupyter Notebook with Python language. Since the students had already taken a C programming course, the introduction to Python programming and Jupyter notebook use was very short in the first weeks of the course. Jupyter notebook allows users to write and execute code, create visualisations, insert text, images and videos, all in one place. In particular, we proposed Google Colab (Colab 2023), a cloud-based platform available to anyone with a Google Account. Thus, students do not have to do any installation-which can sometimes be problematic-and they can easily share the notebook in Google Drive and collaborate in group for data analysis and lab report writings. Jupyter Notebooks were also a valuable resource during data analysis and probability classes, where the teacher used some notebooks to introduce complex concepts and simulation, as well as to provide hands-on examples and exercises for students. An example of this teaching method can be seen in the notebook available at the following link, which was used to propose an application of Bayes' theorem (Jupyter 2023). The use of Jupyter Notebooks addresses the recently emerged need to introduce computational skills as early as possible in the undergraduate courses. Examples of this include the AAPT Conference Report from 2019 (Computational 2019) and the PICUP group (Caballero et al. 2019).

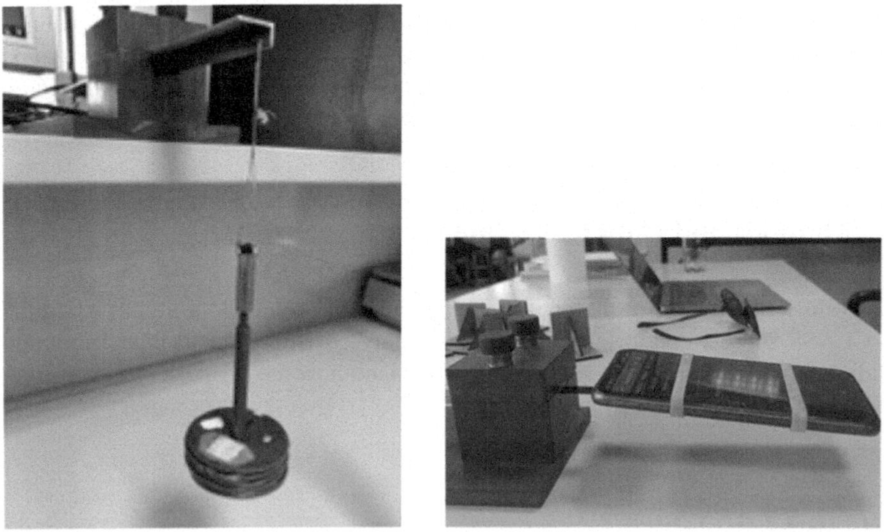

(a) (b)

Fig. 2 **a** An example of an experimental apparatus with Arduino and ultrasound sensor to study the dynamics of a spring. **b** Students' experimental setup for the pendulum experiment with smartphones and Phyphox

2.3 Pedagogical Approach

The lectures typically revolved around explaining the principles of statistical data analysis, introduced first heuristically and then proceeding to a more formal justification of the practices used. This way we were promoting scaffolding. Students were introduced to Bayesian reasoning, fitting data with the least squares method, and the maximum likelihood principle. They learned how to estimate confidence intervals and p-values. One of the intentions of the course is to help students understand that performing a measurement is always likened to drawing a random number with an appropriate probability distribution. We discussed the experiments and the way in which they are conducted only loosely, without too many details, to foster students' engagement in the preparation of them. By utilising Arduino and smartphones instead of a dedicated standard experimental setup, students are encouraged to design the experimental apparatus, take decisions and in case of discrepancies between measurements and the mathematical model, work on improving the apparatus and clarifying the approximations or assumptions of the model. The Arduino-based instrumentation is intended to be a tool for students to explore and understand. The objective is to move beyond treating the data acquisition system as a black box. Students are encouraged to experiment with editing both the data analysis code in the Jupyter notebook on the computer and the data acquisition code on the Arduino IDE. Before each lab session, students were sent a document containing general guidelines about each lab activity; these sheets were designed so as not to specify too many details, not following a step-by-step recipe format, leaving room for student creativity and initiative. Students are also given the autonomy to incorporate their own research question related to the experiment into the final report.

3 Methods and Data Collection

To evaluate the impact of the revised lab, we assessed students' attitudes using the E-CLASS survey, conducted observations, analysed student work and submitted additional questionnaires to students to understand how the proposal was received, including aspects related to group work.

3.1 E-CLASS Survey

The E-CLASS survey is a research-based assessment tool composed of 30 Likert-like questions. Various aspects are probed, including confidence, experimental design, purpose of labs, statistical uncertainty, systematic error, troubleshooting. The E-CLASS assessment is described in this paper, along with a summary of the research conducted using it (Wilcox and Lewandowski 2018). We translated into Italian (available on PhysPort website (ECLASS 2023)) and administered it to students for the first time in the 2021 edition of the course both at the start and end of the semester. Recently, the E-CLASS survey has been translated and used in German laboratory courses (Teichmann et al. 2022). The E-CLASS requires students to provide two responses to the same question: one from their own point of view and one from the point of view of a professional physicist. These responses are labelled in Fig. 3 as YOU (the student viewpoint) and EXPERT (the expert viewpoint) and are analysed separately. The answers provided by the students are then compared with those provided by the experimental physicists, who serve as a reference for the EXPERT (Zwickl et al. 2014). To calculate the score, we applied a two-point scheme, by awarding students 1 point for each question where their response matches the expert-like response and zero point for any other response. This means that the average score for any given question represents the percentage of students who gave a favourable response. The overall E-CLASS score is determined by summing the scores on all 30 questions, which represents the average number of favourable responses from each student.

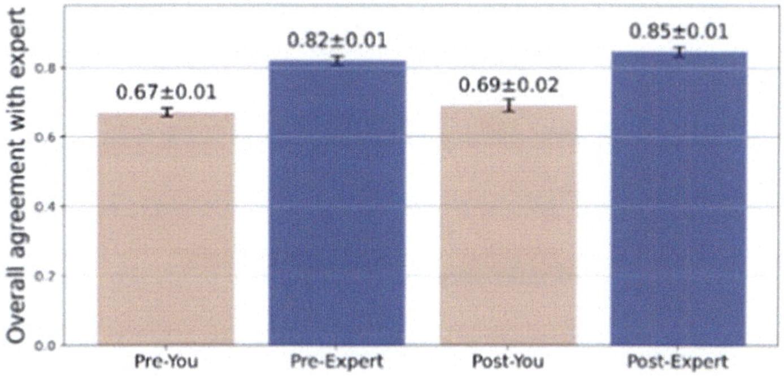

Fig. 3 Average fractions of questions with favourable responses ("You" and "Expert" views) both before and after lab activities for year 2021/22

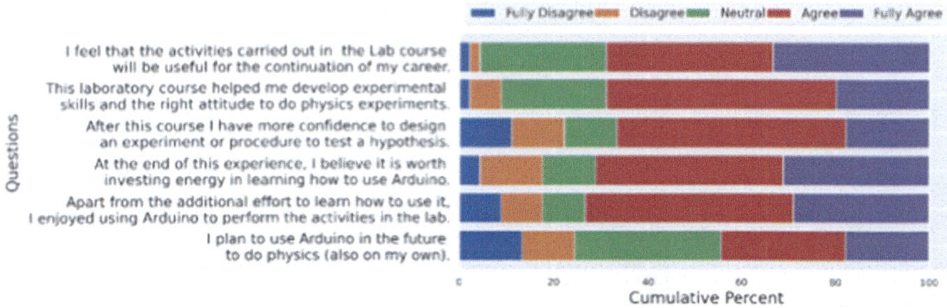

Fig. 4 Stacked bar plots showing student responses to survey administered at the end of the course

3.2 Student Perception of the Lab Experience

In the last week of the course, and before the final exam, we administered another anonymous questionnaire, referred to as the end-of-course questionnaire. This instrument developed following general guidelines outlined for example in (Lovelace and Brickman 2013), and further designed by us, contained a set of questions aimed at obtaining feedback from students about the perceived usefulness of the lab experience and its practical organization. The questionnaire also included questions specifically related to the use of Arduino in the lab. Most of the questions in the questionnaire can be viewed in Figs. 4, 5, and 6. The number of students who responded to the end-of-course questionnaire was 46, which is about 50% of those enrolled in the section of the course lab.

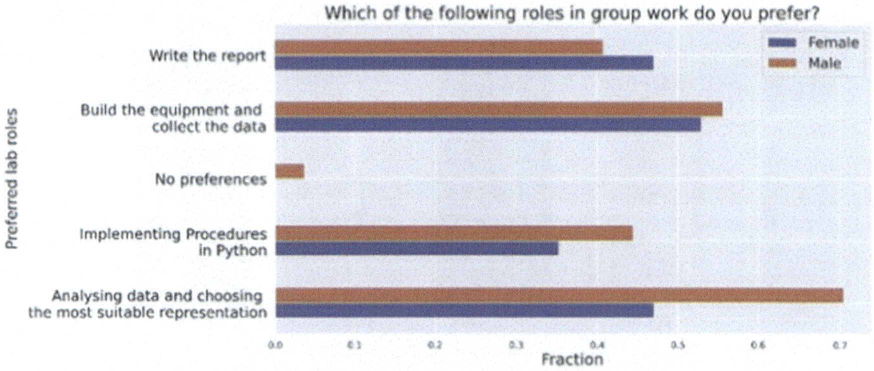

Fig. 5 Student responses to survey items about preferred lab roles

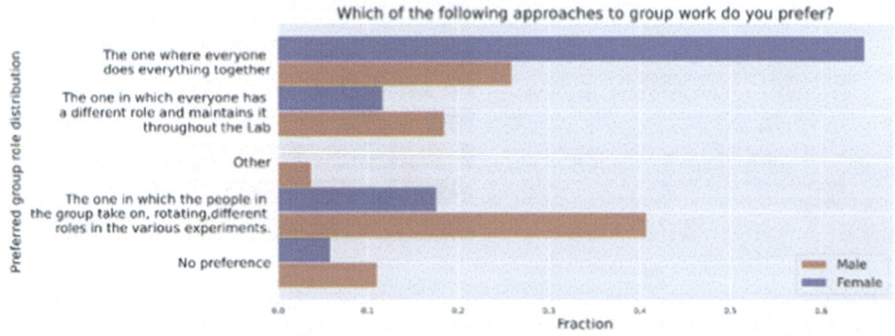

Fig. 6 Student responses to survey items about preferred lab group distribution of work

4 Student Learning: Results

In this section we illustrate the results obtained with the tools described in Sect. 3.

4.1 Analysis of E-CLASS Survey

In Fig. 3 we show the overall class score for the 2021/22 edition. For the YOU (student's point of view) questions, only a slight increase in the overall score, not statistically significant, was observed between the pre- and post-questionnaires. As you can see, the score for the student viewpoint towards EXPERT increased from 0.82 to 0.85. The latter result may indicate that through the laboratory course, knowledge of how physicists work has slightly improved in students. The fact that there is no deterioration in the average E-CLASS score is a good indication of the methodology adopted in the course with the introduction of digital devices into the laboratory.

In overly structured and guided lab courses normally a deterioration of 1.9 point in average is observed (Wilcox and Lewandowski 2016b). Other types of laboratory courses where there are no negative shifts in the average E-CLASS score are ISLE or SCALE-UP type lab courses (Wilcox and Lewandowski 2016a). The E-CLASS survey was administered in presence to students in the lab during the first week and the final week of the semester. The number of participants was 88, and students chose a nickname to maintain their anonymity.

More than for shifts in mean values, the most interesting indications provided to instructors by the questionnaire concern shifts (positive or negative) in individual questions. The questions for which we measured the statistically significant shifts obtained using the Mann-Whitney U-test, are:

- Q1 (When doing an experiment, I try to understand how the experimental setup works), from 0.77 to 0.94.
- Q8 (When doing an experiment, I try to understand the relevant equations) from about 0.73 to about 0.94.
- Q10 (Whenever I use a new measurement tool, I try to understand its performance limitations) from 0.53 to 0.74.

These three results seem to follow from the intentional and continued use of Arduino and smartphone to carry out the experiments. In addition, some questions kept their initial low scores, which means those aspects are not addressed sufficiently during the course to induce change. These include: Q4 (If I am communicating results from an experiment, my main goal is to have the correct sections and formatting), Q17 (When I encounter difficulties in the lab, my first step is to ask an expert, like the instructor) and Q29 (If I don't have clear directions for analysing data, I am not sure how to choose an appropriate analysis method). The result of item Q4 seems to indicate that students misunderstand the demands on the lab report, so in future editions we will try to discuss with them more explicitly the characteristics of a good lab report and what to pay attention to. The results of Q17 and Q19 seem to indicate that low confidence and lack of autonomy in decision-making persist. Changing these convictions likely takes more than just the duration of one course.

4.2 *Student Perception of the Lab Experience*

Referring to the end-of-course questionnaire described in the previous section, our focus was on understanding student perceptions of the reformed lab course. Figure 4 illustrates the responses to six of the questions from this questionnaire that relate to the student's perception and the use of Arduino.

The analysis of student feedback revealed that most students have a favourable perception of their laboratory experience and feel that it has been important for the development of their skills. Students generally responded positively to the introduction of Arduino boards; some of them expressed excitement about the potential for the device and expressed interest in continuing to use Arduino in the future also on their own.

4.2.1 Roles in Lab Activities and Gender Equity

In the two editions of the course, spontaneous formation of group members was not allowed, primarily to accustom students to working with people they may not know, as is common in professional settings. In the same survey we asked about the preferred roles during group lab work. While not mandatory, we recommended that each group rotate roles for each experimental session to give each group member the chance to take on different roles. Therefore, we also asked in the end-of-course questionnaire which role distribution model was preferred. Although the questionnaire was anonymous, we requested that participants indicate their gender. Of the 46 students who responded, 28 were male, 17 female, and one did not respond to the question. Analysing the responses according to gender showed some interesting aspects. Contrary to popular belief, we found that gender is not a significant factor in the selection of roles for those responsible for building the equipment and collecting data. Males show a slight preference for programming compared to females, while, surprisingly, they seem to prefer the role of data analysis and selection of appropriate data representation more than females do, contradicting stereotypes (Fig. 5). In general, about the role distribution scheme, females prefer a collaborative approach where everyone works together, while males tend to prefer a rotating system where roles are assigned among group members, as shown in Fig. 6. The findings align with previous research (Holmes et al. 2022) and provide a valuable foundation for further studies in the field of physics education. Other interesting insights came from the comments written by students (see the following section).

4.2.2 Student Perspectives

The last question in the end-of-course questionnaire asked students to make explicit the aspects of the lectures and laboratory activities that they perceived as problematic, both from the point of view of content and the teaching approaches. Some students highlighted the problem of time management, especially the time allowed for the delivery of lab reports. In the first lab sessions, students were asked to submit the report at the end of the corresponding lab session. Following feedback from students, the teacher adjusted this requirement, providing a 24-h submission window. According to some students' feedback, accommodating their request for more time was greatly appreciated by them and significantly improved their attitude towards the course. This underlines the importance for a teacher to listen to (and ponder) students' requests and show interest in their difficulties. Some students' comments refer to the composition of the group, and remarkably, all these comments come from female students, for example:

> As for the lab activities, by not allowing freedom of choice of groups, I have found in my group and others a great imbalance in the knowledge of their respective members. As for the four hours to accomplish the activity, I don't disagree, only that for groups with only one or two individuals who meet the required knowledge base, it becomes hell (I am referring to the unbalanced groups)(translated).

> My group is very uneven, I'm the only one doing anything, and I don't think staying up all night trying to come up with interesting conclusions would be a plausible option now or in the future. Group exercises only created stress and a desire for me to quit, because I never felt stimulated by my group mates and, rather, just left to myself (literally). (translated)

> The assortment of the group made it impossible to divide roles equally, leading one person to do the work of four people (translated)

Such imbalances in groups, like being left alone to do all the work, seems to have had a more detrimental effect on female students. Perhaps this is due to a lower level of confidence in female students. To promote fairer and more equal group work in future editions we will allow students the freedom to form groups while requiring gender balance. We'll also make an effort to frequently ask students how their group work is progressing. Regarding the teaching methods in theory lessons on probability and data analysis, some students appreciated the approach designed to promote students learning on their own. Two female students pointed out the following:

> However, I think the most important lesson I learned from this course was to "jump in": to formulate hypotheses and try to test them rigorously, in a sense to believe in myself and my abilities (even if we were often wrong). (translated)

> ...Nothing to report except in the fact that bringing all the lessons together in order to construct a proper logical thread was not a simple matter... (translated)

A male student reported:

> ...I really appreciated the professor's approach to teaching the subject, applying and drawing on Physics education sources to propose more effective teaching methods. I also appreciated the beginning of the lectures helpful in contextualizing the purpose of studying physics and what it is for... (translated)

In contrast, three other students expressed a desire for more formal and systematic lectures:

> There is no clear support material to go along with the classroom lectures" or "For the future, I think more step-by-step guidance would be beneficial... (translated)

> During the study of probability, not carrying out problems in class in a progressively related way with the theory disadvantages its consolidation. (translated)

> As for lectures, however, I think they need to be supplemented with PowerPoints or multimedia presentations of different kinds (translated)

The latest comments seem to indicate that students are accustomed to the traditional lecturing mode, where the teacher explains very systematically, and they take notes. As reported in study (Deslauriers et al. 2019) on measuring actual learning vs. feeling of learning in active learning and passive lecture classrooms, students who are exposed to active learning methods may underrate their own knowledge, but actually perform better than their peers who were taught using traditional lecturing methods. In a future edition of the course, when this request for lecturing emerges from some students, we plan to propose a discussion of this article, to encourage a change of expectation.

4.3 Distribution of Students' Final Exam Scores

Analysis of overall student scores on the final exam of the 2020/2021 and 2021/2022 editions (Organtini and Tufino 2022) reveals no measurable difference between students in the Arduino/smartphone lab section and students enrolled in the "traditional" lab sections. This result indicates that although students had to learn how to use Arduino and smartphone and had to spend time making decisions on how to carry out the various proposed experimental activities, the reformed lab format did not put students at a disadvantage in the course. During the final assessment of the 2021/22 academic year, although the distribution of scores was not greatly different from traditional sections, one of the authors, the teacher of the course, noted that students initially struggled with the course demands. However, as the course progressed, they began to fully grasp the tools of their lab work and displayed a deeper understanding than in previous years, resulting in some cases, in excellent outcomes. Students evaluated the limitations and elaborated more critically and extensively on the answers. This aligns with findings on active learning - students need time to adjust to more open-ended teaching methods, and this can be facilitated by providing the appropriate support and guidance.

4.4 Conclusions

In this paper, we have described the transformation of a first-year physics laboratory course, which was achieved by introducing digital acquisition devices such as Arduino and the smartphone and elements of active learning in both theory lessons and laboratory work. This transformation gave students the freedom to design the experimental apparatus and analyse the data. The results obtained from the E-CLASS responses of students show that this transformation is possible without a worsening of students' beliefs and expectations; E-CLASS allowed us to identify aspects in which there was improvement from the beginning of the course and areas in which students' perspectives remained far from those of the experts. The end-of-course questionnaire on students' experience indicated that students enjoyed working with Arduino. The responses to this questionnaire revealed some interesting aspects about teamwork and equity with respect to gender, that is an interesting research topic. For example, it shows that women much more want to work together with group members by sharing all tasks. A subset of students would also benefit from additional support and scaffolding. Considering the examination results, we obtained encouraging outcomes. Our conclusion is that active learning strategies and hand-on activities with Arduino and smartphone are beneficial in enhancing students' soft skills, fostering creativity and critical capabilities. In subsequent editions of the course we will try to take these implications into account.

Competing Interests Author Giovanni Organtini declares to be the author of a textbook about doing physics with Arduino and smartphones.

Ethics Approval The authors declare that an informed consent for participation and publication was obtained from all participants in this study. Consent was also obtained for the use of responses to the questionnaire, which were anonymous. The study was conducted according to the institution data protection rules.

Acknowledgements Giovanni Organtini thanks Guido Fantini and Adriano Frattale Mascioli for their commitment running the laboratory classes. Eugenio Tufino acknowledges Micol Alemani and Michael F.J. Fox for useful discussions and thanks Stefano Oss for his support. We are indebted to retired Franco Meddi, who took this course till 2020/21 who gave us the possibility to test the methods and for his invaluable insights.

References

Caballero MD, Chonacky MN, Lea Engelhardt (2019) Picup: a community of teachers integrating computation into undergraduate physics courses. Phys Teach 57:397

Colab (2023) Google colab. URL https://colab.research.google.com/. Accessed Jan 2023

Computational (2019) Advancing interdisciplinary integration of computational thinking in science. https://www.aapt.org/Resources/upload/Computational_Thinking_Conference_Report_Final-200212.pdf. Accessed Jan 2023

Deslauriers L, McCarty LS, Miller K, Callaghan K, Kestin G (2019) Measuring actual learning versus feeling of learning in response to being actively engaged in the classroom. Proc Natl Acad Sci USA 116:19251–19257

ECLASS (2023) E-class translation on physport website. https://www.physport.org/assessments/assessment.cfm?A=ECLASS. Accessed Jan 2023

Fox MFJ, Werth A, Hoehn JR, Lewandowski HJ (2020) Teaching labs during a pandemic: lessons from spring 2020 and an outlook for the future. arXiv:2007:01271

Holmes NG, Heath G, Hubenig K, Jeon S, Kalender ZY, Stump E, Sayre EC (2022) Evaluating the role of student preference in physics lab group equity. Phys Rev Phys Educ Res 18:010106

Jupyter (2023) An application of the Bayes theorem with google colab. https://colab.research.google.com/drive/1rZ22a6lF51BhB_GM8m7z7sVkdwhAnnEX. Accessed Jan 2023

Lovelace M, Brickman P (2013) Best practices for measuring students' attitudes toward learning science. CBE-Life Sci Educ 12(4):606–617

Organtini G, Tufino E (2022) Effectiveness of a laboratory course with Arduino and smartphones. Educ Sci 12:898

Phyphox (2023) Phyphox app. https://phyphox.org. Accessed Jan 2023

Smith EM, Holmes NG (2021) Best practice for instructional labs. Nat Phys 17:662–663

Teichmann E, Lewandowski HJ, Alemani M (2022) Investigating students' views of experimental physics in German laboratory classes. Phys Rev Phys Educ Res 18:010135

Wieman C, Holmes NG (2015) Measuring the impact of an instructional laboratory on the learning of introductory physics. Am J Phys 83(11):972–978

Wilcox BR, Lewandowski HJ (2016a) Impact of instructional approach on students' epistemologies about experimental physics
Wilcox BR, Lewandowski HJ (2016) Open-ended versus guided laboratory activities: impact on students' beliefs about experimental physics. Phys Rev Phys Educ Res 12:020132
Wilcox BR, Lewandowski HJ (2018) A summary of research-based assessment of students' beliefs about the nature of experimental physics. Am J Phys 86:212
Zwickl BM, Hirokawa T, Finkelstein ND, Lewandowski HJ (2014) Epistemology and expectations survey about experimental physics: development and initial results. Phys Rev ST Phys Educ Res 10:010120

Diagnosing the Cognitive Source of Students' Difficulties Within the Physics-Mathematics Interplay Context

H. Levi, A. Merzel, Y. Lehavi, and B. Schwarz

Abstract Students exhibit difficulties concerning "physmatics"—the interplay between mathematics and physics in the context of physics education, which seldom gets attention in the learning diagnostics literature. Based on previous categorization of students' physmatic difficulties, we present the results of a diagnostic questionnaire that aims to identify possible cognitive sources for some of these difficulties. We describe four possible cognitive sources based on students' paired responses to mathematical questions with and without physics context.

1 Introduction

Difficulties with "physmatics", the interplay between mathematics and physics in the context of physics education (Lehavi et al. 2015), can be defined as difficulties in harnessing knowledge in both disciplines to make sense of physics equations and problem-solving. These physmatic difficulties are described in the literature in several ways—math as a language (Redish and Kuo 2015), general errors using math in physics with unproductive framing (Modir et al. 2019), and misalignment of students' reasoning and what they reason about (Gifford and Finkelstein 2020). Although there is ample research on diagnosing students' difficulties in learning physics, such rarely, if at all, focus on students' physmatic difficulties.

H. Levi (✉) · A. Merzel · B. Schwarz
Hebrew University of Jerusalem, Mount Scopus, 9190501 Jerusalem, Israel
e-mail: hnlevi@gmail.com

A. Merzel
e-mail: avraham.merzel@mail.huji.ac.il

B. Schwarz
e-mail: baruch.schwarz@mail.huji.ac.il

H. Levi
Jerusalem College of Technology, Havaad Haleumi St. 21, Jerusalem, Israel

Y. Lehavi
David Yellin Academic College of Education, Ma'agal Beit HaMidrash St. 7, Jerusalem, Israel

S. Faletič and J. Pavlin (eds.), *Teaching and Learning Physics Effectively in Challenging Times*, Challenges in Physics Education, https://doi.org/10.1007/978-3-031-72541-8_12

In previous work, we analyzed recorded physics lessons in 9th to 12th grade classes and induced a grounded theory categorization of various students' physmatic difficulties (Levi et al. 2020 and Levi et al. 2021). That study found difficulties in three categories—(I) performing arithmetic operations on physics equations,[1] (II) mathematical generalization of a zero value of a product of physical parameters and variables; and (III) sensitivity to the region of applicability of formulas in physics. The latter can be related to reported students' (as well as teachers') unawareness of the borders of validity of fundamental laws such as Galileo's law of free fall (Lehavi and Galili 2009) and to the complexity of constructing and validating a mathematical model for physical phenomena (Uhden et al. 2012 and Karam et al. 2010).

In this present study, we go further and diagnose those identified difficulties in order to underpin the cognitive structures responsible for them. A great challenge in diagnosing physmatic difficulties is distinguishing them from mutually independent difficulties in mathematics or physics. Some works addressed this challenge by comparing students' problem-solving in purely physics contexts and purely mathematical context (Carli et al. 2020 and Macuca et al. 2022). These studies, conducted in middle school and high education levels, found that although there is some correlation between students' performance, they do not necessarily use the same problem-solving strategy in both contexts. Therefore, our current study aims to broaden this line of research by comparing high school students' responses to these two types of problems and diagnosing their specific difficulties.

Our study, thusly, may contribute to identifying students' cognitive sources of physmatic difficulties and, hence, may help teachers and curricula designers to develop tools to respond efficiently when such difficulties appear in class, and adapt instructional procedures in remote settings (e.g. Mishra et al. 2020).

2 Research Question

What are the cognitive sources of the three physmatic difficulties mentioned above (categories I, II, and III)?

3 Method

3.1 Tools

In view of the three categories of difficulties mentioned in the introduction, we designed a diagnostic questionnaire that aims at distinguishing between difficulties that stem from physics from those that stem from mathematics. This questionnaire

[1] Although the use of contextualization in mathematics education is considered beneficial for understanding mathematics (Perin 2011 and Jones 2013), here we employ only the physics education standpoint.

incorporates two paired variations of questions for each physmatic difficulty—one has a physical context (PC), and the other has a purely mathematical context (MC).

We kept both types of questions with about the same (short) length. To avoid the bias of students' prior knowledge, we chose an out-of-curriculum (in the country of research) physical context—pressure in static fluids. All the questions were related to different aspects of the expression for static pressure: $P - P_0 = \rho \cdot g \cdot h$.

The questionnaire contained three pairs of questions, each addressing one of the three difficulties mentioned above.

For example, the multiple-choice PC question that addressed the overuse of mathematical habits when the product of several parameters and variables is zero was:

The formula for pressure in a pool of water as a function of the depth is:
$P - P_0 = \rho \cdot g \cdot h$
Where P represents the pressure at a certain depth, h; P_0 represents the air pressure measured on the water surface; ρ represents the water density; and g is the free fall acceleration
At a certain point, it was found that the value $(P - P_0)$ is zero. Which of the following factors can be zero?
1.ρ, 2.g, 3.h, 4. P, 5. Any of the quantities in answers 1,2,3 can be zero

---

The parallel MC question was:

The formula that describes Y as a function of x is: $Y - Y_0 = a \cdot b \cdot x$
Where Y represents the function value at a certain x. Y_0 represents a parameter; a represents another parameter; and b represents another different parameter
It is given that the value $(Y - Y_0)$ is zero. Which of the following factors can be zero?
1.a, 2. b, 3. x, 4. Y, 5. Any of the quantities in answers 1,2,3 can be zero

---

See Appendix for all the questions in the questionnaire.

We selected these three categories since each of them involves a different level of math-vs-physics focus (Pospiech and Geyer 2022): category I (using arithmetic) is almost blind to physical considerations, category II (zero-product generalization) is concerned with the different meaning of symbols in the two languages despite the similarity of mathematical operation, and category III (applicability region of formula) is concerned with the limits posed by physics on performing a mathematical operation.

The questionnaire was validated by the authors and by think-aloud interviews with three 10th-grade students. We formatted the questionnaire in a Google Form and mixed the order of the questions, so no student answered two questions that represented the same difficulty in the two different contexts (PC and MC) in a row.

3.2 *Population*

The research population was 167 students of 10th-grade who study algebra-based physics from 24 different high schools in the country of research. The students volunteered to answer the questionnaire anonymously and in their free time. All students were not familiar with under-water pressure. Most of them learned kinematics, and some learned geometrical optics during the school year. The research had our IRB's ethical approval. 16 students did not correctly answer any of the MC questions, so we removed their responses from the analysis of the difficulty described as category III, since this category requires arithmetic calculations in addition to the difficulty of the applicability region of formula. (category III).

4 Findings

We describe the findings of students' responses by indicating the success or failure in the MC question (M ±) and the PC question (P ±) for each pair of questions. So, (P + M-), for example, represents a difficulty in math but not in physics. We first describe the analysis of the whole sample of responses (see Fig. 1). We shall then analyze the physics cognitive load relative to the math cognitive load in each pair of questions.

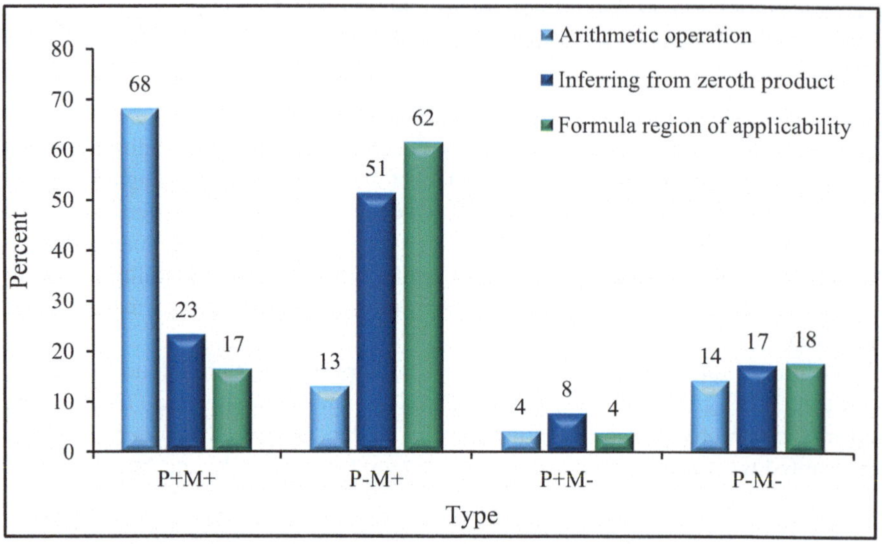

Fig. 1 Responses to all categories of difficulties (N = 167)

4.1 Whole Sample Analysis

Notably, regardless of the specific type of difficulty, only rarely did students who answered correctly in a physical context fail to answer the purely mathematical context questions correctly. From a cognitive perspective, this may imply that most students 'use' their mathematical thinking skills for improving their physics thinking skills, with regard to the three categories studied here.

Students with no difficulties in mathematics (the two left categories in Fig. 1) performed differently in the three cases.

In the arithmetic case, most of the students performed well in both (math and physics) contexts (P + M + ~ 68%). In the mathematical over-generalization case, about quarter of them performed well in both contexts (P + M + ~ 23%). Finally, only a few were sensitive to the region of applicability of a formula (P + M + ~ 17%).

We infer that in each of these three cases, the cognitive challenge posed by physics thinking is different, with the arithmetic challenge being the lowest while the region of applicability of formula is the highest.

4.2 Difficulties in Arithmetic Operations

In the arithmetic case (Q1:PC and Q1:MC), most students (N = 114, ~68%) performed well on the MC question (see Fig. 2). For the few of those students who failed in the PC question (N of P-M + = 22, about 16% from the M + students), the physical context provided an extra barrier. Seemingly, those students differently comprehended pure mathematics and mathematics in a physical context, exhibiting a physmatic difficulty. About 14% of the students failed in both PC and MC questions. These findings suggest that even when physics plays a minor factor, it still provides a cognitive load for a considerable number of students.

4.3 A Mathematical Overgeneralization of the Zero-Product Property

Questions 2:PC and 2:MC addressed a case when the product of several parameters and variables yields a zero result. In mathematics, even though each of the factors in the product can attain a zero value, such a generalization may not hold in the physical context. In the specific conditions of Q2:PC, the water density (ρ), the free-fall acceleration (g), and the pressure in the water (P), cannot attain a zero value.

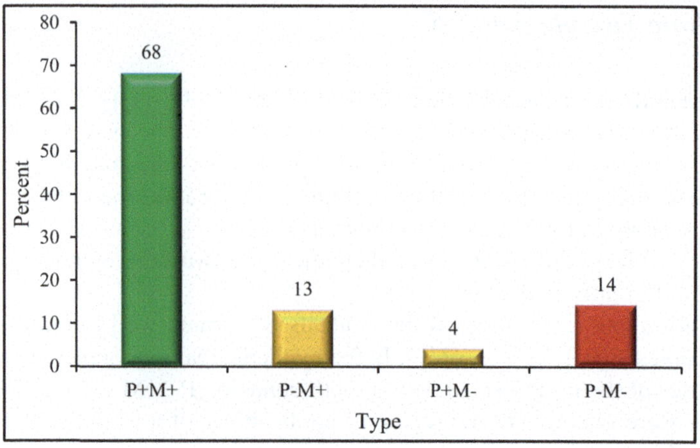

Fig. 2 Responses to question concerning the difficulty of Arithmetic operations (N = 167)

We differentiate here between students who ascribed zero to one of the values of ρ, P and g and students who chose that if the product $\rho \cdot g \cdot h = 0$ then each of the factors in it can be zero, regardless of its physical content (denoted "P-$\rho \cup g \cup h$"). These results are visualized in Fig. 3.

Here, the majority of the students (~75% exhibited no difficulty in the purely mathematical context, but almost 69% of them did not perform well on the physics

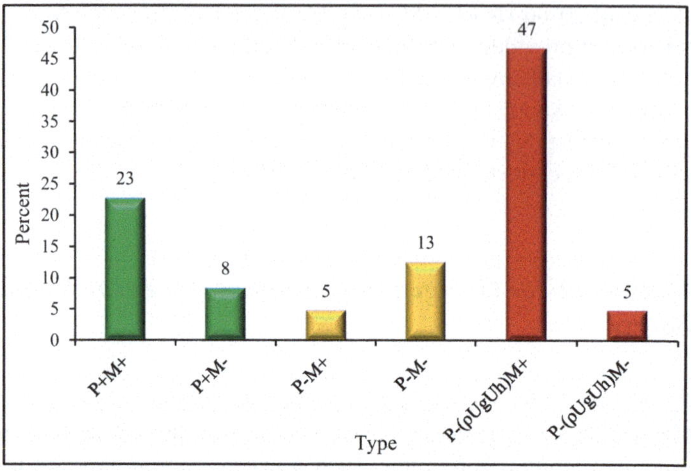

Fig. 3 Responses to questions Q2:PC and Q2:MC, addressing the physmatic difficulty of over-generalization of the zero-product property—P-($\rho \cup g \cup h$) represents students who exhibit this difficulty in Q2(PC), while P- refers specifically to students who chose one wrong value (ρ / P/ g) (N = 167)

question. Apparently, for these students, the challenge posed by the physics context—in this case—was more pronounced than in the arithmetic one. This group of students was composed of two subgroups: those that employed in the physical context an overgeneralization of the mathematical strategy for a zero product and those that ascribed a zero value to one value, *r*/*P*/*g* (~47% and ~5%, respectively, of the M + students). These students in both subgroups seem to disregard the physical content of the symbols in this case. However, one may interpret the answer $P = 0$ as representing a physical view that ignores the meaning of P_0 rather than an automatic application of a mathematical procedure.

4.4 Region of Applicability of a Formula

The constraints imposed by physics on using a mathematical model to describe a phenomenon (the pressure above the water level), which were addressed in the paired questions Q3:PC and Q3:MC, revealed specific difficulties regarding the sensitivity to the region of applicability of a formula. Evidently, the applicability region of a formula posed the greatest challenge to our addressees.

While almost 77% had no trouble with the MC question, only 22% of them (~17% of the whole sample) noticed that the formula was not applicable for calculating the pressure above the water level. Apparently, it was rare to find students in our sample that were able to employ physical thinking regarding the applicability region of a formula without proper mathematical thinking (P + M- ~ 4%).

A little less than a half of the students (P-M+ ~41%) merely plugged the calculated values given in Q3:PC into the pressure formula showing no attention to the fact that the minus sign for the calculated depth means that the formula is not applicable. Another interesting case is those who chose to replace their calculated depth by its absolute value (P-(|x|)M+ ~19% in Fig. 4). Those students may exhibit discomfort with the minus sign of their calculation which they could not resolve being unaware to limitations in applying the formula in the case presented in the question. In this sense, they may possess partial understanding of the borders of applicability of formulas. Their answer may indicate a cognitive dissonance between the belief that formulas are ever applicable in physics and the unreasonable result of using a formula to describe a phenomenon in the world.

One may infer that it might be easier to teach the concept of the region of applicability to these students by the conceptualization of the possibility that a formula might not always be applicable.

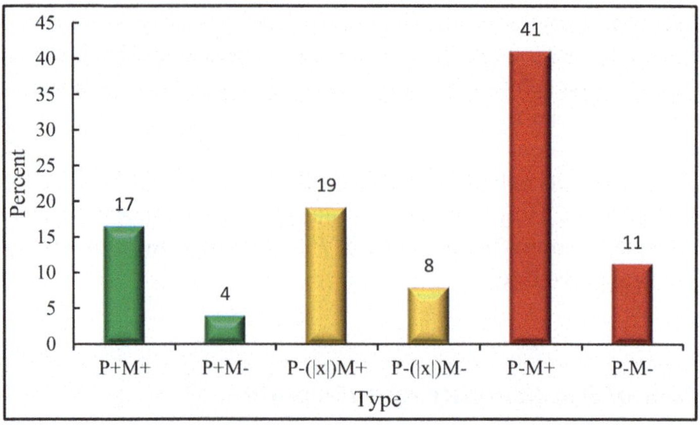

Fig. 4 Different types of responses to questions concerning the physmatic difficulty to identify the region of applicability of a formula (N = 151)

5 Summary and Discussion

Our research enabled us to distinguish within the physmatic domain between difficulties originating from mis-thinking in mathematics and physics with regard to three categories. Moreover, our tool enabled a refined analysis that revealed a different extra-cognitive challenge posed by the physics context in each of these categories. Implicitly, using this tool in other categories has the potential to establish a map of more physmatic difficulties to be considered by physics educators.

The findings indicate that while the majority of our addressees performed very well in using mathematical operations in a purely mathematical context, many of them exhibited difficulties in the physmatic domain (see Fig. 5). We diagnosed three types of such difficulties (which do not stem from mathematical incompetency): type I is a physmatic arithmetic difficulty (~13% of our population, who had difficulty doing arithmetic operations in questions that included physical context, but had no difficulty if the context was pure mathematics), type II is characterized by overgeneralization of the mathematical zero-product property in physical contexts (~51%), and type III reflects unawareness of the region of applicability of a physical formula (~62%). We may infer that there is some hierarchy of physmatic difficulties that are related to the physical context focus. Specifically, as the difficulty has more focus on the physical context in addition to the mathematical one, more students might present this difficulty.

The results of the analysis of the first physmatic difficulty (arithmetic operations) are in line with the correlation between succeeding in solving ‘purely mathematical context’ problems and solving ‘physics context’ problems that were shown in the literature (Carli et al. 2020 and Macuca et al. 2020). Past studies focused on loading context to mathematical problems. In this study, however, the physical meaning is the physical context. This correlation is, notably, found with (Carli et al. 2020) and

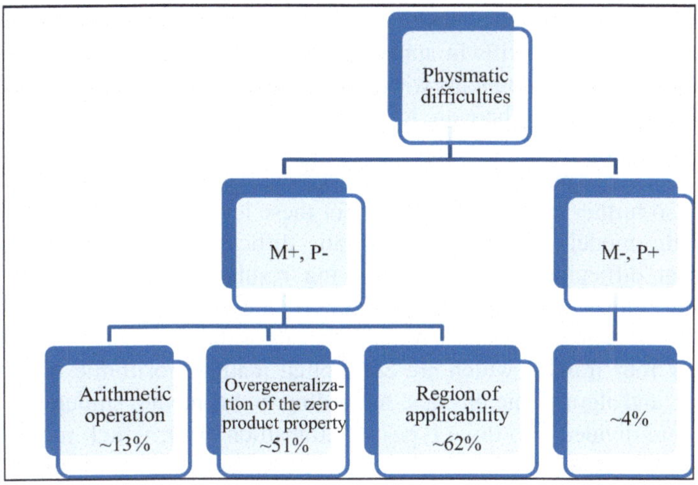

Fig. 5 Categories of physmatic difficulties and the frequencies of their appearance among the 10th-grade physics students

without (Macuca et al. 2020 and here) familiarity with a specific topic, implying that physmatic difficulties probably do not stem from misunderstanding of mathematics. Abilities in mathematics are, however, necessary to solving problems in physics.

The distribution of those who did not answer correctly to Q2:PC is unbalanced (see Fig. 3), as those who answered in a mathematical way (i.e., $\rho \cup g \cup h$) were far more frequent than those who showed some physical misunderstanding by choosing ρ or P or g. The latter might be because of the unfamiliar content (hydrostatic pressure), as the pupils were, presumably, unaware that these quantities cannot have a zero value in a pool, underwater. From the frequent response, therefore, we can see that mathematical technical thinking overcomes the physical thinking. This is in line with past studies (Macuca et al. 2020). We found three levels of cognitive response; two of which are misunderstandings—the first is mathematically-focused ($\rho \cup g \cup h$) and the second is more physically-focused (ρ or P or g). One can interpret these two misunderstandings as cognitive challenge to interpret mathematical expressions in a physical context (Uhden et al. 2012). The third level includes those who did not overgeneralize.

It seems that there are also three cognitive levels of understanding the region of the applicability of a formula. The first level, expressed in choosing the distractor with the minus sign, indicates that students adopted the mathematical technical algorithmic approach and know how to plug numbers into the formula but do not interpret the results to represent a valid description of the world. One can interpret these findings as a difficulty in applying the skill of examining validity (Uhden et al. 2012). The second level is for whom the negative result of the depth did not fit with their understanding of the world ("what does a negative depth mean, given that more positive is deeper?") and therefore, instead of putting aside the formula, they modified their calculations' results and chose the absolute (positive) value of their calculations. Choosing this, they demonstrated some skills in applying validity, but in a technical, algorithmic way. Finally, there are the students who realized that the formula is not applicable in the case described. Just as there are levels in the physmatic difficulty to identify the region of the applicability of the formula, and the difficulty with overgeneralization from a zero value of a product, we assume that there are also levels in other physmatic difficulties, so further studies should look for these levels (not necessarily three).

One of the models, describing physmatic difficulties in quantum mechanics, suggests that difficulties in problem-solving result from inappropriate shifting between four frames, denoted by a 2-by-2 matrix that is defined by two axes (Modir et al. 2019)—the math-physics axis and the algorithmic-conceptual axis. These axes portray four frames, which are conceptual math, algorithmic math, conceptual physics, and algorithmic physics. According to this model, students' physmatic difficulties are divided into three types: (1) transition error, which requires a transition from one frame to another; (2) displacement error, in which the problem is directed to a specific frame, but students place it in another frame; and (3) content error, in which students identify the situation within the correct frame but do not use the right resources to solve it.

Our findings could also be interpreted using this framework. Arithmetic operation difficulty (Q1), for example, fits the algorithmic mathematics frame with its MC version and the algorithmic physics frame with its PC version. Our findings suggest that these two frames are equally difficult for students. As for the physmatic difficulties region of applicability of a formula and the overgeneralization from an equation in which one side is zero, about half of the students present mathematical reasoning, which is in accordance with Modir et al.'s concept of displacement error (Modir et al. 2019). In Q2 and Q3 questions, the PC problem is directed to the frame of conceptual physics, but the students place it in the frame of algorithmic physics. This prevents the students from interpreting their calculations' results concerning the physical context of the problem.

Transfer of knowledge from mathematical to physical contexts was found effective in solving algebraic problems but transfer of knowledge from physical to mathematical contexts was not helpful (Bassok and Holyoak 1989 and Rebello et al. 2017). Our results support these findings and show that most of those who had previous mathematical knowledge (i.e., most of our population) could successfully solve the physical problem (with unfamiliar content) but not vice versa (see P + M + vs. P + M- in Fig. 2). However, we found that transfer from mathematics is not always beneficial, since it might cause overgeneralization, as we showed above (in Q2 and Q3).

The pedagogical content knowledge model for physmatic has four components (Lehavi et al. 2015) and our findings can contribute to all four. Regarding the knowledge of physmatic contents, it brings forth specific aspects of the interplay of physics and mathematics in physics lessons and, regarding teachers' knowledge about students, we highlight the cognitive source of students' difficulties and their possible hierarchy. We expand teachers' knowledge of assessment by offering questionnaires, like what we have developed and presented here that enable to assess physmatic difficulties. Practically, our findings can help teachers better understand their students' physmatic difficulties and better plan their lessons by adjusting to the different levels of each difficulty.

The implementation of remote teaching necessitates the adaptation of instructional procedures (e.g. Mishra et al. 2020) Consequently, those who wish to apply these instructional steps must possess a comprehensive understanding of students' difficulties and the underlying causes of these difficulties, enabling proactive preparation of supportive aids (Stefanidou et al. 2022). Our research can help provide information about cognitive sources of physmatic difficulties of students.

Our study has some limitations. Firstly, the research population consisted of 10th grade students only. Secondly, it is possible that similar research in different physical content would yield different results. Lastly, broader research in both perspectives is needed in order to better generalize our findings.

Appendix

The questions presented in the questionnaire appear below, in Table 1.

Table 1 The data collection tool

Difficulty category	Physics	Mathematics
Q1: Arithmetic operations	Q1 (PC): The formula that describes the pressure as a function of the depth underwater (h) is:$P = \rho \cdot g \cdot h + P_0$. It is also given that the water density is $\rho = 1000kg/m^3$, $g = 10m/s^2$ and the atmospheric pressure is $P_0 = 101000\frac{kg}{m \cdot s^2}$ What should be the depth where the pressure will be $121000Pa$? a. $2m$ b. $18990m$ c. $0.00012m$ d. $22.2m$ e. Other: ________	Q1 (MC): The formula that describes Y as a function of x is: $Y = a \cdot b \cdot x + Y_0$. It is also given that the parameter a is: $a = 10$, the parameter b is $= 1000$ and the parameter Y_0 is $Y_0 = 103000$ What should be the value of x, when Y is 123000? a. 2 b. 18990 c. 0.0001 d. 22.6 e. Other: ________

(continued)

Table 1 (continued)

Difficulty category	Physics	Mathematics
Q2: Overgeneralization from an equation in which one side is zero	Q2 (PC): The formula for pressure in a pool of water as a function of the depth is: $P - P_0 = \rho \cdot g \cdot h$ Where P represents the pressure at a certain depth, h; P_0 represents the air pressure measured on the water surface; ρ represents the water density; and g is the free fall acceleration At a certain point, it was found that the value $(P - P_0)$ is zero. Which of the following factors can be zero? 1. ρ 2. g 3. h 4. P 5. Any of the quantities in answers 1, 2, or 3 can be zero Explain your answer: ____________ ______________	Q2 (MC): The formula that describes Y as a function of x is: $Y - Y_0 = a \cdot b \cdot x$ Where Y represents the function value at a certain x; Y_0 represents a parameter; a represents another parameter; and b represents another different parameter It is given that the value $(Y - Y_0)$ is zero. Which of the following factors can be zero? 1. a 2. b 3. x 4. Y 5. Any of the quantities in answers 1, 2, or 3 can be zero Explain your answer: ____________ ________________
Q3: Region of the applicability of the formula	Q3(PC). The formula that describes the pressure in a pool, as a function of the depth underwater (h) is:$P = \rho \cdot g \cdot h + P_0$. The pressure is measured. It is also given that the water density is $\rho = 1000\ \text{kg/m}^3$, $g = 10\ \text{m/s}^2$ and the atmospheric pressure is $P_0 = 101000 \frac{\text{kg}}{\text{m} \cdot \text{s}^2}$. The measure outcome was $100000 \frac{\text{kg}}{\text{m} \cdot \text{s}^2}$. What can we say about the measurement location in this case? Note: Depth is measured from the surface of the pool, and the positive axis is set downwards a. The depth was −0.1 m b. The depth was 0.1 m c. One cannot infer the water depth by using the formula with the given numerical values d. Other: _____	Q3(MC). The formula that describes Y as a function of x is: $Y = a \cdot b \cdot x + Y_0$. The value of Y is calculated It is also given that the parameter a is: $a = 10$, the parameter b is $b = 1000$ and the parameter Y_0 is $Y_0 = 103000$ The calculated value of Y was 100,000. What can we say about the value of x in this case? a. The value of x is -0.3 b. The value of x is 0.3 c. One cannot infer the value of x by using the formula with the given numerical values d. Other: _____

References

Bassok M, Holyoak K (1989) Interdomain transfer between isomorphic topics in algebra and physics. J Exp Psych: Learn Mem Cogn 15:153

Carli M, Lippiello S, Pantano O, Perona M, Tormen G (2020) Testing students' ability to use derivatives, integrals, and vectors in a purely mathematical context and in a physical context. Phys Rev Phys Educ Res 16:010111

Gifford J, Finkelstein N (2020) Categorical framework for mathematical sense making in physics. Phys Rev Phys Educ Res 16:020121

Jones S (2013) Understanding the integral: Students' symbolic forms. J Math Behav 32:122–141

Karam R, Pospiech G, Pietrocola M (2010) Mathematics in physics lessons: Developing structural skills. In: Proceedings of the GIREP-EPEC & PHEC 2009 international conference; 2009; Leicester, UK. Leicester: University of Leicester, pp 17–21

Lehavi Y, Galili I (2009) The status of Galileo's law of free-fall and its implications for physics education. Am J Phys 77:417–423

Lehavi Y, Bagno E, Eylon BS, Mualem R, Pospiech G, Böhm U, Krey O, Karam R (2015) Towards a PCK of physics and mathematics interplay. In: Proceedings of the GIREP-MPTL 2014 international conference on teaching/learning physics: integrating research into practice; 2014 Sep; Palermo, Italy. Palermo: Università degli Studi, pp 843–852

Levi H, Merzel A, Lehavi Y, Schwarz B (2020) Categorization of physmatic misconceptions. In: Abstracts of the IPS conference; 2020; Rehovot, Israel. Rehovot: Weizmann Institute of Science. Talk

Levi H, Merzel A, Schwarz B, Lehavi Y (2021) Students' reasoning while using mathematics in physics lessons. In: Abstracts of the 3rd world conference of physics education, Hanoi, Vietnam. 13 – 16 December 2021. Online talk

Macuca A, Poljančić B, Jurdana-Šepić R, Lončarić K (2022) Relationship between mathematical competencies and achievements in physics of 14-year-old students. In: Abstracts of the GIREP conference, university of Ljubljana, 4–8 July 2022

Mishra L, Gupta T, Shree A (2020) Online teaching-learning in higher education during lockdown period of COVID-19 pandemic. Int J Educ Res Open 1:100012

Modir B, Thompson J, Sayre E (2019) Framing difficulties in quantum mechanics. Phys Rev Phys Educ Res 15:020146

Perin D (2011) Facilitating student learning through contextualization: a review of evidence. Community College Rev 39:268–295

Pospiech G, Geyer M (2022) Structural skills of students in solving physical–mathematical tasks. In: Michelsen C, Beckmann A, Freiman V, Jankvist U, Savard A (eds) Mathematics and its connections to the arts and sciences: mathematics education in the digital era, vol 19. Springer, Cham. https://doi.org/10.1007/978-3-031-10518-0_16

Rebello N, Cui L, Bennett A, Zollman D, Ozimek D (2017) Transfer of learning in problem solving in the context of mathematics and physics. In: Learning to solve complex scientific problems. Routledge, New York, pp 223–246

Redish E, Kuo E (2015) Language of physics, language of math: disciplinary culture and dynamic epistemology. Sci Educ 24:561–590

Stefanidou C, Kyriakou K, Mandrikas A, Stavrou I, Skordoulis C (2022) Students' views on physics teaching at a distance in the context of COVID-19 pandemic. Eur J Sci Math Educ 10(3):284–297

Uhden O, Karam R, Pietrocola M, Pospiech G (2012) Modelling mathematical reasoning in physics education. Sci Educ 21:485–506

Identification of the Factors Influencing Game-Based Learning in a Game "The Tournament of J. C. Maxwell"

Ladislav Janiga and Viera Haverlíková

Abstract The paper presents educational game "The Tournament of J. C. Maxwell" and results of pedagogical survey realized among 55 students in Slovakia. The goal of the survey was to identify factors influencing successful use of the game as a learning tool. The students expressed their positive attitude to the game they played in questionnaire and interview. Teamwork, the way the team is formed, movement and change of activity, content and difficulty of the game, discussion and competition were identified as factors that might affect attitudes to the game as a learning method.

1 Game-Based Learning

Game is a word we all use in our daily lives. Everyone has played a game at some point. Yet it is difficult to define what a game is. Everyone has their own definition of a game because everyone has their own experiences and memories. According to Kapp (2012), the term game represents a system in which players engage in an abstract challenge, defined by rules, interactivity and feedback, that results in a quantifiable outcome often eliciting an emotional reaction. In the early years of human life, game play takes up most of the time. Their purpose is not just entertainment. Playing games is one of the most natural ways of learning. Game-Based Learning (GBL) is an educational approach that reinforces this aspect of a game and integrates games in a learning process. According to Greipl et al. (2020), GBL catalyzes learning process by focusing on student's cognitive, emotional, and social factors and learning environment. Many studies point to the positive effects of educational games such as: players develop communication and teamwork skills, players evolve in terms of self- and peer-evaluation, games can increase students' engagement in subject, etc. (Cardinot and Fairfield 2019). Most of the research has focused on

L. Janiga (✉) · V. Haverlíková
Faculty of Mathematics, Physics and Informatics, Comenius University in Bratislava, Mlynská dolina, 84248 Bratislava, Slovakia
e-mail: ladislav.janiga@fmph.uniba.sk

V. Haverlíková
e-mail: viera.haverlikova@fmph.uniba.sk

S. Faletič and J. Pavlin (eds.), *Teaching and Learning Physics Effectively in Challenging Times*, Challenges in Physics Education, https://doi.org/10.1007/978-3-031-72541-8_13

virtual games (Naik 2014). Although there are many papers concerning GBL and educational games, most of them refer to specific game and the experience with it. There is a lack of a theoretical framework of the aspects that make educational games successful in the learning process (Greipl et al. 2020)

1.1 Game-Based Learning in Slovakia

Although there is a significant increase in educational games research worldwide, educational games research in Slovakia is only at the beginning. Many games aimed at repetition and consolidation of knowledge are used in Slovak schools, but educational games aimed at knowledge acquisition, cognitive development and creativity advancement are not so commonly used. Specifically in physics, research from 2012, showed that Slovak teachers do not know enough about physics educational games and do not have experience in using them. Also, the number of educational games used in school decreases as the age of students increases (Horváthová and Haverlíková 2012). Our experience and recent discussions with teachers show that the interest in educational games in primary education is still prevalent. At higher levels of education, games are most often used in pedagogical-psychological diagnostics and in language teaching. This is also confirmed by the appearance of the keywords game and education in diploma (bachelor, master or doctoral) theses. Although physics teachers are interested in using educational games in their lessons, they do not know any, and do not have enough time to design and prepare their own new educational game.

1.2 Research goals

In long-term research, our first goal is to identify the key factors influencing successful use of games in physics education and to map the types of games capable of achieving selected learning objectives using a grounded theory approach.

Based on the minimal number of educational games aimed at upper secondary physics education in Slovakia, our second long-term goal is to design a set of educational games suitable for physics education that could be offered to Slovak teachers in the future. The focus of our research is on games in real environments that offer students the opportunity to interact, play, discover and learn in real world, and to manipulate with real objects.

In the first part of this paper, the goals and structure of one of our own games "The Tournament of J. C. Maxwell" is introduced. The second part describes and interprets our observations and data obtained from testing this game.

2 The Tournament of J. C. Maxwell

2.1 Educational Goal of the Game and It's Physics Content

The Tournament of J. C. Maxwell is an interactive educational game in real (non-digital) environment focused on the topic of magnetism and electromagnetism for upper secondary school students. The game is intended as a revision tool prior to summative assessment. We have designed all the tasks in such a way that students can gain some new knowledge, new skills or a new perspective on the topic.

The game covers the content of Slovak National Curriculum in the topic of magnetism and electromagnetism (representation of magnetic field by magnetic field lines, to describe magnetic field of a straight wire and coil in which an electric current passes, Ampere's right-hand rule, the effect of magnetic field on a straight wire in which an electric current passes, electromagnetic induction, Lenz's law, principle of electric motor, principle of the transformer). In addition to knowledge, the game aims on the development of specific skills, such as communication skills, teamwork skills, problem-solving skills, equipment-manipulation skills, experimental skills, analytical skills, etc. The game mechanics allow natural competitiveness and emotions of the students to be harnessed to engage them more deeply in the learning process.

2.2 Game Mechanics

The game mechanics are inspired by the game Diver (Haverlíková 2005). The game consists of 15 stands located all over the school building. Each stand has a different task. By answering the task, students get the location of the next stand. If the answer is correct, the next stand is on the main path to the finish. If the answer is incorrect, the next stand offers a clue to the previous task and students must go back and find the correct answer. Students work in groups of 2–4 people and aim to pass the stands in the shortest possible time and with the correct answers. Figure 1 shows the path of the stands. All stands are marked with letters. The purple ones (S and F) mark the start and the finish. The orange ones are stands on the main path (task stands) and blue ones are the clue stands. Green arrows mark correct answers and red arrows mark incorrect answers. If a team answers all the tasks correctly, they will only visit the orange stands, and the word MAXWELL can be formed from the letters on their path. For each team, the order of the main path should be mixed up to avoid accumulation of the teams at the stands, communication between teams, distractions from the game, and a drop in attention span. During the first testing, we used the Edupage platform (an integrated information system for schools) to deliver some of the instruction on the stands. The reason for this was to provide the teams with different navigation instructions according to their own path. We later evaluated that this was not an optimal solution (see Sect. 3.5.1 for more details). We used the game

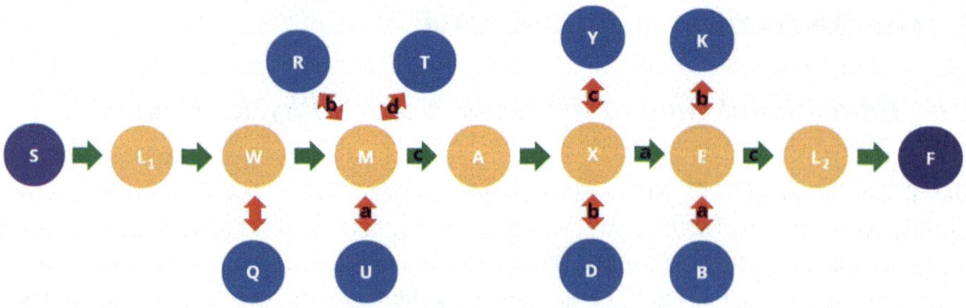

Fig. 1 The path of the stands

during a physics lesson to revise knowledge and skills before summative assessment of the learning outcomes.

2.3 *Tasks of the Game*

We designed the tasks based on the physics content requirements and skill development requirements described in Sect. 2.1 and listed in Table 1.

The tasks of the main path are described in the following paragraphs with a brief commentary.

Table 1 Tasks requirements

Task	Physics content requirements	Skill development requirements
L_1	Usage of electromagnetism transformation ratio	Experimental skills Equipment-manipulation skills
W	Magnetic field of a wire Right-hand rule for a wire	Picture analysis
M	Magnetic field of permanent magnets Magnetic field lines	Picture analysis
A	Earth's magnetic field	Compass usage
X	Right-hand rule for a coil Lenz's law	Problem-solving skills Graph analysis Analytic skills
E	Magnetic field of a coil Right-hand rule for a coil	Problem-solving skills Equipment-manipulation skills Analytic skills
L_2	Usage of electromagnetism	Knowledge application in a new situation

Fig. 2 Task L_1—equipment

2.3.1 Task L_1

In front of you, there are several coils, an AC power source and a voltmeter. Construct a transformer that changes the voltage from 10 V to 1 V. What is the ratio between the number of threads of the primary and secondary coils?

Continue to W.

Students know the concept of electromagnetic induction, they know how do coil parameters influence the induced voltage, but they do not know what the transformation ratio is. The task requires working with real tools (Fig. 2). Students try to achieve a voltage change from 10 V to 1 V by trying different combinations of coils. In this way, they find the relationship between the input and output voltages and the number of threads of primary and secondary coils. Thus, the aim of this task is to experimentally discover transformation ratio. This is the only stand where a teacher was present. His role was to observe and guide (if necessary) the students' activities and, after successful completion of the task, to navigate the students to the next task.

2.3.2 Task W

Direction of current in straight wire on the diagram shows the way to the next stand.

Figure 3 (located at station W) shows schematically the conductor and its magnetic field. By understanding the diagram and using the right-hand rule for the conductor, pupils will see that the next station is on the right (analogous diagrams can be used to guide pupils left, up or down).

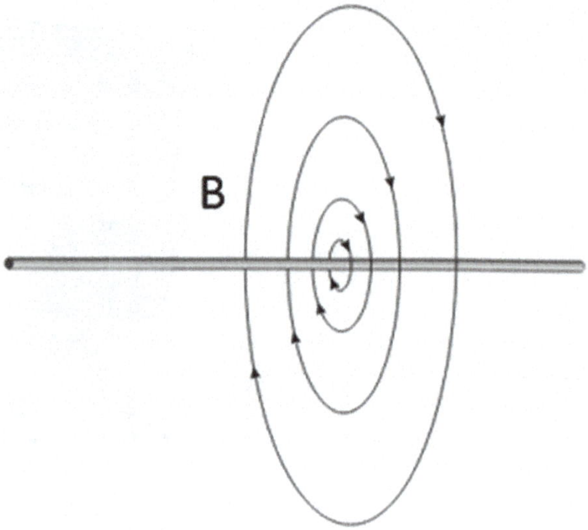

Fig. 3 Task W—diagram of a straight wire and its magnetic field

2.3.3 Task M

In the picture in front of you, the magnetic field of three magnets is visualised using iron filings. Select the configuration of the magnets that was used to create the magnetic field image from the options provided.

The magnetic field of two magnets is a well-known situation for students. In this task, they need to solve non-standard situation—a magnetic field of three magnets (Fig. 4). To do so, they need to analyse the situation, simplify it by reducing it to a two-magnet situation, use their knowledge of the magnetic field of two magnets and solve it in parts.

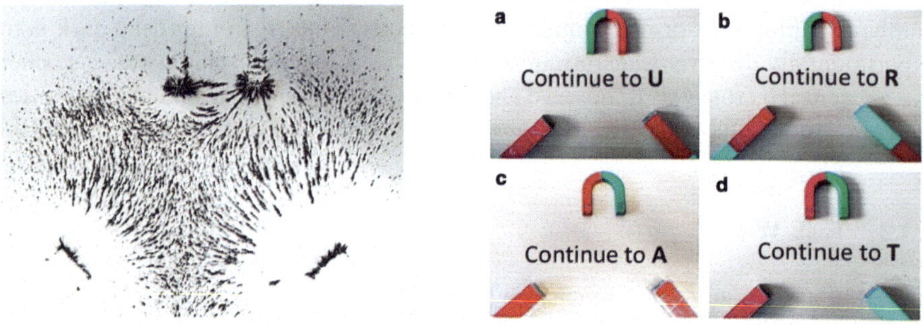

Fig. 4 Task M—images of the investigated magnetic field and configurations of the magnets

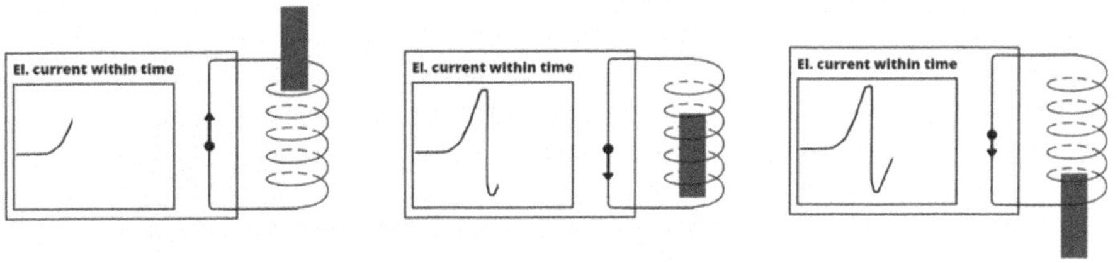

Fig. 5 Task X—sequence of pictures representing the fall of the magnet through the coil cavity

2.3.4 Task A

The next task is in a south-easterly direction.

Every team was given a compass at the beginning of the game. Using a compass is not a common skill for today's students. Students must use the compass correctly to find the next stand.

2.3.5 Task X

The pictures show the magnet falling through the coil cavity. The coil is connected to an electric current sensor that displays a current-time diagram. What was the orientation of the magnet during the fall? Which pole of the magnet entered the coil first?

(a) South pole—continue to E
(b) North pole—continue to D
(c) Both could be correct—continue to Y.

The situation presented at stand X (Fig. 5) is not entirely new to the students. In previous lessons, they have solved a situation in which a magnet falls through a copper tube. What is new to them is the combination of examining the direction of current in the coil and the current-time diagram. As in the previous problem, the students have to use a variety of knowledge and skills (understanding the situation, graph analysis, Lenz's law and the right-hand rule) to solve this problem.

2.3.6 Task E

The box in front of you hides DC power source. You cannot unpack the box. Plug the power source in. The coil is connected to the power source by blue and black wires. Which wire is connected to the positive pole of the power source?

(a) Blue wire—continue to B
(b) I cannot define—continue to K
(c) Black wire—continue to L_2.

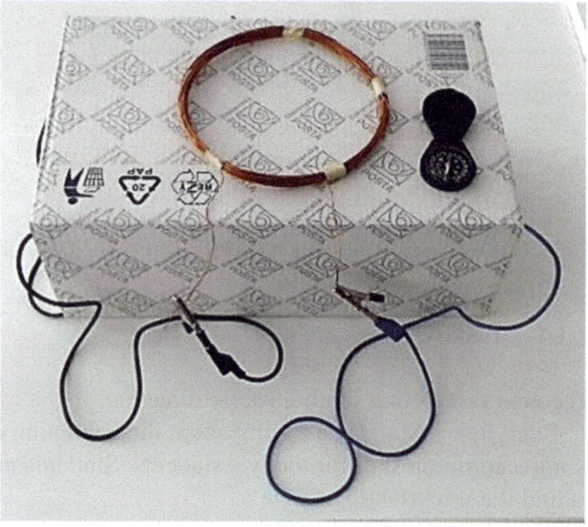

Fig. 6 Task E—equipment

In most of the problems involving the right-hand rule that students have solved so far, the direction of the current was known, and they needed to investigate the magnetic field. This time it is the other way around. In addition, students have to use a variety of knowledge on this topic (determining magnetic poles using a compass, Ampere's right-hand rule, conventional direction of current). The complexity and reversibility of this task increases its difficulty (Fig. 6).

2.3.7 Task L_2

QR code links to the video. Watch part of it (3:40–5:10) and answer the questions. MagLev propulsion is maintained by propelling coils located at the edges of the track. Propelling coils are:

(a) electromagnets.
(b) permanent magnets.

Train speed is controlled by:

(a) Magnetic field intensity of the propelling coils.
(b) Frequency of changes of magnetic field orientation of propelling coils.

The QR code (Fig. 7) is linked to a video explaining the principles of electromagnetic trains (MagLevs). Students are asked to watch only the part related to the MagLevs propulsion. This task is aimed at acquiring new knowledge, but it also requires understanding the audio-visual information and using prior knowledge to understand new principle. We chose the topic of MagLevs because the students were interested in it.

Fig. 7 Task L_2—QR code

3 Research

3.1 Goal

Our research focused on two objectives:

1. To get feedback for the game design. Since this was the first realization of this game, it was expected that some improvements would be necessary considering students' reactions (formulation of tasks, understandability of instructions, tasks and images, estimated time, etc.).
2. To map which factors have main influence on successful use of the game as a learning tool.

3.2 Sample

The research was conducted on 55 upper-secondary students from 4 different classes of a grammar school in Bratislava. Students played the game during a 45-min physics lesson (10 min introduction to the game mechanics and rules, 30 min game play and 5 min for discussion). In individual classes, teams consisted of 2-4 students and were formed in 3 different ways (randomly, by student request, or by researcher intention). The game took place in the school building which is well known environment for the students. The game was used as a revision of knowledge and skills before the summative assessment of learning outcomes. Students were informed that game was not part of the assessment process and that the tasks of the game could help them to prepare for the assessment. The researcher was also the physics teacher in all 4 classes.

3.3 *Methods*

During the game students' behaviour and task handling were observed. Observations included emotions, engagement, teamwork, task handling and competitiveness. Unfortunately, only one researcher was present during the game, so it was not possible to observe all teams for the whole time.

One week after the game, the online questionnaire was administered to the students. It focused on the difficulty of the game, understandability of the tasks, evaluation of teamwork skills, the benefits of the game and the students' attitude to the game. All 55 students were invited to complete the questionnaire, and 37 students completed it voluntarily.

Students whose answers were interesting in some way (in a conflict with observation, extreme in comparison with other students, etc.) were invited to the interview. The interview was semi-structured, some questions were the same for all respondents and others were based on the student's answers in the questionnaire. The interview was conducted with 8 respondents.

3.4 *Findings*

3.4.1 Observation

Observations showed that the most students were well engaged in the game, as soon as the game started, they ran out of the classroom and started solving problems. There was also the high level of competitiveness and desire to win. Some students showed frustration when they got stuck in some tasks. No problems with teamwork were observed. Task handling was directly observed only in task L_1. Students tried to solve it by thinking only and not by using experimental method. Most teams used experimental method only after a little intervention by the researcher. The best team completed the tasks in 20 mins. Some teams were not able to complete the game in the time limited by the end of the lesson.

After completing the game, the need for discussion about the tasks, students' emotions and experiences from the game was observed both among students and from students to teacher. The teams naturally started to compare their answers and solutions and to share their experiences. They showed positive emotions and attitudes towards the game. Teams that did not complete the game showed frustration and dissatisfaction with their performance through the body language as well as verbally.

3.4.2 Questionnaires

The online questionnaire was administered one week after the game and it was completed by 37 students.

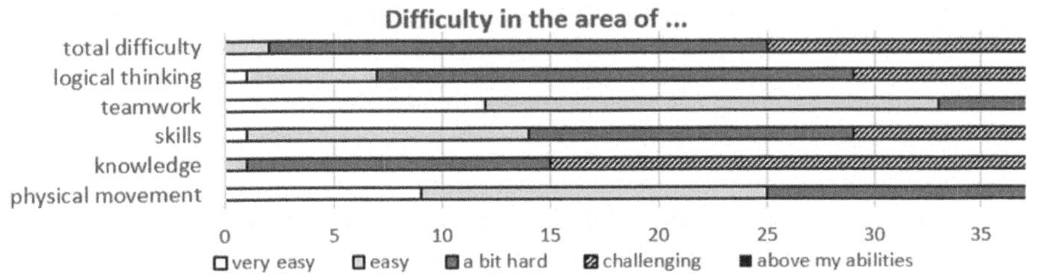

Fig. 8 Absolute numbers of students' evaluation of the game difficulty

The first part focused on the perceived difficulty of the game. Students were asked to rate the difficulty of selected aspects of the game. The results are shown in Fig. 8. The game was designed to be challenging for the students, especially in terms of logical thinking, skills, and knowledge. Collected data showed that students mostly perceived the game as a bit hard or a challenging in these aspects. The game is not intended to be difficult in teamwork (in fact, it should be possible to complete the game individually) or physically, which was also reported by the students in the questionnaires. Overall, the difficulty level of the game seems to be set properly, as it is not easy, but also not beyond students' abilities. It needs to be a bit hard or a challenging to encourage the development of students.

The understandability of the game was the second aspect under investigation. The students were asked to rate the degree of understandability of selected parts of the game. Findings are displayed in the Fig. 9.

The students understood the instructions and rules of the game very well. Some students stated that they did not know what to do because they had not paid attention during the instructions. The understandability of the tasks was rated high as only 1 student reported many problems, and 3 students rated it as acceptable. The majority of students had only a few or no problems with the understandability of the tasks. Regarding the understandability of clues, 2 students reported many problems, but this had no effect on their performance in the game. Another 2 students did not understand the clues at all. Further investigation showed that they actually did not visit the stands with clues, as they answered all the tasks correctly. Consequently, they did not know what the clues were in this game when they were completing the questionnaire. The rest of the students understood the clues very well. The lowest

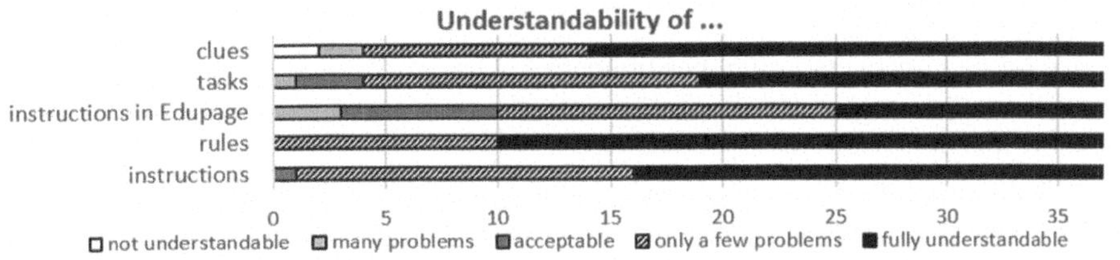

Fig. 9 Absolute numbers of students' evaluation of the game understandability

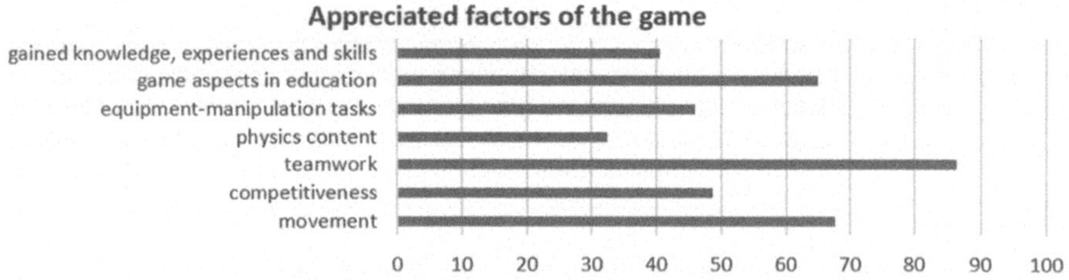

Fig. 10 Percentage of students appreciating selected factors of the game

score in this category was for the instructions in Edupage. Students reported that many of them were confused by many different communication channels (oral instructions, Edupage, written instructions, pictures, equipment).

In the third part of the questionnaire, students were asked to reflect on whether and how the game enriched them. Students mentioned applying knowledge (71.8%), gaining new physics knowledge (61.5%), discovering new views on the topic (56.4%), gaining new experiences (41.0%) and developing skills (23.1%) as benefits of the game. Only one student answered that the game had no benefits for them.

The fourth part of the questionnaire focused on teamwork (strengths and weaknesses of the team, leadership, roles in the team). As strengths of their teams, the students identified teamwork, communication, peer-learning, work distribution, decision making, speed and peer-motivation. Weaknesses included lack of knowledge, communication, tiredness, and poor work distribution. In total, 9 students did not identify any strength and 14 students did not identify any weakness. The most reported team roles were related to task solving (solving, explaining). Few students reported roles as navigator, motivator or writer. Students with these roles saw themselves as less valuable members of the team. One student identified herself as leader.

The fifth part of the questionnaire focused on the students' attitudes towards the game. Students appreciations of selected aspects of the game are displayed in the figure 10. Students also reported that after the game, 81% of them discussed the game with other players and 18% discussed with students who did not play the game. Only 8% reported that they did not return to the game in their thoughts.

3.4.3 Interviews

Based on their answers in the questionnaires, 11 students were invited for an interview, 8 students responded. For this article, the answers of 2 students were chosen to show the game from their perspective.

Student A is very active and proactive and shows interest in physics. Her team was randomly formed, 2 other girls (S and T) were her teammates. Student A mentioned that she was not in good health during the game, and it was difficult to stay motivated. According to her: "*At the beginning of the game there was chaos in our team, all of us wanted to read instructions at once. Then, after the first task we distributed the team-roles, and it got better. … I was a leader and girls were ok with that. I had theoretical background, and I was able to apply it on the tasks. S was a writer and T was using compass if the task required. … We were able to help each other and explain solutions."* To the question, how was your game experience affected by the fact, that you cannot choose your teammates, student A responded: *"I had no problem to work with them. Actually, it was beneficial. I would choose L (her best mate in class) and we would chat a lot and did not focus on the game."* She appreciated the game as a good revision tool for the test. She liked game mechanics, teamwork, and support from other teammates. Also, choice and design of the tasks she assessed as very good.

Student B is not interested in physics and her grades are below class average. Her team was assigned intentionally by researcher, the best student in physics in a class (K) was her teammate. She described the game: *"K is the best in physics from our class. Firstly, we take our roles, K was solver, and I was navigator and writer. The game was a positive change in learning. It's better for me to think during the movement or manipulation. Different types of thinking were included in the game."* Considering difficulty and teamwork, she said: *"Tasks were very hard for me. I couldn't solve them by myself. I managed to think about them with my previous knowledge. I understood how K was thinking. He was thinking out loud. … K always asked for my opinion. And if I didn't understand his solution, K explained it to me. I was valuable member of the team."* At task E, they reported some confusion and difficulty: *"At task with a coil, we met a few teams and some students from other class. Little discussion started there. It was confusing, because of too many opinions. K solved it, at the end."* To the question, how was your game experience affected by the fact, that you cannot choose your teammates, student B responded: *"It had large impact. If I can, I would choose R, but both of us are bad in physics. So, it would be a very long game for us. It was unusual to have my mate assigned."*

3.5 Interpretation

Collected data is interpreted in two ways:

1. What improvements should be made to the game?
2. What were the key factors influencing successful use of the game as a learning tool?

3.5.1 Feedback to the Game-Design

First tests of the game have shown that the game needs some improvements. The difficulty level seems to be set properly. The students' evaluation of the difficulty of the game is in line with the researchers' intentions and expectations in all investigated aspects of the game. The tasks were well designed and appreciated by the students.

As expected, the most difficult tasks were E and X. Students also appreciated the clues. They were useful and helped not only to find the correct answer but also to understand why it was correct.

The biggest confusion was caused by using too many different communication channels. In the initial version of the game, the need to give the teams various navigation instructions resulted in the use of Edupage. In further version, this problem was solved by using colour differentiated task instructions at the stands—each team was identified by its colour.

After the game, the students showed a need for discussion, not only about the physics content, but also about their emotions and experiences from the game. It is important to give the players a chance to express these feelings immediately after the game. Nevertheless, the core of the discussion should be focused on the physics content (task solution, way of thinking, knowledge and skills gained, problems) and on the game experience (teamwork, benefits of the game, problems, ...).

The tests also showed that one lesson (45 mins) is not enough for the instructions, the game and the follow-up discussion. At least one team in each class was unable to complete the game in time. Ideally, the full time for the game (including instructions and explanation of the rules, the game, discussion and reward) is 2 lessons (90 mins).

3.5.2 Key Factors Influencing Successful Use of the Game

Factors identified as having a major influence on the use of the game as a learning method were teamwork, the way the team is formed, movement and change of activity, content and difficulty, discussion and competition.

Teamwork positively influences the game in many ways. Teamwork reduces the difficulty of the game, especially for students who are not so good at physics. Students can help each other in the team. For students who are good at physics, it can be challenging to explain their thinking and solution to others. In this way, teamwork enables peer-learning.

From the preliminary results, it seems that the most effective way to achieve the goals of the game is the creation of teams by the researcher. Teams created by students or randomly can lead to situations that negatively affect the game. An uneven distribution of students interested in physics means that some teams struggle with tasks and lose motivation, while others solve tasks easily and win. Good friends in the team can be distracted by chatting and lose their attention. Bad relationships can cause problems in the teamwork and be distracting. Therefore, team members should be a mixture of students who are good at physics and students who are not so good at physics. Also, relationships between students should be considered.

Students appreciated the inclusion of movement in the game and some of them reported that change of the manner of learning and thinking helped them to concentrate, learn and engage. Different types of tasks require different skills and different ways of thinking to solve them, which helped students to stay focused, develop different skills and develop their understanding of the topic.

Content and tasks seem to be well selected. The tasks cover all important topics of magnetism and electromagnetism and focus on specific skills. The tasks challenge the students, their difficulty is not too low, but also not beyond the students' abilities. Students reported that it was a good way to prepare for the test. Most of them reported that they gained some new knowledge, skill or experience which was in line with the learning objectives of the game.

The competitive aspect of the game increased students' motivation and engagement in the learning process. They wanted to win or at least not to be the last. At the end, all teams were rewarded, and winners were awarded.

Discussion is an important aspect of the game in several ways. The majority of the students showed a need for discussion about the tasks and their solutions. Discussion can reinforce the learning aspect of the game. It provides an opportunity to revisit the tasks, summarise the solution process, answer the questions and explain some misunderstandings.

4 Conclusion

The research showed the students' positive attitude towards the game and some positive effects on the learning process, such as increased motivation and engagement, need for discussion on the topic, and acquisition of new knowledge and skills. It was the first educational game these students had played in physics. It is planned to continue the research with more games that the students will play. With more games, there could be some effect of losing the enjoyment of this new method. Although the research is only in its early stages, its results show that the games could help to build students' positive attitudes towards physics and science and could be an effective learning tool in physics at upper-secondary school. Factors influencing the successful use of the game The Tournament of J. C. Maxwell in the learning process were identified as teamwork, the way the team is formed, movement and change of activity, content and difficulty, discussion and competition.

Acknowledgments This work has been supported by Slovak KEGA grant 059UK-4/2022 "Preparation of physics teachers for the use of non-formal and informal education" and by Granty mladých UK UK/233/2022 "Mapping key factors influencing inclusion of games into educational process."

References

Cardinot A, Fairfield JA (2019) Game-based learning to engage students with physics and astronomy using a board game. Int J Game-Based Learning 9(1). https://doi.org/10.4018/IJGBL.2019010104

Greipl S, Moeller K, Ninaus M (2020) Potential and limits of game-based learning. Int J Technology Enhanced Learning 12(4):363–389. https://doi.org/10.1504/IJTEL.2020.10028417

Haverlíková V (2005) Educational games in physics. In: Planinsic G, Mohoric A (eds) Informal learning and public understanding of physics: 3rd International GIREP Seminar. University of Ljubljana, Ljubljana, 5–9 September 2005
Horváthová J, Haverlíková V (2012) Výskum používania hier vo vyučovaní fyziky. Scientia in Educatione 3(1):3–17. https://doi.org/10.14712/18047106.26
Kapp KM (2012) The gamification of learning and instruction: game-based methods and strategies for training and education. Pfeiffer, San Francisco
Naik N (2014) A comparative evaluation of game-based learning: Digital or non-digital games? In: Busche C (ed) Proc of the European conf on games-based learning, vol 2, pp 437–445

Contemporary and Complex Topics

Introducing Quantum Physics with Low-Cost Quantitative Experiments

Marco Di Mauro, Luigi M. Gratton, and Pasquale Onorato

Abstract We present three quantitative optical low-cost experiments aimed at introducing students to quantum physics from a phenomenological point of view. The first experiment, focussed on light diffraction, introduces the uncertainty principle while the others could be used to discuss quantum states and measurements and the collapse of the wave-function, first analyzing light transmission through three polarizers, and then using the latter to set-up a classical analogue of a quantum eraser. The experiments do not require expensive materials or specific laboratory instruments and employ the sensors and camera of common mobile devices. The experiments were tested with master students in mathematics and physics, who aim at becoming high school teachers.

1 Introduction

In the last years, the teaching of quantum physics (QP) at the pre-university level has been one of the greatest challenges to Physics Education Research (PER). In the past, research on the teaching and learning of QP focused mainly on the comparison between two different approaches: in the first, students meet QP through the study of continuous-variable systems; in the second, they are introduced to the postulates of QP through a spin-first approach, in which 2-level systems are preferred (Sadagyani 2016). In recent years many studies appeared, which advocate the idea of an approach oriented towards quantum technologies (cf. the Quantum Technology Education Project (QTedu n.d.)). In teaching quantum computing, researchers in PER often prefer the second approach, using the Stern-Gerlach experiment to motivate a

M. Di Mauro (✉) · L. M. Gratton · P. Onorato
Physics Department, University of Trento, Via Sommarive 14, 38123 Povo, Trento, Italy
e-mail: marco.dimauro@unitn.it

L. M. Gratton
e-mail: luigi.gratton@unitn.it

P. Onorato
e-mail: pasquale.onorato@unitn.it

S. Faletič and J. Pavlin (eds.), *Teaching and Learning Physics Effectively in Challenging Times*, Challenges in Physics Education, https://doi.org/10.1007/978-3-031-72541-8_14

quantum description of the world, often making use of simulations (cf. the Quantum Mechanics Visualization Project of the University of St. Andrews (QuVis n.d.)). Others have instead proposed a phenomenological approach aiming at building a theoretical framework starting from what is observed through an analysis of experimental data. A famous proposal employs the polarization of light obtained through polaroid filters and bi-reflective crystals (Ghirardi 2007). A second approach was proposed by Richard P. Feynman in the renowned book "QED" and developed by some researchers in PER (Taylor et al. 1998; Malgieri 2015). This approach can be supported by various observations of real experiments.

In this work, we propose and exhibit some experimental activities based on physical optics, which introduce some features of QP, realizable with low-cost materials, that can be presented to students in the classroom or performed at home. They are based on the use of smartphones as measuring instruments and consist of quantitative experiments, where students can in addition achieve the typical learning goals of a laboratory course. The materials used for the experiment consist in six carefully calibrated copper wires of different thickness (the thickness was accurately measured by the manufacturer to within 0.01 mm), one commercially available LED powered by a common button battery, three small polaroid film cutouts measuring 2 cm $\times$ 2 cm (these are easily retrievable online), and finally one low-power (less than 1 mW) laser diode.

2 A Lab in the Pocket: Experiments with the Smartphone

Recently, numerous smartphone-based physics experiments have been proposed in the literature (Monteiro 2022; Onorato et al. 2021; Hawley et al. 2018; Löfstrand and Pendrill 2016; Monteiro et al. 2017; Vieyra et al. 2015). These experiments take advantage of the built-in smartphone sensors as the accelerometer, gyroscope (angular velocity sensor), magnetometer, proximity sensor, luxmeter (ambient light), pressure (barometer) among others. Some other experiments employ the cell phone camera to acquire photos or videos, which are then analyzed with dedicated software. The apps employed to use the sensors as a measurement tool in various experiments are Phyphox (Phyphox n.d.), Physics Toolbox Suite (PhysicsToolbox n.d.) and AndroSensor (Androsensor n.d.) and they are available for the main operating systems on their respective websites.

Tablet apps such as Vernier Video Physics (Vernier n.d.) or laptop software such as Tracker (Physlets n.d.) can be used to analyze motion in the video clip and generate a range of graphs to represent different aspects of motion. Moreover, using Tracker or ImageJ (ImageJ n.d.) students can also work on static pictures to analyze spectra and perform optic experiments (Onorato and Gratton 2020).

In the next sections we discuss how these two approaches, the one based on the ambient light sensor and the one employing the camera, can be applied to low-cost experiments designed to introduce some topics in quantum physics.

3 Introducing the Uncertainty Principle Using Diffraction of LED Light

One of the most didactically simple ways to present the uncertainty principle is based on diffraction (Muiño 2000) starting from a collimated beam of monochromatic photons—hence with defined momentum—whose position is measured through a slit of width b. Let us consider a collimated beam of ideal monochromatic photons (the beam width is S, much larger than b), all with momentum p along the $x-$axis defined with infinite accuracy ($p = h/\lambda$). We want to determine their position along the $y-$axis using a slit placed along the $z-$axis (cf. Fig. 1, left): the passage through the slit represents an observation of the photon, which allows to delimit its position along the $y-$axis, with an uncertainty given by the width of the slit b (i.e., $\Delta y \approx b$). Classically, we expect to observe the image of the slit on the screen, with exactly the width b. This is indeed what happens if the slit has a very large width, so the geometric optics approximation is valid. On the contrary, what is observed for generic b is that a diffraction pattern is formed on the screen, with a more intense central maximum that indicates where the photon is most likely to arrive. The image which is formed on the screen in this case is larger than both the beam width S and the slit width b. This means that the beam has widened to form the diffraction pattern, and for this to happen the momenta of photons must have acquired a y component. Moreover, the image gets wider as b gets smaller. Thus, the diffraction caused by the obstacle indicates that the photons acquire a component of the momentum transverse to their motion, with a magnitude that we can estimate based on the deflection.

The experiment consists in measuring the interval between the two lines of darkness (minima) closest to the centre, 2δ, to approximate the value of $\Delta p_y/p \approx \delta/L$; this is the minimum value for the uncertainty in the momentum. Since, as a standard

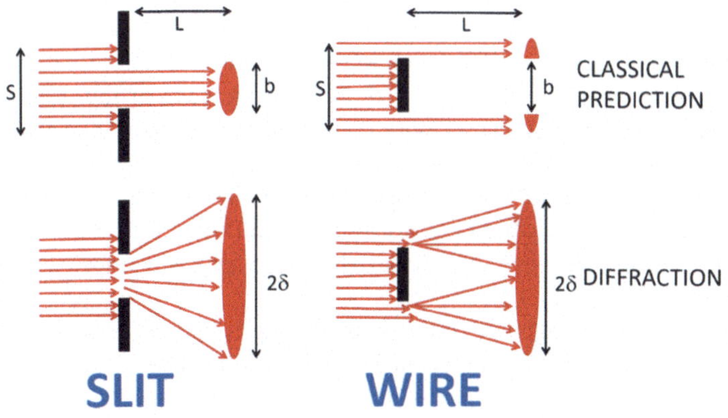

Fig. 1 Classical (geometric optics) and "quantum" (wave optics) outcomes for the experiment, for the slit (left) and the wire (right) cases

computation shows (Born and Wolf 2019), $\delta \approx \lambda L/b$, it is easy to obtain a weak uncertainty principle in the form:

$$\Delta p_y \Delta y \approx p \frac{\delta}{L} b \approx \frac{h}{\lambda} \frac{\lambda}{b} b = h. \tag{1}$$

While it seems that the photons acquire the transverse momentum component because of the disturbing effect due to the presence of the slit, a more correct interpretation is that the presence of the slit gives information on the position of the photons along the $y-$axis, thus reducing the sharp information that was available on the complementary variable p_y.

Students typically measure the dependence of δ on the slit width. The experiment is usually performed using a laser diode and slits of different widths, all material which is available only in well-quipped labs. In order to make this experiment safer and cheaper, we replaced the potentially hazardous laser and the expensive calibrated slits, and we propose to investigate light diffraction with a thin copper wire (cf. Fig. 1, right) and a LED (Onorato et al. 2022). The advantage is that copper wires of different thicknesses which are perfectly suitable for this purpose are much easier to obtain than slits. The theoretical basis of our proposal exploits Babinet's principle (Born and Wolf 2019; Babinet 1837), according to which a thin wire produces a diffraction pattern which is the same as that produced by a slit with the same width. Also in this case, the presence of the wire implies that the momenta of the photons acquire a $y-$component. Heuristically, in fact, the wire makes available the complementary information concerning the position along y of the photon with respect to the slit case, and this results in the same indeterminacy on the complementary variable p_y. A more rigorous, if still somewhat heuristic way to understand the relation between the two cases involves the Fourier transform of the wave functions of the photon. Let us consider the case of the slit first (see Nikolic and Nesic 2011 for details). Suppose the incoming photons are described by a plane wave. After passing through the slit, the wave function is restricted, along the $y-$axis, to a region of the same width b. Its Fourier transform, i.e., the wave function in momentum space, has therefore a width of $\Delta p_y = h/b$, from which $\Delta p_y \Delta y \approx b\frac{h}{b} = h$. In the case of the wire, the wave function along y is complementary with respect to the slit case, i.e., the incoming plane wave minus the wave function of the slit case. Hence the Fourier transform in this case will be equal to a term proportional to a Dirac delta (the transform of the incoming plane wave), which has no spread, minus the Fourier transform of the wave function in the slit case, which has the same spread. Hence, we find again $\Delta p_y \Delta y \approx b\frac{h}{b} = h$.

The experimental arrangement consists in placing copper wires of known widths on the cell phone camera, as shown in Fig. 2 (top left). The camera lens ensures the validity of the Fraunhofer approximation, so that the results obtained in that approximation are expected to be valid. In order to evaluate the diffraction patterns achieved with wires of different widths, we employ a reference wire, which is required to be present in all the photos. Then, the LEDs are framed from a distance of three meters, with a black cardboard sheet put behind the LED to enhance contrast. All

Fig. 2 (Top left) Positioning of the wires of different widths on the camera. (Centre) The experimental apparatus. (Right) Diffraction patterns obtained with a red LED and using wires of width: (Top) 0.05, 0.08, 0.1, 0.2*mm* and (Bottom) 0.05, 0.325, 0.5 *mm*

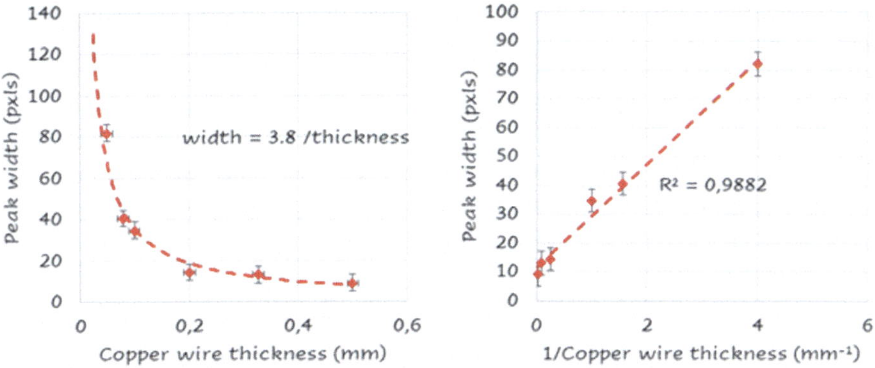

Fig. 3 Distance between the dark fringes as a function of the thickness of the wire (left) and of its inverse (right)

measurements are best performed with the room lights off, but we noted that excellent results are in fact possible even if the lights are on. The setup of the experiment, including the cardboard, is shown in Fig. 2 (centre). In the experiment, we frame a red LED and photograph the diffraction patterns thus obtained, as shown in Fig. 2 (right). The results of the measurements, obtained by analyzing the photos with Tracker, are reported in Fig. 3 and are in good agreement with the expected dependence of the distance between minima on the inverse of the width.

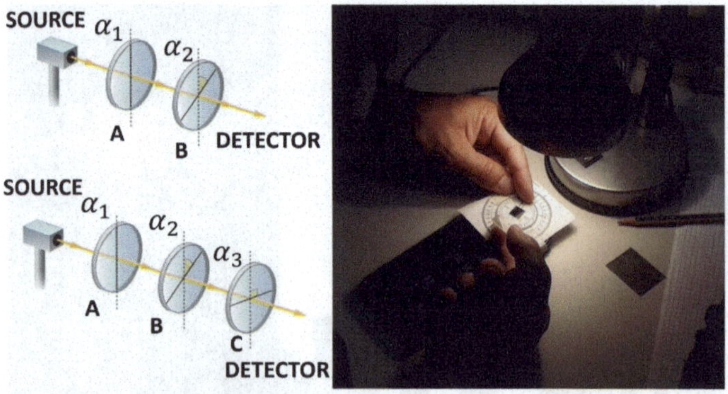

Fig. 4 Setup of the polarizer experiment

4 Introducing the Notion of Quantum States: The Malus Law and the Three-Polaroid Experiment

In his famous textbook on quantum mechanics (Dirac 1958), Dirac introduced the purely quantum concepts of superposition and uncertainty referring to polarized photons. Dirac's examples suggest a very simple quantitative experiment to demonstrate the notions of quantum states, the projection of a given state vector in another basis, the preparation of quantum systems in states with known properties, and the measurement of the various properties with state vector reduction.

Consider two linear polarizers A and C, and a light beam that is shone through both, as shown in the upper left part of Fig. 4. It is well-known by students that, if the polarization directions of A and C are orthogonal to each other, light does not pass through. Indeed, the intensity of the light passing through the polarizers depends on the relative angle between them through Malus' law:

$$I = I_0 \cos^2(\alpha_2 - \alpha_1) = I_0 \cos^2(\alpha_{12}), \tag{2}$$

where I_0 is the incoming intensity, α_1 and α_2 are the angles of the polarization directions of the two polarizers with respect to a reference axis, and α_{12} is their difference, i.e., the relative angle. Things are not changed if a third polarizer B is put before A or after C. If B is positioned between A and C, instead, light transmission is recovered (this corresponds to noncommutativity of the measurements of polarization); the situation is described by the equation (where the first polarizer sets the reference axis)

$$I = I_0 \cos^2(\alpha_2) \cos^2(\alpha_3 - \alpha_2). \tag{3}$$

The experiment we propose thus employs just three low-cost polarizers, a light bulb, and the ambient light sensor of a smartphone. We use some printed goniometers to determine the polarization angle of each polaroid, and we put the polaroid filters

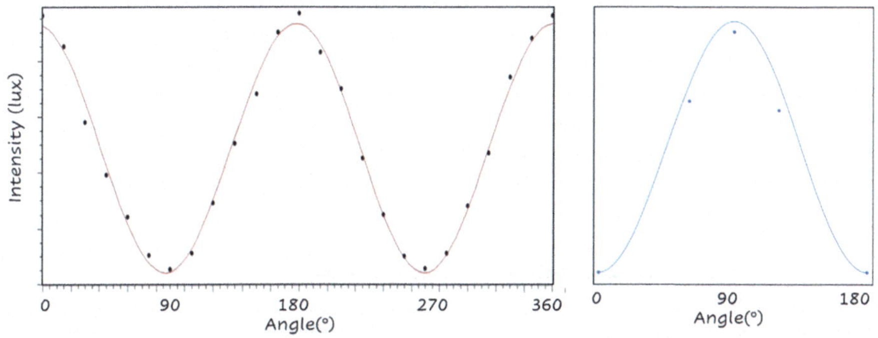

Fig. 5 Results of the polarizer experiment obtained by students

directly on the sensor and rotate them. Students start with a simple verification of the Malus law (2), using just two polarizers, then they perform the measurements with three polarizers. The results they obtain are usually in good agreement with Eqs. (2) and (3), respectively, as in the instance reported in Fig. 5.

5 A Classical Analogue of the Quantum Eraser

In this section we describe a simple experimental apparatus which allows the quantitative study of the classical analogue of a quantum eraser. The quantum eraser is one of the most famous and apparently paradoxical experiments of quantum physics (Scully and Drühl 1982; Scully et al. 1991). In this experiment, a basic feature of quantum physics is tested. While a measurement, i.e., the acquisition of information about a system, can destroy a quantum superposition, this effect can be undone by an appropriate action on the system, which erases that information. A typical realization involves the classic double-slit experiment. If information about which slit the quantum particle has gone through (Walborn et al. 2000)—the which-way information—is acquired, the interference pattern is destroyed. However, if the which-way information is erased, the interference pattern is recovered.

The experiment we consider (Hillmer and Kwiat 2007; Khandewal et al. 2021) can be regarded as a classical analogue of the quantum eraser we described above. In fact, we consider a classical system which can be described by the same formalism which describes the above-discussed quantum system and the measurements one performs on it. The experimental set-up is that depicted in Fig. 6, where also a photo of the real apparatus is provided. The apparatus consists in a laser diode, two polarizers providing the which-way information, and a third polarizer which erases that information. The double-slit is replaced by a thin copper wire, which, as already mentioned, is much easier to retrieve. The experimental results are analyzed with the aid of a digital camera and the Tracker application and are compared with the result of a classical computation of the interference patterns. Using wave optics

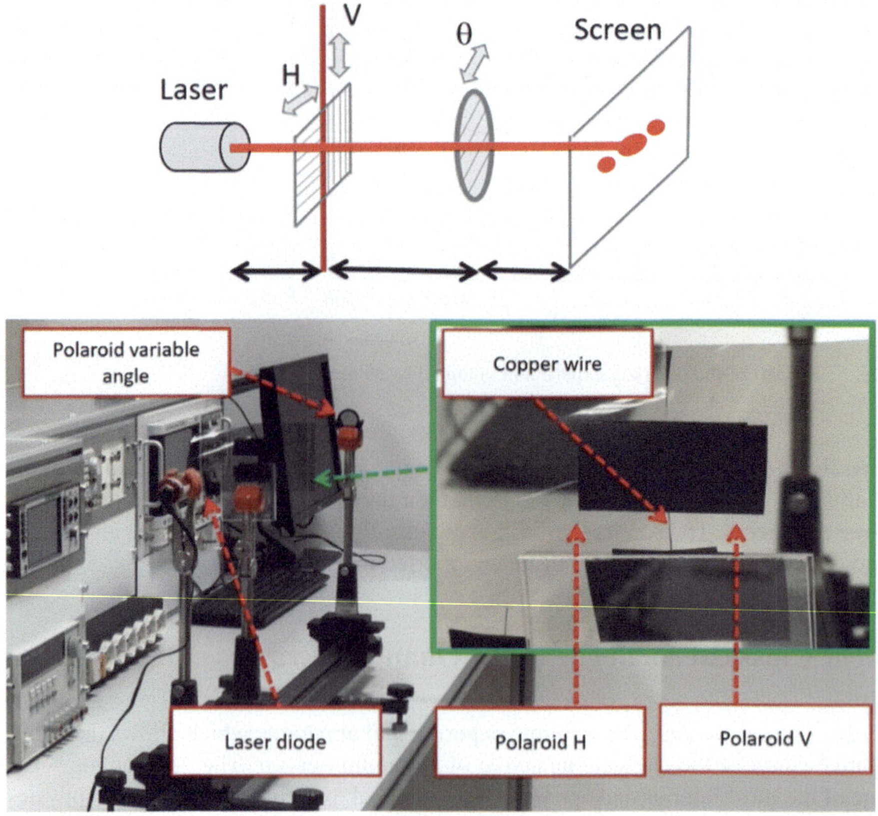

Fig. 6 The experimental apparatus: a low-power, $650nm$ laser beam, is directed towards a thin copper wire, which is put at a distance of about 8 cm from the laser diode. The width of the wire is $d = (0.20 \pm 0.01)$ mm. Then the laser beam reaches a screen at a distance D of about $260cm$ from the wire. Two linear polarizers are put on the two sides of the wire, one being orthogonal to the wire and the other being parallel to it. A third polarizer is inserted before the screen (at a distance of about 65 cm), tilted by an angle θ with respect to the horizontal polarizer

theory, including the Babinet principle and polarization (Born and Wolf 2019; Babinet 1837), the experiment can be studied from a theoretical point of view, and its outcome can be predicted. The result for the intensity pattern on the screen in the case with polarizers (Di Mauro et al. 2023) is:

$$I_\theta = \frac{1 + \cos(kd \sin\gamma)\sin 2\theta}{k^2 \sin^2 \gamma} \tag{4}$$

where $k = 2\pi/\lambda$ and the angle γ is the one which subtends the points on the screen where light arrives. This angle is always very small, thus in first approximation we

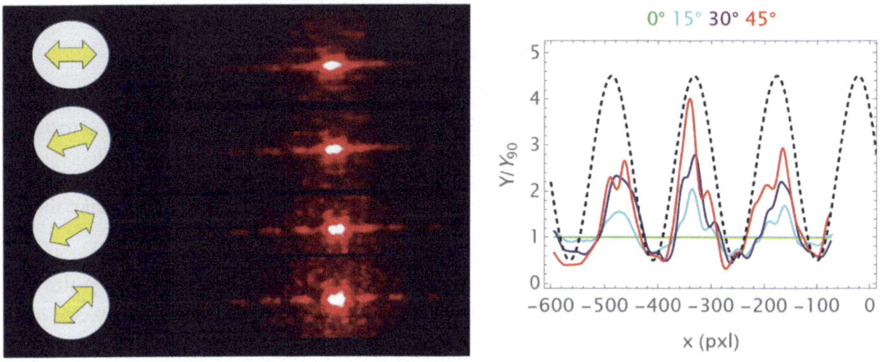

Fig. 7 (Left) Diffraction patterns for different angles. (Right) Tracker analysis of the diffraction pattern pictures obtained at different angles θ. The plots show the measured intensities for the various angles as functions of the position. The intensities are normalized with that measured for $\theta = 0$, and compared with the theoretical curve (dashed line), evidencing that they have the same periodicity

have that $\gamma \simeq x/D$, where x is the position on the screen and D the distance between the screen and the scattering wire.

The experimental results we obtained with our setup are reported in Fig. 7, where the images of the diffraction pattern which were obtained for different polarizing angles are shown (left). The intensities of the light patterns on the screen were measured by acquiring digital photos with the digital camera of a cell phone, in .jpeg format. The pictures were analyzed with Tracker, which allows to extract the relative luminance (which is directly proportional to the intensity of the light, cf. Rosi et al. 2016; Onorato et al. 2021) as a function of the position. The intensities of the normalized patterns, measured with Tracker, are shown in Fig. 7, and closely resemble the theoretical expectations.

6 Evaluation and Results

The experiments described in the previous sections were proposed to a group of 16 university students of mathematics and physics who attended a course oriented to the training of future physics teachers. During the final exam, students were asked to autonomously replicate the experiments, interpret the results in the context of a corpuscular model of electromagnetic radiation, and provide a pedagogical commentary on this "inductive" approach to quantum physics. All students were able to perform measurements using the experimental apparatuses discussed in Sects. 3 and 4. The obtained results consistently matched the theoretical predictions, and students were capable of offering adequate interpretations in quantum terms. In particular, concerning the diffraction experiments, students successfully derived the uncertainty relations (see Eq. 1) and discussed the utility of this experiment in explaining the uncertainty principle. One student described the experiment as follows:

> The particles (photons) are all launched in a direction perpendicular to the screen, so there is an evident attempt to measure the position of the particles directed towards the screen. The diffraction pattern is wider than the slit, even though photons are shot in a direction perpendicular to the screens. The attempt to define the particle's position more precisely results in a loss of knowledge of its momentum along the direction in which we make the measurement.

Regarding the experiment in Sect. 4, when discussing the interpretation of polarization with the emergence of a corpuscular idea of radiation, the students were able to interpret the phenomenon in terms of classical and quantum probabilities, highlighting the aspects of non-commutativity and emphasizing "the fact that the product rule of probability does not yield correct results." In particular, one student identified some strengths of this pedagogical approach:

> The introduction of Quantum Physics through the three-polaroid experiment has the primary advantage of being easily replicable in the classroom, allowing students to observe the effect in real-time. [...] This experiment introduces quantum physics by naturally incorporating concepts of probability and statistics... and allows us to compare classical and quantum probability, where commutativity of addition and multiplication does not hold.

Regarding the experiment in Sect. 5, one student stated:

> If we introduce an observer, the interference pattern disappears. It's as if we were blocking one of the slits without actually doing so... Obtaining 'which way' information reduces the possible indistinguishable paths to only those that pass through the analyzed slit, destroying interference. If it's possible to know which path the photon took, there will be no interference; if I have information, a piece of reality, that allows us to distinguish which paths the quantum object couldn't have taken, those paths are no longer counted among the possible ones.

Another student reiterated the idea of the seemingly magical nature of Quantum Physics and discussed how polarization experiments are useful in dispelling some misconceptions about the theory as "a niche discipline, relevant only in academic contexts. It's abstract. It provides indeterministic and probabilistic results."

7 Conclusions

In this contribution, we have described three simple experiments, which can be easily and cheaply performed in school and university laboratories, employing poor materials and free software, but giving quantitative results which are in good agreement with theoretical analyses. The experiments can help students grasp some fundamental concepts in quantum physics, such as the uncertainty principle, quantum states and measurements, and quantum erasure. These experiments can also be fruitfully included in an experiment-based teaching-learning sequence centred on quantum physics. The quantitative character of the experiments will allow students to achieve, in addition, the typical learning goals of a laboratory course, e.g., data analysis, elaboration and visualization. The experiments are also perfectly suited to classes focussing on electromagnetism and wave optics, since they involve such key concepts as diffraction, interference, polarization, and the Babinet principle.

The experiments were tested with master students in physics and mathematics at the University of Trento, aiming at becoming high school teachers. Despite the students had already had an exposure to quantum physics, the results obtained in the final exam indicated that they achieved a better understanding of the relevant principles of quantum physics.

References

https://imagej.nih.gov/ij/. Accessed 11 Jan 2023

https://phyphox.org/. Accessed 11 Jan 2023

https://physlets.org/tracker/. Accessed 11 Jan 2023

https://play.google.com/store/apps/details?id=com.chrystianvieyra.physicstoolboxsuite (Android); https://apps.apple.com/it/app/physics-toolbox-sensor-suite/id1128914250 (Apple). Accessed 11 Jan 2023

https://play.google.com/store/apps/details?id=com.fivasim.androsensor (Android). Accessed 11 Jan 2023

https://qtedu.eu/. Accessed 11 Jan 2023

https://www.st-andrews.ac.uk/physics/quvis/. Accessed 11 Jan 2023

https://www.vernier.com/physics/vernier-video-analysis/. Accessed 11 Jan 2023

Babinet J (1837) Mémoires d'optique météorologique. C R Acad Sci 4:638

Born M, Wolf E (2019) Principles of optics. Cambridge University Press, Cambridge

Di Mauro M, Gratton LM, Onorato P, Oss S (2023) A classical experiment for the quantum eraser: a comparison between model and experimental results. Phys Educ 05(01):2350002

Dirac PAM (1958) The principles of quantum mechanics, 4th edn. Clarendon Press, Oxford

Ghirardi G (2007) Sneaking a look at god's cards: unraveling the mysteries of quantum mechanics. Princeton University Press, Princeton

Hawley SH, McClain RE Jr (2018) Visualizing sound directivity via smartphone sensor. Phys Teach 56:72

Hillmer R, Kwiat P (2007) A do-it-yourself quantum eraser. Sci Am

Khandewal A, Tan JBJ, Leong TK, Yang Y, Venkatesan T, Jani H (2021) A cost-effective quantum eraser demonstration. Phys Educ 56:033007

Löfstrand M, Pendrill A-M (2016) Aerodynamics in the amusement park: interpreting sensor data for acceleration and rotation. Phys Educ 51:055015

Malgieri M (2015) Teaching quantum physics at introductory level: a sum over paths approach. Ph.D. Thesis, University of Pavia, 2015

Monteiro M (2022) Resource letter MDS-1: mobile devices and sensors for physics teaching. Am J Phys 90:328

Monteiro M, Stari C, Cabeza C, Marti AC (2017) The polarization of light and malus law using smartphones. Phys Teach 55:264

Muiño PL (2000) Introducing the uncertainty principle using diffraction of light waves. J Chem Educ 77(8):1025

Nikolic D, Nesic L (2011) Verification of the uncertainty principle by using diffraction of light waves. Eur J Phys 32:467

Onorato P, Gratton LM (2021) Measuring the Raman spectrum of water with a smartphone, laser diodes and diffraction gratring. Eur J Phys 41(025304):1

Onorato P, Rosi T, Tufino E, Caprara C, Malgieri M (2021) Using smartphone cameras and ambient light sensors in distance learning: the attenuation law as experimental determination of gamma correction Phys Educ 56:045007

Onorato P, Rosi T, Tufino E, Caprara C, Malgieri M (2021) Quantitative experiments in a distance lab: studying blackbody radiation with a smartphone. Eur J Phys 42(0545103):1

Onorato P, Gratton LM, Di Mauro M, Rosi T (2022) A simple experimental setup for investigating light diffraction with a thin copper wire and a LED. Phys Educ 57:063003

Rosi T, Malgieri M, Onorato P, Oss S (2016) What are we looking at when we say magenta? Quantitative measurements of RGV and CMYK colours with a homemade spectrophotometer. Eur J Phys 37(6):065301

Sadagyani HR (2016) Spin first vs. position first instructional approaches to teaching introductory quantum mechanics. PERC Proceedings 2016:292–295

Scully MO, Drühl K (1982) Quantum eraser: a proposed photon correlation experiment concerning observation and "delayed choice" in quantum mechanics. Phys Rev A 25:2208

Scully MO, Englert B-G, Walther H (1991) Quantum optical tests of complementarity. Nature 351:111

Taylor EF, Vokos S, O'Meara JM, Thornber NS (1998) Teaching Feynman's sum-over-paths quantum theory. Comput Phys 12:190

Vieyra R, Vieyra C, Jeanjacquot P, Marti A, Monteiro M (2015) Turn your smartphone into a science laboratory. Sci Teach 082(09):32

Walborn SP, Terra Cunha MO, Pádua S, Minken CH (2000) Double-slit quantum eraser. Phys Rev A 65:033818

Contributions from Pilot Projects in Quantum Technology Education as Support Action to Quantum Flagship

S. Faletic, P. Bitzenbauer, M. Bondani, M. Chiofalo, S. R. Goorney, K. Krijtenburg-Lewerissa, O. Mishina, R. Müller, G. Pospiech, İ. Ercan, M. Malgieri, A. Merzel, M. Michelini, P. Onorato, H. Pol, L. Santi, Z. C. Seskir, J. Sherson, H. K. E. Stadermann, A. Stefanel, E. Surer, K. Tóth, J. Yago Malo, and O. Zabello

Abstract The GIREP community on teaching and learning quantum physics and the Education section of the Quantum flagship project of the European Union (QTEdu) have brought together different stakeholders in the field of teaching quantum physics on all levels, including outreach. The goal of QTEdu is to pave the way for the

S. Faletic (✉)
Faculty of Mathematics and Physics, University of Ljubljana, Jadranska Ulica 19, 1000 Ljubljana, Slovenia
e-mail: sergej.faletic@fmf.uni-lj.si

P. Bitzenbauer
Department of Physics, Physics Education Research, Institute for Physics Education, Leipzig University, Vor dem Hospitaltore 1, 04103 Leipzig, Germany

M. Bondani
CNR-Institute for Photonics and Nanotechnology, Via Valleggio 11, Como, Italy

M. Chiofalo · J. Yago Malo
Department of Physics 'Enrico Fermi' and INFN, University of Pisa, Largo B. Pontecorvo 3, 56127 Pisa, Italy

S. R. Goorney · J. Sherson
Department of Management, School of Business and Social Sciences, Aarhus University, Aarhus, Denmark

Niels Bohr Institute, Copenhagen University, Copenhagen, Denmark

K. Krijtenburg-Lewerissa
Freudenthal Institute, Utrecht University, P.O. Box 85170, 3508 AD Utrecht, The Netherlands

O. Mishina
Istituto Nazionale di Ottica (CNR-INO), Area Science Park, Basovizza 34149 Trieste, Italy

R. Müller
Physikdidaktik TU Braunschweig, Bienroder Weg 82, 38106 Braunschweig, Germany

G. Pospiech
TU Dresden, Department of Physics, Research Group Physics Education, Helmholtzstr. 10, 01069 Dresden, Germany

S. Faletič and J. Pavlin (eds.), *Teaching and Learning Physics Effectively in Challenging Times*, Challenges in Physics Education, https://doi.org/10.1007/978-3-031-72541-8_15

training of the future quantum workforce. To this end, it is necessary to understand the needs of the quantum technology (QT) field, make the general public aware of the existence and importance of QT, and introduce quantum physics already in high school, so that high school students can choose QT as their field of study and career. Finally, new university courses need to be established to support emerging specific profiles such as a "quantum engineer". In this symposium, four QTEdu pilot projects were brought together to demonstrate how their complementary approaches have worked towards realising the above goals.

1 Introduction

Quantum technology (QT) is a rapidly developing field which has the potential to become an extensive source of employment with relevant economic impact in the near future. Therefore, there is a growing demand for a quantum workforce. One hindrance

İ. Ercan
Department of Microelectronics, Delft University of Technology (TU Delft), Mekelweg 4, 2628 CD Delft, the Netherlands

M. Malgieri
Department of Physics, University of Pavia, Via Bassi 6, 27100 Pavia, Italy

A. Merzel
The Seymour Fox School of Education, The Hebrew University of Jerusalem, Mount Scopus, 9190501 Jerusalem, Israel

M. Michelini · L. Santi · A. Stefanel
Dipartimento di Scienze Matematiche, Informatiche e Fisiche, Unità di Ricerca in Didattica Della Fisica, University of Udine, via Delle Scienze 206, 33100 Udine, Italy

P. Onorato
Physical Science Communication Laboratory, Department of Physics, University of Trento, 38123 Povo, TN, Italy

H. Pol · H. K. E. Stadermann
ELAN, Department of Teacher Development, University of Twente, P.O.Box 217, NL-7500 AE Enschede, The Netherlands

Z. C. Seskir
Institute for Technology Assessment and Systems Analysis (ITAS), Karlsruhe Institute of Technology, Karlsruhe, Germany

E. Surer
Department of Modeling and Simulation, Graduate School of Informatics, Middle East Technical University, Ankara 06800, Turkey

K. Tóth
Eötvös Loránd University, Budapest, Hungary

O. Zabello
Offenburg University of Applied Sciences, Badstraße 24, 77652 Offenburg, Germany

for the development of a future quantum workforce is that students often become familiar with quantum physics only at university level, and even that not necessarily in the first year. By then, their career paths may have already been chosen. The goal of the outreach activities described below is, therefore, to bring the quantum world to students before university so they may consider choosing it as their career path.

However, QT education is not only about training skilled workers for occupational fields in quantum technologies. It is also crucial that the population as a whole gains a basic understanding of the underlying physical principles in order to support extensive funding efforts, regulation and adoption of future quantum products. Therefore, it is important to develop and evaluate suitable concepts and approaches for a wide range of recipients, including high school students in formal education. Physics education research has been investigating student conceptions and learning difficulties in quantum physics for decades. In 2019, the different approaches were discussed at a GIREP symposium (Michelini et al. 2021). A GIREP Thematic Group was formed to discuss ways and means to efficiently communicate quantum physics at pre-university level (Michelini et al. 2021).

In QTs, the industrial applications of basic principles of quantum physics and the general education in quantum physics meet each other: quantum physics leaves the research laboratories and develops novel technological products for society and industry, some of which are still future perspectives, but are nevertheless extensively covered in the media.

To support this development, the European Quantum Technology Flagship was founded (Quantum Flagship). Within the Flagship, Quantum technologies are divided into four areas:

- *Quantum computing*: Single quantum objects (like ions or superconducting states) are used as qubits. Unlike the bits of classical computing, which can be either in state 0 or in state 1, qubits can be brought into quantum mechanical superposition and entangled states. Computations can then be performed in some kind of parallel manner. However, quantum computers are currently expected to only outperform classical computers on very specific tasks.
- *Quantum sensors*: In this field, quantum physical effects are used to measure quantities such as magnetic fields much more precisely than is possible today. This can be used for diagnostic purposes, for example, to measure the magnetic activity of the human brain (Hill et al. 2019).
- *Quantum communication*: With single photons, one can develop protocols that promise secure ways of communication – a reason to further develop the corresponding technologies.
- *Quantum simulation*: By using one easier-to-control system (i.e., that of cold atoms in optical lattices), one can simulate other quantum mechanical systems, e.g., the behaviour of electrons in ordered systems. This concept, known as *analog quantum simulation* is not universal (unlike quantum computing), but it remains a promising perspective for the development of new chemical catalysts or medical drugs based on molecular simulations, especially in the nearer-term as quantum computers are being developed.

Quantum technologies have the potential to be disruptive not only economically, but also in terms of teaching quantum physics. They enable a new view of quantum physics: the perspective of applications in everyday life and technology. In the language of physics education: Quantum technologies make context-related teaching and learning of quantum physics possible, as is already practised in many other fields of physics.

By being able to focus on concrete applications of quantum physics in practical contexts, quantum physics possibly loses some of its abstractness. In addition, quantum technological applications can be quite motivating: the encryption and decryption of secret messages in quantum cryptography is probably seen by learners as an interesting and highly relevant topic. Moreover, new experimental possibilities are opened up, for example on quantum sensors, on the transmission of information, or on the use of two-state systems as qubits.

In the European Quantum Technology Flagship, the topics of education, outreach and training were pursued within the Education section's QTEdu Coordination and Support Action (CSA) (Quantum Flagship) until 2022 and as part of the new Qucats CSA (Consolidating the Course of QT) since then. Within QTEdu, 11 different pilot projects on Education, Outreach and Training were established. Four of them were chosen for the 2022 GIREP symposium to discuss the new perspectives on teaching and learning quantum physics arising from the new quantum technologies. The first to react to this new opportunity were outreach programs. The quantum technology context proved a fertile ground to bring quantum physics closer to the general public, because now it had immediate meaning to them. Outreach also has the benefit of not being constrained by formal curricula, which are slow to change.

But, be it formal education programs or outreach, the goal is to achieve quantum literacy: the ability of the general person to understand quantum terminology and the impact of quantum physics on technology and society. To this end, several projects have been launched to develop a modular and as universal as possible assessment instrument that will enable measurement of the effectiveness of each approach.

The contributors to the symposium discussed how outreach and assessment go hand in hand towards the common goal of building the foundation for a quantum workforce.

2 Outreach Programs

In this section, we reflect on the role of outreach, which we consider to be non-formal education, in the field of QTs. Through inspiration, outreach is a possible route to the long-term development of a quantum-ready workforce, capable of researching, developing, applying, and industrialising quantum computers, sensors, simulators, and communication infrastructure. Increasing efforts on training at the university level (Kaur and Venegas-Gomez 2022) are a short-term solution which address the urgent need for a workforce, but do not help to inspire future generations of Quantum

Physicists. We believe that outreach serves an important purpose in helping all citizens to navigate and contribute to modern society. Widespread quantum awareness may be of essential importance as the technologies are disseminated into everyday life to overall transform our economy, in many jobs areas such as the legal & ethical (Kop 2021), financial (Hogg et al. 2010), medical and environmental (Liu et al. 2022) among many others. These transformations require new patterns of thought and skills among citizens of the public in our present "society of acceleration and uncertainty" (Rosa 2013). In order to be aware and valuable citizens of the public, we now require many more and different skills than we have in previous times. Exactly which are required, and how we may develop them through outreach, we discuss in the following examples of two projects.

The *Quantum Technologies Education for Everyone* (QuTE4E) project has spearheaded novel research-based approaches to outreach and developed practical guidelines for outreach. The *Italian Quantum Weeks* (IQW) project brought together a wide range of scientists and communicators for a nation-wide coordinated effort that reached thousands of people (Quantum weeks). Combining knowledge from both projects can bring us a step closer to effective outreach for achieving quantum literacy.

2.1 Quantum Technologies Education for Everyone

The pilot project Quantum Technologies Education for Everyone (QuTE4E) (Quantum Technologies) has addressed several questions around the what, why, and how of outreach in a research-based approach we call *Physics Outreach Research* (POR). The ultimate aim of QuTE4E has been to develop practical guidelines for conducting outreach activities, given limited time and opportunities available to practitioners.

First we consider the "why" of outreach: for what purpose do we want to communicate Quantum Technologies to the public? We believe that with a suitably designed storytelling, we may both inspire and educate citizens of the public with the skills necessary to navigate the "society of acceleration and uncertainty" in an informal and enjoyable manner that is inclusive, accessible and manageable by the practitioners.

2.1.1 Outreach for Modern Thinking Skills

An inspiration for us is the collective works of Howard Gardner, in particular his Five Minds for the Future (Gardner 2009). Therein, Gardner promotes the value of the disciplined, synthesising, creative, respectful, and ethical minds. We believe that outreach may be an engine to develop these "minds". In particular, Gardner's minds are inherently addressed in the everyday activities of scientists through the so-called inquiry cycle (Kuhn 2010). Thus presenting the public with activities which enable them to make use of inquiry, in a kind of scientific role-play, can enable outreach to have a dual role in promoting quantum awareness and inspiration, as well as helping to equip participants to navigate the "society of acceleration".

Engaging in scientific thought is precisely the disciplinary thinking to which Gardner refers. The aspects of experimentation, conceptualisation, and theory-building develop the synthesising and creative minds (Kuhn 2010). To address the respectful and ethical minds, one must consider *Responsible Research and Innovation* (RRI) (Coenen and Grunwald 2017), the need for scientific research to be conducted in a manner which respects potential societal and environmental impact. Significant work within the pilot has gone into developing a narrative which addresses these minds, leading to the development of a storytelling framework Goorney et al. named *culturo-scientific storytelling* (CSS) (Goorney et al. 2022). The CSS (see Fig. 1) combines the theoretical framework of the discipline-culture (Tseitlin and Galili 2005) with the inquiry cycle to address Gardner's minds, developing skills such as disciplinary thinking, creativity, and awareness of the ever-changing nature of scientific knowledge.

The CSS was developed for outreach of QT, but it is not specific to it. It could be applied in other rapidly emerging fields, such as Artificial Intelligence. However, what is unique to our field is, perhaps more than any other, the need for creative thought. For example, how can one understand the concept of quantum superposition or entanglement or spin, when no such examples of this exist in daily life? Even more challenging may be those concepts to which the public can relate due to their presence in popular science, such as Quantum Teleportation. QM forces us to "see" what cannot be seen with eyes and experienced with other senses, to connect abstraction and concrete, and to create visualisation forms that would otherwise be unnecessary. The creative mind is of essential importance here, and we believe it to be one of the ways that the CSS can make the greatest impact in our field.

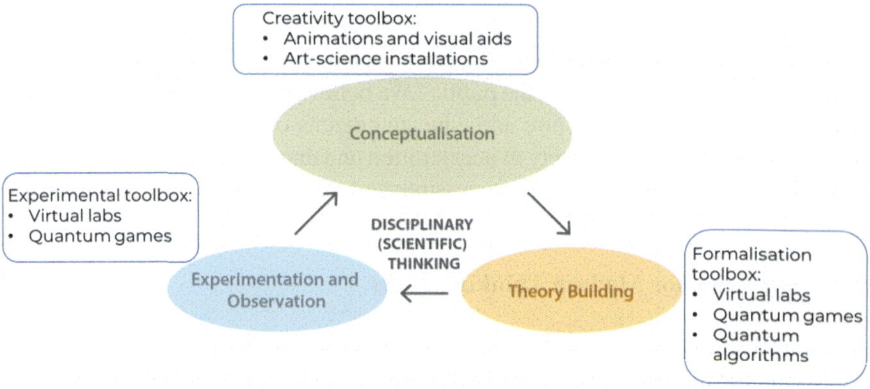

Fig. 1 The toolbox of community-developed resources available for quantum technologies outreach, able to support scientific thinking and thus implement the culturo-scientific storytelling. Reproduced from Goorney et al. (Goorney et al. 2022)

2.1.2 Community Resources—The "How" of Outreach

The CSS provides a theoretical framework for the narrative approach, but it would be abstract without a means through which to implement it. The QuTE4E pilot has been a fertile ground for the development, testing, and validation of resources aiming to support different aspects of the scientific thinking cycle. This has led to the construction of a community toolbox (Seskir et al. 2022), consisting of aids for creativity, experimentation, and formalisation. Use of these tools has proved highly successful, for example in several events during *Italian Quantum Weeks* (see next section), and in *Europe Quantum Day 2022* (Europe Day), a large-scale outreach exhibition taking place in the headquarters of the European Commission in Brussels (estimated 15,000 public participants). In these cases, the events would not have been practical to hold without such resources freely available. Other use cases of these tools, such as a series of workshops based around the art-science exhibition *Quantum Jungle* in Pisa (Baumgarten et al. 2022), and extracurricular courses for high school students (Bondani et al. 2022a, b), (Quantum Moves 2), have been investigated quantitatively with research instruments, demonstrating the potential value of these tools. Some examples of these include *Quantum Games*, such as *Quantum Moves* (Quantum Moves 2) and *Quantum Odyssey* (Quarks Interactive), both with their own engaging narratives for participants (citizen science and integration with *Qiskit*, respectively). Visualisation tools, such as *Quantum Composer* (Quantum Composer) and the *Quantum Mechanics Visualisation Project* (QuVis) (University of St. Andrews), are particularly valuable as a means of fostering creativity whilst remaining grounded in the real workings of quantum mechanics. There are many more such tools available, and they are consistently being developed by groups around the world. Several of these are represented in the pilot, including *ScienceAtHome* (ScienceAtHome), *Quarks Interactive* (Quarks Interactive), and *QPlayLearn* (QPlayLearn).

Quantum Technologies are complex applications of core concepts in QST. Thus our storytelling, through the CSS, allows us to build these conceptual micro-stories into the macro-story of QT. In so doing, educational paths can be developed on conceptual knowledge in the former case, and on procedural knowledge in the latter. The produced resources could be tailored to the beneficiaries needs, and validated using the work of the QCI pilot described here. Finally, we note that the same approach can be extended from non-formal to formal educational activities, by just changing the level of formal description suited to the audience.

2.2 *Italian Quantum Weeks*

In the framework of the *EU Quantum Flagship*, on the occasion of the first *World Quantum Day* (WQD, April 14, 2022) (World quantum day) the Italian scientific and business community have decided to propose the *Italian Quantum Weeks* (IQWs) (Quantum weeks), several weeks from March 26 to May 31 dedicated to the exploration of the foundations and applications of quantum mechanics with the aim to raise awareness of the importance of quantum technologies in our daily lives. This can help promote informed discussions and decisions about how these technologies are developed and used. In addition, teaching quantum technologies to a diverse group of people can help break down barriers and increase engagement with science, which will benefit society as a whole.

More than 130 researchers, technicians, disseminators, communicators, teachers, belonging to more than 40 research institutions, universities, scientific societies, in 17 Italian cities were involved in the project and set up a bundle of initiatives to be proposed to students and general public, with the aim of offering an interactive experience of the peculiar behaviour of the quantum world and of its immense application potential. The activities have been organised with the support of the QTEdu, QTE4E and are officially inserted in the WQD project.

Two national conferences broadcast on the IQWs *YouTube* channel (Italian quantum weeks). On April 8, the conference for the general public "From bit to qubit: the advantages of quantum" was held, focusing on the advantages of quantum technologies and the opportunities offered by the Second Quantum Revolution. The conference featured high-level talks by quantum experts from Italy and abroad. Then, the "Quantum@School" conference titled "Stories of Quantum Technologies" took place on April 12. The conference was structured on the talks of five young Italian scientists, who told their personal success stories to inspire students to pursue studies and careers in the quantum field.

Local lectures, guided tours of research laboratories, interactive workshops, quantum game sessions and "Quantum aperitifs" were organised in the different cities, exploiting local expertise and research.

To foster awareness and make connections with personal experience, the "Quantum Suggestions" creativity contest was proposed, aimed at developing a "quantum-inspired" creative project. Participation in the contest is open to people of all nationalities, individually or in groups, and the involvement of schools and students is strongly encouraged. Eligible creative projects may relate to any aspect of quantum mechanics and include any type of artwork. We received 31 artworks of different nature and complexity that were presented to the public during the *European Researchers' Night*. The national winners will be announced at the next *World Quantum Day* on April 14, 2023, and the submitted projects will be included in a national exhibition to be organised in 2024.

The final event of the IQWs 2022 initiatives was held in Palermo on May 22 and consisted of an open conference complemented by a "quantum-inspired" music and dance performance.

2.2.1 Speaking the Unspeakable

The most complex and challenging activity of IQWs was the development and organisation of the interactive exhibition "Speaking the Unspeakable-Quantum Superposition," set up in 8 different Italian venues (Catania, Como, Milan, Florence, Modena, Naples, Padua, Rome (two locations)).

The ambitious goal of the exhibition was to introduce the general public to one of the unique properties of quantum states, namely quantum superposition, which, having no classical analogue, is extremely difficult to explain using categories and concepts derived from everyday experience and without resorting to the language of mathematics. The effort of the exhibition was precisely to try to "speak the unspeakable" without misrepresenting the genuine quantum message. To this end, the superposition state was introduced using a quantum experiment that has no classical analogues, namely the single-electron interference experiment proposed by Richard Feynman (Feynman et al. 1965). A recording of the experiment performed in Italy at the laboratories of the University of Modena and Reggio Emilia was used in the exhibit. In the description of the experiment, we used only the concept of quantum state and decided never to talk about wave-particle dualism so as not to suggest classical analogies that confuse rather than clarify. To support the intuition of the superposition properties of quantum states, we used the analogy with the perception of bistable images, that is, images composed of two different parts that coexist and cannot be seen at the same time.

The heart of the exhibition focused on quantum mechanical description of the world, represented by the axioms of the theory describing preparation, evolution and measurement of quantum states. The probabilistic nature of quantum measurements, the difference between quantum superposition states and classical mixed states, and the noncommutativity of some quantum observables were represented through simple interactive games with coloured cubes, disks and squares of different colours. In addition, the entire logic of the axioms was illustrated with sequences of polarizers.

After discussing the fundamentals of quantum description of the properties of states, the importance of state superposition for applications of quantum technologies, such as computer science and quantum cryptography, was described. First, the concept of "qubit" (minimum amount of quantum information) and "quantum gates" (logical operations on qubits) were introduced, and the advantage of using quantum resources was exemplified through the description and simulation of the classical and quantum "coin-flip" game (Meyer 1999).

Finally, a description of the physical realisation of qubits and the different architectures of quantum computers was presented, along with examples of quantum algorithms implemented on the free IBM *Quantum Experience* platform for quantum computing (IBM quantum computing).

The exhibition was supported by written panels, installations, and interactive exhibits. Two original videos on the history of quantum mechanics were placed at the beginning and in the middle of the exhibition, while a video on the experimental research in Italy in the field of quantum technologies closed the exhibition.

2.2.2 Outcome

Visitors, both organised school groups and interested individuals, were guided by physics researchers, Ph.D. and University students. The exhibition in the different cities lasted an average of one week and attracted more than 5,000 visitors. Evaluation of the exhibition and other events was carried out in the different locations through a voluntary and anonymous satisfaction questionnaire administered at the end of each event. The questionnaire included items on the level of interest, satisfaction, perceived usefulness, difficulty, and sharing the experience with family members. We obtained 564 responses, 449 from visitors to the exhibition at the different venues and 115 from the other events. The results of a preliminary data analysis are summarised in Fig. 2 and show that the scores are reasonably high for all items and not too low for the "difficulty" item. Figure 2a shows that females (51% of participants), males (37%) and those who chose not to answer the question on gender (21%) score similarly, which is quite good considering the known difficulties in involving girls in STEM activities. In contrast, Fig. 2b shows that teachers and common visitors, who are likely to be interested and educated, score better than students in all categories, a result that indicates the need to increase student literacy and awareness by teaching quantum physics in high schools. Finally, Fig. 2c shows that the different types of in-person activities scored the same, while online conferences were much less effective, probably due to a different level of personal involvement and a gradual disengagement with online activities due to the saturation of the use of this mode during the COVID-19 pandemic.

3 Assessment

The two examples of outreach presented in this paper do not stand alone. Over the last decades, teaching proposals for quantum physics taking multiple approaches (e.g., history-based, experiment-based, two-state approaches amongst others) with a focus on different key topics (Stadermann et al. 2019) have been developed and implemented to make quantum physics (QP) accessible both for the secondary school level (for example, see (Müller and Wiesner 2002); (Weissman et al. 2022); (Michelini and Stefanel 2021); (Bondani et al. 2022a, b); (Bitzenbauer 2021a, b); (Pospiech 2000); (Malgieri and Onorato 2022); (Faletič, 2020); (Bitzenbauer and Zenger 2022)) and the general public. To be able to evaluate and refine the developed teaching and outreach proposals, it is important to have instruments available which can be used to assess students' conceptual understanding within a broad spectrum of approaches, but applicable to the framework of the individual approach. Since there is no such instrument available (Krijtenburg-Lewerissa et al. 2017), two initiatives for developing such an instrument were started within the QTEdu pilot projects.

The first initiative is the project *Community-based development of the Quantum Concept Inventory (QCI)*. The aim of this project is to develop a modular quantum concept inventory, the QCI, based on community input which allows for the assessment of students' understanding of quantum physics' key concepts in different contexts. The QCI will be useful for evaluating the numerous teaching concepts and for comparing their impact on student learning of quantum physics despite their different emphasis. Hence, the QCI will serve as an important resource for

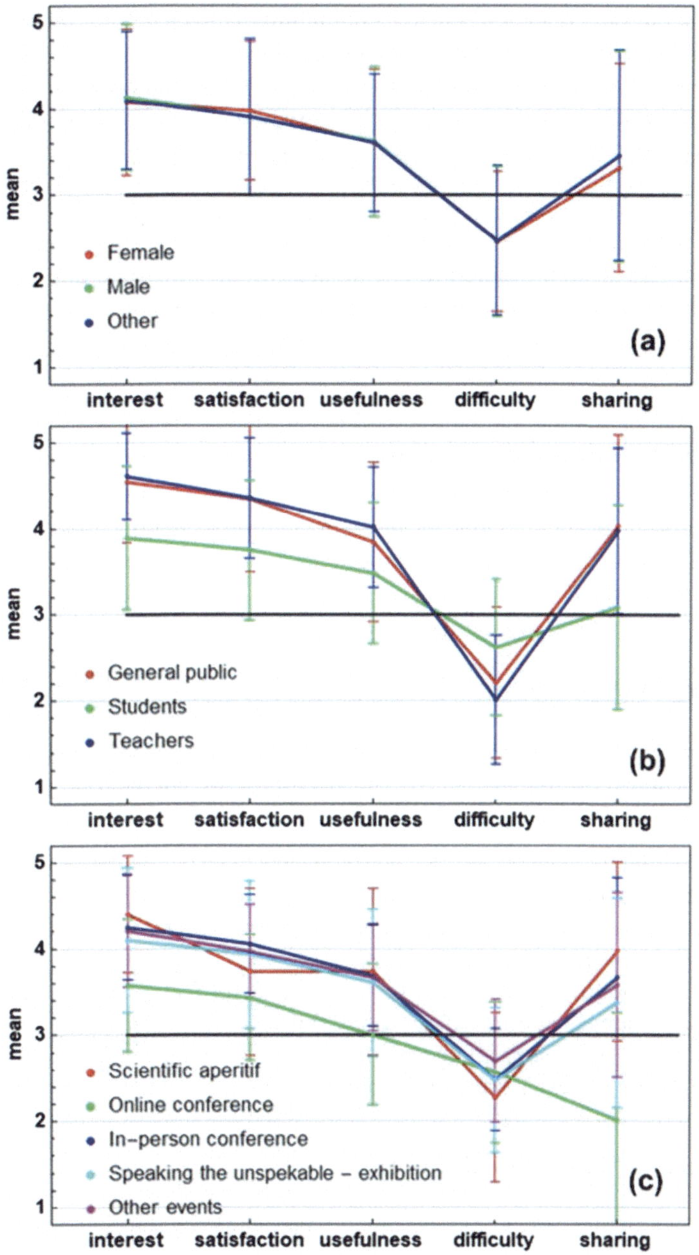

Fig. 2 Results of the satisfaction questionnaire administered at the end of IQWs outreach events. The panels show the mean score value according to the different items: interest in the activity, satisfaction, perceived usefulness, difficulty, and level of sharing the experience with family and friends. Panel **a** compares the results by gender, panel **b** by category of participants, and finally panel **c** by type of event

the quantum physics education research community with regards to robust empirical research into teaching and learning of quantum physics.

The second initiative is the project *Development of quantum concepts* via *different two-state approaches* (DQC-2stap). The ultimate goal of the project DQC-2stap is to give recommendations for secondary school teachers or educators at university on how two-state approaches and appropriate teaching material can be optimised in the sense that students best can get a grasp of the quantum concepts. To this end, steps have been taken to develop a questionnaire that is intended to be an assessment module that can be administered to any student of any two-state approach in any setting with only minor modifications to concretize particular questions to the context of the approach.

3.1 The Quantum Concept Inventory

The QCI project (Quantum Concept Inventory) aims to create a concept inventory that is based on the communities' perspectives and that is applicable for different contexts and key topics.

3.1.1 Project Objectives

To create a concept inventory that is based on the communities' perspectives and that is applicable for different contexts and key topics, we pursue four main research objectives in this project:

1. Identification of quantum physics' key concepts to be taught to secondary school students from a community perspective,
2. Collection of existing test instruments to create a platform of existing test items on the key concepts of quantum physics,
3. Creation of additional test items in various task formats (e.g., multiple choice, likert scale items, open-ended questions, concept cartoons), and
4. Evaluation of test items, both qualitatively (e.g., think-aloud interviews) and quantitatively in a later stage of the project.

3.1.2 Current Status of the Project

So far, the project members have collected test instruments published in the literature to assess students' understanding of quantum physics topics and have categorised the existing items in pillars with respect to the content aspects covered in an online community event (research objective 2). This categorization helps to identify content domains for which there is:

- a sufficient set of empirically tested items that allow for the assessment of learners' conceptual understanding at secondary school level;
- a lack of empirically evaluated test items to assess students' understanding.

From our categorization, it became apparent that:

- Most of the items published in the literature cover the quantum formalism (Cataloglu and Robinett 2002), (Goldhaber et al. 2009), topics associated with wave-particle dualism (Wuttiprom et al. 2009), (McKagan et al. 2010), matter waves (di Uccio et al. 2019) or time evolution (Singh 2001).
- Most of the items published in the literature have been developed for the use with university students and, hence, are predominantly not suitable for the use in the secondary school setting (Bitzenbauer 2021a, b).
- There are no items available in the literature yet that are suitable for evaluating secondary school students' understanding of quantum concepts which are seen as particularly relevant with respect to understanding modern applications of quantum technologies (for example superposition, entanglement, or measurement).

To close the above-mentioned gap of empirically validated test items suitable for the secondary school setting, it is important to understand which content domains are considered relevant for teaching and learning in this educational context today—and more importantly—in the future. Hence, in the sense of research objective 1, the authors decided to conduct a Delphi study to identify the key concepts that are foreseen to be relevant in future quantum physics curricula in secondary schools across Europe, and that consequently should be covered in the QCI. This procedure ensures content validity of our instrument in an early stage of its development.

3.1.3 The Future Quantum Physics Curriculum at Secondary Schools: A Delphi Study

A Delphi study is currently being conducted to identify key concepts for secondary school teaching of QP that represent that community-based view. In the first round of this Delphi study, physicists, physics teachers, and physics education researchers were asked why secondary school students should learn quantum physics and what topics should be addressed in classroom teaching. In the second round, a questionnaire was distributed, in which the QP community was asked to select which of the topics that emerged from round 1 (including various specifications considering QTEdu's Competence Framework (European Commission, 2021)) should be part of the secondary school curriculum (see Fig. 3).

Lastly, a third round is planned: The most frequently selected topics will be ranked by a group of selected experts. Additionally, focus group discussions will help to derive consensus among the experts and will, finally, lead to an adopted version of the European Competence Framework for Quantum Technologies with respect to the secondary school level.

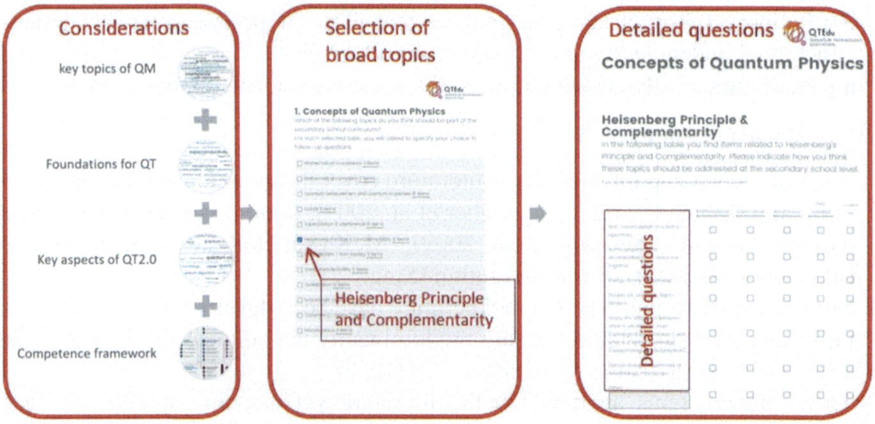

Fig. 3 The experts' answers in round 1 questionnaire were categorised in three pillars "key topics of QM", "foundations for QT" and "key aspects of QT 2.0" and were, together with input from the European Competence Framework for Quantum Technologies, used to develop the questionnaire of Delphi round 2. In this round, the experts were asked to rate whether the topics provided should be part of future secondary school curricula on which level (not needed, awareness, conceptual, mathematical)

3.1.4 Outlook

Based on the results of the Delphi study, the QP topics that should be addressed by the QCI will be determined. Subsequently, new test items will be developed (research objective 3) before the QCI will be evaluated empirically, both using qualitative and quantitative methods (research objective 4). An overview of the QCI development is provided in Fig. 4.

3.2 Development of Quantum Concepts *via* Different Two-State Approaches

A two-state approach to quantum physics seems to be a suitable and flexible means for teaching and learning quantum physics. In two-state systems systems not only are the basic quantum concepts clearly visible but also discoverable by students themselves while a comprehensible mathematical formalism can be used. Such an approach can be taught in different contexts, e.g. with polarisation, spin, a double-well system or with still other systems with two basis states, including several instances of quantum technology.

However, it could be that different approaches have different strengths and weaknesses when it comes to teaching the most important quantum concepts such as quantum state, superposition, uncertainty, entanglement and measuring process. In order to find hints in this direction in the project *Development of quantum concepts*

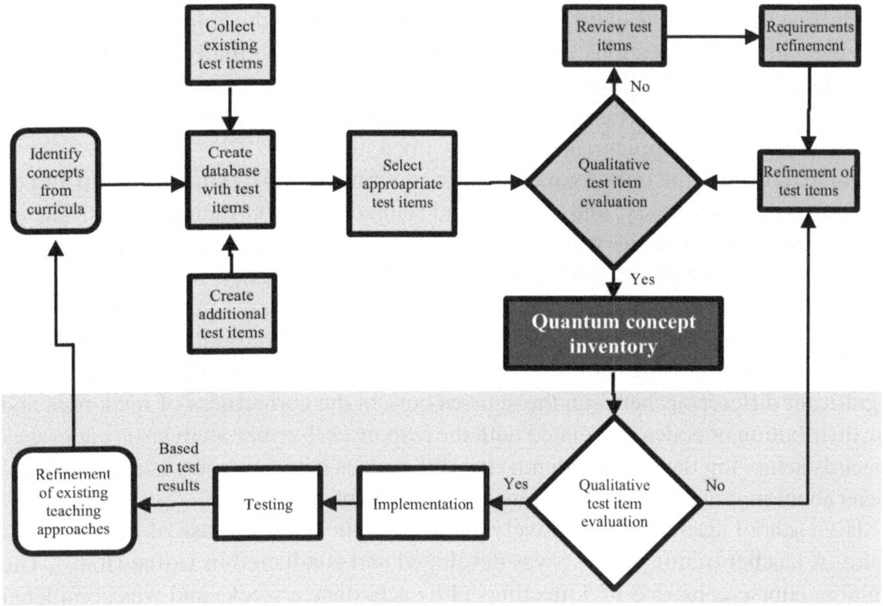

Fig. 4 Overview of the development of the QCI

via *different two-state approaches* (DQC-2stap) (Deveopment of quantum concepts) a questionnaire is being developed that sheds light onto this question. This process is accompanied by corresponding teacher education activities.

The questionnaire is designed to be useful within different approaches and settings. It will also help identify which of the two-state contexts is most suitable for teaching which quantum physics concept at the secondary school level. Especially specific learning difficulties of students should be identified.

The gained insights into students' learning difficulties in two-state approaches will be used to develop or improve secondary school or outreach teaching proposals and material based on the two-state approach within a design based research process. This research goal needs more time and will profit from cooperation with other projects in the QTEdu initiative.

3.2.1 Development of Questionnaire and Preliminary Results

In order to limit the complexity of the questionnaire development, we initially restricted ourselves to the concept of measurement in quantum physics. This concept was also chosen because the measurement process lies at the core of the peculiarity of quantum physics. So the research question is: "Which understanding and learning difficulties do students show with respect to the measuring process?".

The first version of the questionnaire was developed with the help of existing materials and our own suggestions which have been graded individually by each member of the team, discussed and then approved or rejected by a consensus. The items were mostly composed of two tiers: closed single choice items and an explanation of the choice. These were supplemented by a few open items.

In the next step, the questionnaire was tried with a group of students, who agreed to participate in the study, and adapted and refined based on their answers. Then a pilot with teacher students was done. The resulting final version was administered to students of existing two-state-approach courses in Hungary, Slovenia and Italy. The results were analysed with a coding scheme developed iteratively from the data. The codes were selected to represent students' ideas about the measurement process in quantum mechanics. Preliminary results seem to indicate statistically significant differences between the courses both in the correctness of responses and the distribution of codes associated with the responses. Further analysis is under way towards achieving the goal of identifying differences between courses and students' ideas about measurement as they emerge in different courses.

High school teachers were actively involved in the testing phase of the questionnaire. A teacher training course was developed and conducted in Udine (Italy). The training course consisted of 5 meetings (4 h each, once a week) and was completed in March 2022 preparing teachers for the implementation of a field study into teaching/learning quantum physics via a polarisation approach (one of the two-state approaches included in our pilot project). The teachers attending the course implemented the two-state course in their classes, collected data on the questionnaire and contributed in the coding discussion. More teacher training courses are intended to be developed in the project, benefitting from the results of the research conducted with questionnaires.

3.2.2 Outlook

Using the questionnaire, we will conduct field studies in order to compare learning outcomes from students who are introduced to measurement in quantum physics via different two-state approaches. We will use the outcomes of the questionnaire for (a) a refinement of the different teaching sequences, and (b) for answering the question about the suitability of different two-state-approaches to introduce students to measurement in quantum mechanics. Should the instrument prove valid, reliable and easily implementable, it could be part or an inspiration for the measurement module of the QCI.

4 Discussion and Conclusion

The pilot projects launched under the *European Quantum Flagship Educational Coordination and Support Action* aim to address all aspects of making the general public aware of quantum physics and technologies and prepare the ground for a future quantum workforce. This consists of identifying core concepts to teach, developing methods to successfully communicate the concepts to the recipient, and evaluating their understanding of said concepts.

The QCI project with its Delphi study provided basic concepts that experts in the field deem fundamental. These concepts need to be efficiently communicated to the general public in a timely manner to achieve quantum literacy for responsible decision-making about quantum technologies.

Outreach, as non-formal education, can reach large numbers of people and develop in its recipients the minds of aware and capable citizens of the future who understand that scientific ideas can evolve and popular beliefs can change. For this reason, it is essential that non-formal education is engaging and rigorous, independently of the recipients' age and the educational context.

The QUTE4E pilot has designed an original, research-based approach, (*Physics Outreach Research*) and theoretically framed their guidelines in the *culturo-scientific story-telling* (Hogg et al. 2010), which resulted in a toolbox of innovative educational experiences (Seskir et al. 2022), (Baumgarten et al. 2022), (Quantum Moves 2), (Quarks Interactive), (Quantum Composer), (University of St. Andrews), (ScienceAtHome), (Quarks Interactive), (QPlayLearn), which aim to complement the limited experimental and mathematical literacy typical of outreach participants.

The IQW project was a massive initiative by more than 130 content creators and thousands of visitors aiming to educate the general public about the quantum world and quantum technologies. Many types of outreach activities were prepared. The *Speaking the unspeakable* exhibition endeavoured to present the concept of superposition to the general public. In the spirit of physics outreach research, it collected surveys from participants, the results of which show that the topics are equally scored by both genders, that people with more background knowledge consider the topics less difficult and that in-person activities were scored better than on-line ones.

To assess how well participants learn in the various formal and informal programs, a standardised instrument to measure quantum concept knowledge is needed. The QCI project approaches this with ambitious goals. A Delphi study is currently underway to determine the topics to be evaluated. The next steps will develop a modular instrument applicable to all pre-university quantum physics programs. The DQC2-stap project starts with only two-state approaches and only the topic of measurement, but adds teacher education based on its findings. If successful, it could provide a proven stepping stone for the QCI.

The four presentations of the symposium thus cohere very well in the overarching goal of the QTEdu of educating the public about quantum technologies enough for them to consider a career in quantum engineering. Moreover, while significant steps forward have been realised with the QTEdu efforts to qualify and foster non-formal education on quantum technology, we believe that a long way is still to go and the presented pilots may be catalysts to enhance physics education research and accelerate physics outreach research.

Acknowledgements We would like to thank all members of all pilots, listed in web pages of the respective projects (Quantum Technologies), (Quantum weeks), (Quantum Concept Inventory), (Deveopment of quantum concepts) for the dedicated work that they put into each pilot. MB would especially like to thank Italo Testa for his help in the creation and analysis of the questionnaire.

References

Baumgarten R, Garcìa-Perez G, Rossi M, Sokolov B, Foti C, Yago Malo J, Maniscalco S, Chiofalo M (2022) Quantum jungle. https://qplaylearn.com/quantum-jungle. Accessed 18 July 2022
Bitzenbauer P (2021a) Development of a test instrument to investigate secondary school students' declarative knowledge of quantum optics. Eur J Sci Math Educ 9(3):57–79
Bitzenbauer P (2021b) Effect of an introductory quantum physics course using experiments with heralded photons on preuniversity students' conceptions about quantum physics. Phys Rev Phys Educ Res 17:20103
Bitzenbauer P, Zenger T (2022) Online Course on Quantum Physics: Reading Primary Sources with Secondary School Students. The Physics Teacher 60:572
Bondani M et al (2022a) Introducing quantum technologies at secondary school level: Challenges and potential impact of an online extracurricular course. Physics 4:1150–1167. https://doi.org/10.3390/physics4040075
Bondani M et al (2022b) Introducing quantum technologies at secondary school level: challenges and potential impact of an online extracurricular course. Physics 4(4):1150–1167
Cataloglu E, Robinett RW (2002) Testing the development of student conceptual and visualization understanding in quantum mechanics through the undergraduate career. Am J Phys 70:238–251
Coenen C, Grunwald A (2017) Responsible research and innovation (RRI) in quantum technology. Ethics Inf Technol 19:277–294. https://doi.org/10.1007/s10676-017-9432-6
Consolidating the Course of Quantum Technologies. https://qt.eu/about-quantum-flagship/newsroom/consolidating-the-course-of-quantum-technologies/. Accessed 14 July 2024
Deveopment of quantum concepts in different two-states approaches. https://qtedu.eu/project/development-quantum-concepts-different-two-state-approaches. Accessed 14 July 2024
di Uccio S, Colantonio A, Galano S, Marzoli I, Trani F, Testa I (2019) Design and validation of a two-tier questionnaire on basic aspects in quantum mechanics. Phys Rev Phys Educ Res 15:10137
European Commission, Directorate-General for Communications Networks, Content and Technology, Müller R and Greinert F (2021) Competence framework for quantum technologies: methodology and version history. Publications Office of the European Union, Luxembourg https://data.europa.eu/doi/10.2759/347451
Europe Day. https://europeday.europa.eu/index_en. Accessed 14 July 2024
Faletič S (2020) A double well on-line simulation and activities for active learning of introductory quantum mechanics. Eur J Phys 41:45706. https://doi.org/10.1088/1361-6404/ab90db
Feynman RP, Leighton RB, Sands M (1965) The Feynman lectures on physics, vol 3, Quantum mechanics. Addison-Wesley, Reading
Gardner H (2009) Five minds for the future. Harvard Business School, Cambridge
Goldhaber S, Pollock SJ, Dubson M, Beale P, Perkins KK (2009) Transforming upper-division quantum mechanics: Learning goals and assessment. Phys Educ Res Conf 2009:145–148

Goorney S, Foti C, Santi L, Sherson J, Yago Malo J, Chiofalo ML (2022) Culturo-scientific storytelling. Educa Sci 12(7):474. https://doi.org/10.3390/educsci12070474
Hill RM et al (2019) A tool for functional brain imaging with lifespan compliance. Nat Commun 10:4785. https://doi.org/10.1038/s41467-019-12486-x
Hogg T, Fattal DA, Chen K-Y, Guha S (2010) Economic applications of quantum information processing. In: Basu B, Chakravarty SR, Chakrabarti BK, Gangopadhyay K (eds) Econophysics and economics of games, social choices and quantitative techniques. Springer, Milano, pp 32–43
IBM quantum computing. https://quantum-computing.ibm.com/. Accessed 14 July 2024
Italian quantum weeks. https://www.youtube.com/c/ItalianQuantumWeeks. Accessed 14 July 2024
Kaur M, Venegas-Gomez A (2022) Defining the quantum workforce landscape: A review of global quantum education initiatives. Opt Eng 61(8):81806. https://doi.org/10.1117/1.OE.61.8.081806
Kop M (2021) Establishing a legal-ethical framework for quantum technology. Social Science Research Network, Rochester. https://papers.ssrn.com/sol3/papers.cfm?abstract_id=3814422
Krijtenburg-Lewerissa K, Pol HJ, Brinkman A, van Joolingen WR (2017) Insights into teaching quantum mechanics in secondary and lower undergraduate education. Phys Rev Phys Educ Res 13:10109
Kuhn D (2010) What is scientific thinking and how does it develop? In: Goswami U (ed) The wiley-blackwell handbook of childhood cognitive development. Wiley-Blackwell, Oxford
Liu H, Low GH, Steiger DS et al (2022) Prospects of quantum computing for molecular sciences. Mater Theory 6:11. https://doi.org/10.1186/s41313-021-00039-z
Malgieri M, Onorato P (2022) Recent progress on the sum over paths approach in quantum mechanics education. Physics 4(4):1012–1025
McKagan SB, Perkins KK, Wieman CE (2010) Design and validation of the quantum mechanics conceptual survey. Phys Rev Phys Educ Res 6(2):20121
Meyer DA (1999) Quantum strategies. Phys Rev Lett 82:1052
Michelini M, Pospiech G, Faletič S, Stefanel A (2021) GIREP Community on teaching/learning quantum physics in secondary school. J Phys.: Conf Ser 1929:012044. https://iopscience.iop.org/article/10.1088/1742-6596/1929/1/012044
Michelini M, Stefanel A (2021) A path to build basic Quantum Mechanics ideas in the context of light polarization and learning outcomes of secondary students. J Phys.: Conf Ser 1929:12052
Quarks interactive. Quantum odyssey. https://www.quarksinteractive.com/quantum-odyssey-presentation-at-ieee/. Accessed 18 July 2022
Quarks Interactive. https://quarksinteractive.com. Accessed 14 July 2024
Quantum Composer. https://quatomic.com. Accessed 14 July 2024
Quantum Flagship. https://qt.eu/. Accessed 14 July 2024
Quantum Moves 2. Science at Home. https://www.scienceathome.org/games/quantum-moves-2/. Accessed 18 July 2022
Müller R, Wiesner H (2002) Teaching quantum mechanics on an introductory level. Am J Phys 70:200
Pospiech G (2000) Uncertainty and complementarity: the heart of quantum physics. Phys Educ 35(6):393
QPlayLearn. https://qplaylearn.com/. Accessed 14 July 2024
Quantum Concept Inventory. https://qtedu.eu/project/community-based-development-quantum-concept-inventory. Accessed 14 July 2024
Quantum Technologies Education for Everyone. https://qtedu.eu/project/quantum-technologies-education-everyone. Accessed 14 July 2024
Quantum weeks. https://www.quantumweeks.it. Accessed 14 July 2024
University of St. Andrews, The quantum mechanics visualisation project, QuVis. https://www.st-andrews.ac.uk/physics/quvis/. Accessed 18 July 2022
Rosa H (2013) Beschleunigung und Entfremdung—Entwurf einer Kritischen Theorie Spätmoderner Zeitlichkeit. Suhrkamp Verlag, Berlin
ScienceAtHome. https://scienceathome.org. Accessed 14 July 2024

Seskir ZC et al (2022) Quantum games and interactive tools for quantum technologies outreach and education. Opt Eng 61(8):81809. https://doi.org/10.1117/1.OE.61.8.081809
Singh C (2001) Student understanding of quantum mechanics. Am J Phys 69:885–895
Stadermann HKE, van den Berg E, Goedhart MJ (2019) Analysis of secondary school quantum physics curricula of 15 different countries: Different perspectives on a challenging topic. Phys Rev Phys Educ Res 15(1):10130
Tseitlin M, Galili I (2005) Physics teaching in the search for its self. From physics as a discipline to physics as a discipline-culture. Sci Educ 14:235–261
Weissman EY, Merzel A, Katz N, Galili I (2022) Phenomena and principles: Presenting quantum physics in a high school curriculum. Physics 4(4):1299–1317
World quantum day. https://worldquantumday.org/. Accessed 14 July 2024
Wuttiprom S, Sharma MD, Johnston ID, Chitaree R, Soankwan C (2009) Development and use of a conceptual survey in introductory quantum physics. Int J Sci Educ 31(5):631–654

Addressing Complexity in Primary School Through Basic Physics and Metaphorical Narrative: Learning Circuits

Leonardo Colletti, Soufiane Krik, Paolo Lugli, and Federico Corni

Abstract Is it possible to introduce some aspects of the physical science approach as early as the first years of school? Which ones and with what results? Here we describe the first outcomes of the initial stage of a two-year project we have been developing for elementary school. It is conceived as a gentle introduction to complex and systemic thought, and it aims to foster in children the development of powerful basic concepts through analogical skills which we believe are an essential part of the scientific approach. The project—designed so that it can also be carried out independently by teachers—exploits the use of the small set of primary metaphors already available in children's language in order to apply them to the description of the characteristics of various circuits, from marble circuits to water and air circuits, and, finally, to electric circuits. Children's response has been investigated through a single case study with a quantitative methodology. A preliminary look at the results shows a lively and effective engagement of the children and a wide range of ability on their part to grasp and develop analogies on the topic.

1 Introduction

Physics—the paradigmatic representative of the scientific method—has proved particularly effective in devising recurring and versatile models of explanation that are powerful in bringing out the aspects that truly play a major role in a system.

In this research, we ask ourselves whether and how it is possible to teach the rudiments of physics approach—rather than its achievements—already from the early years of education. Is it possible to lead children in identifying the variables that matter and in recognizing similarities between seemingly different situations? For

L. Colletti (✉) · F. Corni
Faculty of Education, Free University of Bozen-Bolzano, Viale Ratisbona, 16, 39042 Bressanone, Italy
e-mail: lcolletti@unibz.it

S. Krik · P. Lugli
Faculty of Engineering, Free University of Bozen-Bolzano, via Buozzi, 1, 39100 Bolzano, Italy

S. Faletič and J. Pavlin (eds.), *Teaching and Learning Physics Effectively in Challenging Times*, Challenges in Physics Education, https://doi.org/10.1007/978-3-031-72541-8_16

children to grow and gradually approach this way of engaging effectively and synthetically with reality, it is necessary to offer them pathways to increasingly refined levels of abstraction.

The *AT-NE-ST* three-year project (the abbreviation stays for *Discovering complexity: Advanced technologies for an education in storytelling and systemic thinking*) at the Free University of Bozen-Bolzan aims at bringing K-8 students (i.e., kindergarten, primary and middle school pupils) closer to the theme of complexity. Presently, we are in the second year of the project development and this paper represents a first partial recapitulation of what has been realized so far.

"Complexity" is a varied and partially ambiguous area. What do we therefore mean by this term? In the first phase of the project—outlined here and addressed to primary school pupils up to third grade—complexity is intended as a multiplicity of phenomena that can be reduced to a core unity, in particular through the use of analogy and the development of synthetic concepts. To this aim, we designed and experimented an activity based upon observation and hands-on practice, in order to lead schoolchildren to identify similarities between circuits of different natures, that is, not only electrical circuits but also mechanical and hydraulic ones.

Concerning the teaching of electricity in the kindergarten and primary school, several studies have addressed children's ability to appropriate the idea of electric current (Osborne 1981; Asoko 1996), to become aware of connections between light bulb and battery and of polarity (Osborne 1983; Osborne et al. 1991, Shepardson and Moje 1994; Glauert 2009; Métioui and Trudel 2020), to recognize what has to do with electricity and describe its operation and properties (Osborne et al. 1991; Solomonidou and Kakana 2000), to understand the relationship between battery and circuit (Shipstone 1984; Kada and Ravanis 2016). Moreover, some of us researched and discussed the preparation of prospective primary teachers on electric circuits, in particular for what concerns the use of basic conceptual metaphors (Corni et al. 2022).

Nevertheless, as explained above, here we wish to approach the subject of circuit from a broader point of view, that is, both with regard to electricity, but also to other types of circulating agents. We aim to emphasize the general structure that circuits of different physical nature have in common, for instance the fact that an acquisition and distribution of energy takes place along the path, whether it is transported by a mechanical, hydraulic or electric vector.

Therefore, the objective of the didactic intervention we have designed and tested, and which we present here, is not primarily aimed at gaining specific skills in electricity but at using electricity, along with a mechanical and a hydraulic system, to refine an interpretative model—that of circuit—which is much more general than its use in the electrical context.

After summarizing the philosophical assumptions and the pedagogical framework behind our study in the second section, in the third one we explain the methods and materials designed and used; in the fourth section, we share a partial preview of the results obtained thus far and the related discussion; finally, we present some conclusions.

2 Theoretical Framework and Goals

2.1 On Abstraction and Image-Schemata

Children who offer themselves to a teacher's listening are not empty vessels to be filled, but are already laden with image-schemata developed from the very beginning of their physical experience with the world (Johnson 1987; Varela et al. 1991; Hampe 2005). These schemata are a source of elementary conceptual metaphors that enable the acquisition of the first rudiments of a cultural organization of physical reality. Categories such as SUBSTANCE, CONTENT/CONTAINER, AGENT/PATIENT, spatial organizers such as PATH, VERTICALITY and CENTRE/PERIPHERY, or conceptual organizers such as POLARITY (hot–cold, high-low, good-bad, etc.) are samples of elementary abstractions that children continuously project onto the world around them, testing and refining them more and more.

The existence and use of these categories allows for an important reflection on the possibility of engaging in experimental scientific discourse with children. The narrative approach typical of primary education is not necessarily a poor one in terms of scientific possibilities (Fuchs 2015). In fact, even if one cannot rely on the unambiguous relationships provided by formal mathematics, every-day language does possess a logical structure. In particular, it includes a whole series of universal metaphorical references, stemming precisely from the above-mentioned image-schemata. Hence, the latter are by their very nature basic tools of synthesis. It is precisely because these image-schemata exist, functioning as a kind of ubiquitous invariant of cognition, that it is possible to work out analogies and, with them, to set up abstract models for physical systems (Fuchs 2011).

Within this set of image-schemata one can also identify that of CIRCUIT. This is truly ubiquitous in everyday experience. In life as in school, there are many occasions when children are confronted with situations that are described, explicitly or implicitly, through a more or less developed concept of circuit. In this way, such experiences are endowed with an identity and a relationship with the cognitive universe that the child carries within him or her and that expands every day. Elementary scientific experiences should leverage these basic image-schemata to lead children to grasp parts of natural reality and, at the same time, enrich them with further attributes. The activity we have designed and tested, and which we would like to become a pathway available to elementary school teachers, aims to facilitate this cognitive transition.

2.2 Analogies at Different Levels

Recognizing and utilizing analogies is a skill that significantly enhances cognitive effectiveness (Hesse 1963; Gentner and Jeziorski 1989; Halford 1992; Hofstadter and Sander 2013; Newton and Newton 1995; Richland et al. 2006; Vendetti et al. 2015;

Gray and Holyoak 2021; Manrique et al. 2022). In particular, this may concern scientific learning (Weller 1970; Gentner 1983; Glynn and Takahashi 1998; Duit 1991) or a strategy to communicate the nature of science (Colletti 2018) or, specifically, in the teaching of the principles of electricity (Black and Solomon 1987; Dupin and Joshua 1989; Schwedes and Dudeck 1996; Hesti 2021) and also in teacher training (Treagust et al. 1998). Regarding physics education, the concept of analogy is intriguing not just in terms of the skills students must cultivate, but also as a tool teachers can hone and employ to enhance the appeal of their instruction. This may involve drawing parallels between physics principles and themes or scenarios commonly found in other disciplines (Colletti 2021, 2022).

The search for analogies is a key cognitive process that fuels scientific abstraction. In establishing an analogy between objects, an abstraction is used, and a model structure is generated.

Vendetti et al. (2015) claim that the analogical ability is "critical for success in education", though it rarely develops spontaneously (Gick and Holyoak 1983) and is difficult to acquire and use for children until the late adolescence. This research suggests that primary school children need structured guidance when looking at areas between which to make relational comparisons.

The analogies we wanted to elicit in these children through our intervention are of two different levels.

The first level, which we expected to be easier to induce, is based on observing the different states in which a given system can be found when a given process occurs. In the system that we first introduce the children to, a certain amount of matter (marbles, water, air) is released and allowed to flow in a tube, causing a wheel to rotate and then it is put back by hand on top of the tube to restart the process again. The system is marked by input quantities—the inclination of the tube, the sort and quantity of matter—and an output quantity, that is, the rotation speed of the wheel. Input quantities may be varied, and the output quantity may change accordingly. The first analogy we expected the children would be capable of making is simply that which occurs among all the possible pairs of input values and the corresponding output values, e.g. (high tube's slope, high wheel's speed) is analogous to (medium tube's slope, medium wheel's speed). While these pairs may vary quantitatively, they are essentially different manifestations of the same singular physical occurrence (which, with older students, would be described through a single formula): In every case, an item descends through the tube, prompting the wheel to spin, and the process can be reinitiated once the initial conditions are reset, regardless of the specific numerical values assigned to the involved quantities (and the substance utilized).

This kind of analogy can be seen simply as a similitude or, as Hofstadter and Sander (2013) jokingly christened it, a *banalogy*: one sees the same structure in the same arrangement of matter. It might be trite, but like all true analogies, it possesses nevertheless the characteristic—typical of analogies (Podolefsky and Finkelstein 2006)—of triggering the formation of inferences. Indeed, the observer may speculate what output (i.e., speed value) will correspond to an input that has not yet been experienced, and such a question may prefigure itself as a true conjecture with exploratory value of the actual value of the 'matter going down-rotational speed' model that has been developed in the observer's mind.

This level of analogy is expected also from the observation of what happens in the second system presented to the attention of the pupils. This system comprises a simple electrical circuit in which one or more batteries connected in series allow electricity to activate one or more light bulbs also connected in series. Similar to the previous case, adjustments can be made to the specific input values (such as the number of batteries and lights) and observe various specific results (high or low light brightness), realizing, however, that all these states of the system are analogous to each other, i.e., they are all manifestations of a general relationship that links input and output in a certain way.

The experiments conducted using the tube always involve the flow of something into it, causing the wheel to rotate. Each trial varies slightly from the others, such as having a steeper slope in the tube or a change in the nature of the object flowing into it. This variation is akin to a collection of toy cars, where there are cars of different sizes, makes, colors, etc. As mentioned earlier, even in this context, the concept of an abstract model summarizing the common characteristics shared by all elements of the collection naturally emerges in the child's cognitive understanding. Similarly, experiments involving a small electrical circuit can be adjusted in length, with varying numbers of light bulbs or batteries, yet they represent the same set of instances. The key image-schemata, including CIRCUIT, FLUID SUBSTANCE, LEVEL DIFFERENCE, HIGH or LOW INTENSITY, INTERACTION, AGENT and PATIENT help the child enormously to grasp the analogy between the different states of the system, both in the case of the tube and that of electricity, separately.

More challenging appears the second-level analogy. Here we aim to elicit an analogy between systems that present themselves radically different to sense perception: on one side, we have the previously established category of 'matter sliding into a tube' and on the other side we have the as well previously established category of 'electric circuit'. In short, concerning the tube, one has a circuit where materials and observable characteristics are completely different from what happens in the case of the electric circuit, but the two systems host processes that share some important aspects with each other. As these aspects emerge, all differences are reduced to second-order details. If this analogical recognition between the two types of system takes place in the children, then their image-schemata (CIRCUIT, FLUID SUBSTANCE, LEVEL DIFFERENCE, HIGH or LOW INTENSITY, INTERACTION, AGENT and PATIENT) acquire a higher level of abstraction and, at the same time, a greater cognitive potential, ready to be taken to a further level of abstraction and generalization through further experiences (e.g., the water cycle, or the production of goods and the market, etc.).

2.3 Choice of the Age of the Pupils and Investigated Cognitive Aspects

According to Egan (1990), the period between 5 and 10 years of age marks the onset of an individual's inclination to inquire about the reasons behind events they experience, to explore potential actions and scenarios, and to discern which interpretations of a phenomenon are feasible and which are not. Egan labels this stage as one of *romantic understanding*, wherein children begin to comprehend the constraints of reality while also engaging in abstract thinking as they develop objective methods of representing the world, primarily through literacy, which involves the acquisition of letters, numbers, and their interrelationships.

This developmental phase is often evident in the inclination of children of this age to collect items, which essentially involves documenting various iterations of a theme and forming an abstract concept that encapsulates these variations, not necessarily reliant on direct perceptual experience.

2.4 Overall and Specific Goals of the Intervention

Summarizing what has been said, the objectives of the structured pathway presented to the children are as follows.

There is an objective on a general level, which is to introduce children to the scientific process of discovery, through the identification of certain characteristics of a system that can vary causing other characteristics to vary. In this sense, the experiment with the tube focused on the CIRCUIT aspect lends itself very well because, in its simplicity, it is paradigmatic of an enormous class of phenomena in which something possesses energy and transfers this energy to something else (one easily recognises here the principles of causality and locality at work).

The second objective is to research some peculiarities of children's cognitive abilities, in particular for what concerns the drawing and using of analogies. The research questions we ask are:

(RQ 1) To what extent are children able to create the first-level analogy between different states of the same system, abstracting from them to produce a general model? Are they able to refer to that model to formulate conjectures on possible outcomes of future experiments and answer questions?

(RQ 2) To what extent are children able to create a second-level analogy between different systems (different materials, i.e., between the mechanical/hydraulic circuit and the electric one), to draw parallels between the parts of one system and those of the other system (e.g., role of slope and role of battery; role of the wheel and role of the lightbulb; role of the tube and role of the electric wires)?

Furthermore, another (pragmatic) goal is to produce and test a prototype suitcase containing teaching materials that can be used by primary school teachers autonomously with their classes.

3 Materials and Methods

3.1 Materials

We used a combination of commercially available materials and elements we designed and produced by our own. In the first stage, we started by proposing the observation of phenomena concerning a transparent tube in which glass marbles can slide. The tube has a variable inclination. The inclination can be measured by the height point from which the marbles are allowed to slide. Along their run, the marbles hit the blades of a wheel whose rotation speed is measured by an IR speed sensor located behind the wheel and displayed both as a number (RPM) shown on an LCD and as a light vertical bar consisting of 4 LEDs for younger children unable to read numbers, with each color corresponding to a preset range of speeds: 0–50 yellow, 50–100 green, etc. The data acquisition is performed using an Arduino UNO board powered by a commercial powerbank. The same kind of observation was also designed in such a way to use, instead of marbles, water or air (the latter by placing an inflated balloon at one extreme of the tube). The spinning wheels were designed and 3D printed specifically for this project.

For the second stage, we built some wooden modules that pupils had to hold in their hands while forming a circle, so that both the number of light bulbs (in series) and the number of batteries (in series) could be easily changed. Each module consists in a wooden stick with ends wrapped in aluminum foil and electric wire connecting these ends with the element taped in the middle of the stick. In particular, we had battery sticks (that is, a wooden stick with a 4.5 V rectangular battery taped in the middle), light sticks (i.e., with a small light bulb taped in the middle) and pairs of conductors handles (that is, two shorter wooden sticks connected to each other by a 40/50 cm long electric wire and with ends wrapped in aluminum foil).

In the third stage, pupils had at their disposal electric wire, pens with conductive ink, conductive tape, LEDs of various colours and coin cells. There were also sheets available in which some sections of the circuits were already screen printed using conductive ink, i.e., silver ink.

We also developed two questionnaires with a series of open-ended questions and small tasks. Questions and exercises were of different types (completing sentences, linking sentences, multiple-choice or true/false type questions, as well as drawing and narrating in one's own words), so as to diversify the analysis and reduce the possibility of some form of bias (e.g., children used to a certain type of exercise but not another).

3.2 The Three Stages of the Intervention

We conducted our activity with five classes of children aged 7 or 8 (a very few of age 9), for a total number of 99 pupils. We visited the first two classes (third graders) two times, as a first activity's trial and assessment, thus proposing only the first two stages of the project, then we went on with the other three classes to whom we offered the whole three-stage project. Each visit lasted approximately one hour and a half. The activity took place during the COVID-19 pandemic period and therefore both children and adults wore face masks.

In all stages pupils were stimulated to get involved and invited to ask and answer questions.

In order not to interfere too much with the teachers' annual planning, we left it up to them to decide whether to have the two questionnaires filled in at home or in class after the visit. In any case, it was recommended that the children carry out their task autonomously and independently from one another.

In particular, the three stages were organized and conducted as described below:

First meeting: Introductory talk citing and explaining words such as observation, measurement, movement, science, making examples and urging the children to refer to their own experience; description of the apparatus (the tubes); experiment with marbles varying the tube's slope and then varying the quantity of marbles and writing the results in the form of a table at the chalkboard; before each attempt pupils were invited to express a conjecture about what was going to happen; experiment with the second tube and water, varying the slope and then varying the quantity; experiment with air (not for all classes), by varying the pressure of air at one end of the tube via an inflated balloon. Questionnaire 1 asked the pupil to draw the apparatus by naming its parts and explaining what was happening in it (with marbles or water), and to answer a question.

Second meeting (a week later): A comprehensive review of the topics discussed in the preceding meeting; a discussion on electric current and its practical applications based on the pupils' experiences; a straightforward elucidation of the modules; in small groups (4–5 groups per class, each group with 4–5 children) circuit construction: each group was asked to form a circle by connecting the various elements held in their hands, firstly only with one battery and one light plus the conductors, then one battery and two lights (in series), two batteries (in series) and one light, two batteries (in series) and two lights (in series); the children were asked to observe any differences in bulb brightness obtained in the different configurations and take note of them. A qualitative table of these outcomes was written on the chalkboard. In the dialogue about what happens in the circuit when the number of batteries and lights varies, the instructor explicitly offered a reminder of what happens in the tube when the inclination varies. The instructor explicitly drew parallels between what happens in the two different systems and between the quantities involved. Children were then asked to observe what happened when the circuit was not closed or was closed with certain materials such as pieces of wood or plastic pens or metal scissors. Questionnaire 2 asked the pupil to draw an electric circuit by naming its parts and explaining what was happening in it; then, several other questions and exercises were proposed on both the new experience (electric circuit) and the experiment with the tube.

Third meeting (after another week): Review of the observations made in both previous meetings; description of the materials to be used (wires, coin cells, LEDs, conductive ink and tape) and explanation of the tasks to be carried out in small groups (two or three pupils), i.e., the construction of four circuits like the macroscopic ones kinesthetically experienced the stage before—this time with LEDs and coin cells glued on paper—i.e., changing the number of lights and batteries (both in series). Finally, a conversation about what was seen in the whole path, particularly going back to what was asked in the second questionnaire and offering a confirming view on the analogy between the systems seen and a parallel between the roles played between their parts and between the processes that happen in them.

4 Results and Discussion

Starting from the observation of the children during the meetings, we can definitely say that the activity met with strong interest, great attention and participation. The dialogue was incessant, the responses to requests numerous, as were the questions, observations, comments and even proposals of new trials. From the qualitative observation of the children's behavior, from the way they quickly learnt to fill in an ordered table, we can certainly conclude that the overall objective of setting up an experiment that would introduce the children to observation and measurement, leading them to correlate two measures, was certainly successful.

The activity was also much appreciated by the teachers, who collaborated very actively and competently, especially during the more demanding stages (second and third meetings). Equally positive is the impression that the class teacher can independently and competently guide this type of investigation. Hence, the tilting tube with the wheel can certainly be an excellent tool to introduce children in a simple and direct way to the systematic study of a regularity of nature.

Limiting ourselves to what we observed in the dialogues, we can argue that children were able to generate inferences from observing the offered attempts, both in the case of the tube (with the different materials) and in the case of the circuit. Correct or incorrect, the emergence of such inferences is, as mentioned, symptomatic of the production of an analogical process.

Let us now turn to the first outcomes we obtained from the analysis of the questionnaires. Leaving aside the first two classes that served as a sort of calibration of the project, we collected 48 Questionnaires 1 and 48 Questionnaires 2 of the 60 of each kind we distributed in the remaining classes.

As explained in section two, we analyzed the results by classifying them according to whether they refer to the individual systems, i.e., the tube (with the flow of marbles and water) or the electric circuit (first-level analogy), or whether they refer to a parallelism between these two systems (second-level analogy).

As for the first questionnaire, after the request for a drawing (see samples in Fig. 1) and explanation of the experiment, we focused our attention on the question which aimed to find out whether schoolchildren were able to extrapolate, on the basis of the experiment conducted in the classroom (i.e., the configurations with input/output as actually implemented and observed), the outcome of a limit case (i.e., an imaginary

Fig. 1 **a** The experiment with the tube and marbles in the drawing of a third-grade pupil. **b** The making of a large electrical circuit in the drawing of a third-grade pupil.

instance that in principle belongs to the same collection of configurations). A good majority of the children answered correctly and demonstrated a sound understanding of what was happening in the single system (the tube circuit or the electric circuit) as the variables took different values. In particular, they understood what the role of the tube inclination was, and how this affected the speed of the wheel. They similarly understood what was happening in the electric circuit, understanding for example that more batteries would have a greater effect, i.e., a greater intensity of light.

These results show that, to a large extent (from 70 to 90% of the participants), pupils appear to be able to imagine the system in general, i.e., even without specifying one of its directly experienced configurations (e.g., specifying how many marbles and what slope; or how many batteries and how many lights). This implies that the children have employed an abstract category to which all possible states of the system ideally converge, seen as essentially analogous to each other. In their imagination, the situations described in the sentences to be analyzed do not only correspond to the four slopes actually experienced in the specific cases of four or eight marbles, but are some of the many possible particular cases in which an entirely general model can be found embodied. Hence, we may say that, for each one of the two systems, these children have grasped an analogy that runs among all the possible configurations that the system could be found embodied in, generalizing the process that occurs in each of them.

As might be expected, the analogy between the two systems—a second-level analogy as we called it—was more difficult for the children to grasp. Only about one out of four children provided sufficiently convincing evidence that they had assimilated one system to the other, thus elaborating the two into a single category, that of a generalized circuit, with all image-schemata involved. This, it should be noted, despite the fact that this category is supported by image-schemata such as CIRCUIT, LEVEL DIFFERENCE, SUBSTANCE, INTERACTION, AGENT and PATIENT, HIGH or LOW INTENSITY that children already possess among their cognitive tools. In our opinion, this confirms our view that a teachers' task is that of helping the children in the process of deepening and refining these cognitive tools through the supply of ever new experiences.

5 Conclusions

We developed a path centered on children's awareness of structuring analogies between observations concerning different systems. The activity was intended as a first opportunity to approach the typical way of proceeding of physics, namely that of extrapolating from a complex multiplicity of phenomena a common structure capable of conferring a unitary and effective meaning. For this, we have referred to a very simple but also very fundamental experiment because it outlines the basic idea of a physical process (there is something that moves and sets something else in motion) and its prerequisites (the idea of a cause-effect principle and that of contiguity between cause and effect).

The interest of this experimentation lies first and foremost in the fact that it aims not so much at the acquisition of specific content concerning nature (things like the force of gravity that makes water fall, or the concept of conductor or insulator) but rather at the development of the very broad and general concept of CIRCUIT. This was done through the fostering of an analogical way of thinking, based on (1) extrapolating the truly salient characteristics of a system and (2) through these characteristics comparing different systems, identifying similar things and dissimilar things. This is made possible by bringing out those image-schemata that are already present in the language (and thus in the cognitive system) of children. The development and stimulation of this type of thinking is fundamental in order to actively and hierarchically set up the assimilation of the contents of science that will be offered to them in the subsequent school years.

We have experimented this path with five elementary school classes, obtaining encouraging results. Children, at an age when they are beginning to explore the various possibilities in which a given system—any kind of system—can present itself, if appropriately guided, know how to classify simple experimental data. On the basis of them, they start to elaborate an abstract model of the system in which physically important quantities are selected and correlated, and from which they are able to formulate inferences about further experimental outcomes. Less frequent—about a quarter of the third-grade children—is the ability to draw analogies between different systems, i.e., systems made up of different materials and in which different forces of nature operate but nevertheless characterized by a similar architecture and functional correspondences between different elements.

Acknowledgements We acknowledge the precious help given by the classroom teachers Corradina Bonaccio, Nives Fois and Irene D'Agostino. We also acknowledge the staff at the BITZ unibz fablab for their availability, support, and assistance in developing the mechanical setup used in this work.

Ethics Approval The study was conducted in accordance with the Declaration of Helsinki and approved by the Ethics Committee of the Free University of Bozen-Bolzano (protocol code ATNEST_Cod_2022_05 dd 21.03.2022) for studies involving humans.

References

Asoko H (1996) Developing scientific concepts in the primary classroom: Teaching about electric circuits. In: Welford G, Osborne J, Scott P (eds) Research in science education in Europe. Falmer Press, London, Current issues and themes, pp 36–49

Black D, Solomon J (1987) Can pupils use taught analogies for electric current? School Sci Rev 69:249–254

Colletti L (2018) Teaching the nature of physics through art: A new art of teaching. Phys Educ 53:015004. https://doi.org/10.1088/1361-6552/aa8a8c

Colletti L (2021) The Italian secondary-school graduation exam: Connecting physics with the humanities. Phys Educ 56:015016. https://doi.org/10.1088/1361-6552/abc8fc

Colletti L (2022) Making physics teaching inclusive through a humanistic approach. The Phys Educator 4:2250016. https://doi.org/10.1142/S2661339522500160

Corni F, Michelini M, Colletti L (2022) Training prospective primary and kindergarten teachers on electric circuits using conceptual metaphors. Educ Sci 12(7):457. https://doi.org/10.3390/educsci12070457

Duit R (1991) On the role of analogies and metaphors in learning science. Sci Educ 75:649–672

Dupin JJ, Joshua S (1989) Analogies and 'modeling analogies' in teaching: some examples in basic electricity. Sci Ed 73:207–224

Egan K (1990) Romantic understanding. The development of rationality and imagination, ages 8–15. Chapman & Hall, New York

Fuchs HU (2015) From stories to scientific models and back: Narrative framing in modern macroscopic physics. Int J Sci Educ 37:934–957

Fuchs HU (2011) Force Dynamic Gestalt, Metaphor, and Scientific Thought. In: Corni F, Mariani C, Laurenti E (eds) Innovazione nella didattica delle scienze nella scuola primaria: al crocevia fra discipline scientifiche e umanistiche, Modena e Reggio Emilia, Novembre 2010. Artestampa, Modena

Gentner D (1983) Structure-mapping: A theoretical framework for analogy. Cognitive Sci 7:155–170

Gentner D, Jeziorski M (1989) Historical shifts in the use of analogy in science. In: Gholson BB, Shadish Jr WR, Neimeyer RA, Houts AC (eds) Psychology of science: contributions to metascience. Cambridge University Press, Cambridge, pp 296–325

Gick ML, Holyoak KJ (1983) Schema induction and analogical transfer. Cognitive Psychol 15:1–38

Glauert EB (2009) How young children understand electric circuits: Prediction, explanation and exploration. Int J Sci Educ 31:1025–1047

Glynn SM, Takahashi T (1998) Learning from analogy-enhanced science text. J Res Sci Teach 35:1129–1149

Good R (ed) (1993) Special issue: The role of analogy in science and science teaching. J Res Sci Teach 30:1211–1350

Gray ME, Holyoak KJ (2021) Teaching by analogy: From theory to practice. Mind Brain Educ 15:250–263

Halford GS (1992) Analogical reasoning and conceptual complexity in cognitive development. Hum Dev 35:193–217

Hampe B (ed) (2005) From perception to meaning. Image schemas in cognitive linguistics. Mouton de Gruyter, Berlin

Hesse MB (1963) Models and analogies in science. Sheed & Ward, London

Hesti R (2021) Analogy educational comics to overcome students' misconception on simple electricity circuit material. J Phys Conf Ser 1957:012036
Hofstadter D, Sander E (2013) Surfaces and essences: Analogy as the fuel and fire of thinking. Basic Books, New York
Johnson M (1987) The Body in the Mind. University of Chicago Press, Chicago
Kada V, Ravanis K (2016) Creating a simple electric circuit with children between the ages of five and six. S Afr J Educ 36:1–9
Manrique HM, Marín A, Zeidler H (2022) When is it advisable to introduce analogies to foster creativity in education? Cult Educ 34:403–423
Métioui A, Trudel L (2020) Conceptions about electrical circuits of English and French pupils from Nova Scotia in Canada. Edureview 8:73–82
Newton D, Newton L (1995) Using analogy to help young children understand. Educ Stud 21:379–393
Osborne RJ (1981) Children's ideas about electric current. New Zeal Sci Teach 29:12–19
Osborne RJ (1983) Modifying children's ideas about electric current. Res Sci Technol Educ 1:73–82
Osborne RJ, Black P, Smith M, Meadows J (1991) Primary SPACE project research report: Electricity. Liverpool University Press, Liverpool
Podolefsky NS, Finkelstein ND (2006) Use of analogy in learning physics: The role of representation. Phys Rev Spec Top-Ph 2:020101
Richland LE, Morrison RG, Holyoak KJ (2006) Children's development of analogical reasoning: Insights from scene analogy problems. J Exp Child Psychol 94:249–273
Schwedes H, Dudeck W (1996) Teaching electricity by help of a water analogy (how to cope with the need for conceptual change). In: Welford G, Osborne J, Scott P (eds) Research in science education in Europe. Routledge, London, pp 50–63
Shepardson DP, Moje EB (1994) The nature of fourth graders' understandings of electric circuits. Sci Educ 78:489–514
Shipstone D (1984) A study of children's understanding of electricity in simple circuits. Eur J Sci Educ 6:185–198
Solomonidou C, Kakana D (2000) Preschool children's conceptions about the electric current and the functioning of electric appliances. Eur Early Child Educ 8:95–111
Treagust DF, Harrison AG, Venville GJ (1998) Teaching science effectively with analogies: An approach for preservice and inservice teacher education. J Sci Technol Educ 9:85–101
Varela FJ, Thompson E, Rosch E (1991) The embodied mind. MIT Press, Cambridge
Vendetti MS, Matlen BJ, Richland LE, Bunge SA (2015) Analogical reasoning in the classroom: Insights from cognitive science. Mind Brain Educ 9:100–106
Weller C (1970) The role of analogy in teaching science. J Res Sci Teach 7:113–119

Analysing the Storyline Approach's Competence-Developing Potential for Climate Change in Science Education

Lorenzo Miani and Olivia Levrini

Abstract Climate Change is the greatest threat to mankind and the biosphere. It challenges Science Education from many points of view, including searching for effective ways to enable citizens to make decisions, embrace complexity, deal with uncertainty and envision possible futures. Two approaches are used to assess the influence of Anthropogenic Climate Change in extreme weather events: the so-called risk-based approach and the storyline one. In this study, we compared these two approaches from a Science Education point of view to understand if they have the same potential to develop sustainability competences in secondary school students.

1 Introduction

Climate Change (CC) is considered the greatest challenge to humans and the biosphere (IPCC 2021), and in the need to prevent planet Earth's ultimate destruction (DeBoer 1991) scientific literacy (Yacoubian 2018) can be seen as a strong tool to promote sustainability and sustainable development for citizens, policymakers and stakeholders (UNESCO 2017). Between CC researchers, IPCC (Intergovernmental Panel on Climate Change) has become in the last thirty years the most important organ for delivering information and forming about the science of CC and its impacts. In the last reports, however, there has been a change in the approach they used to present data for researchers and policymakers (IPCC 2014, 2021). This change regards the approach used to treat uncertainties and develop future scenarios and also originates from a debate that formed in the detection and attribution [D&A] of extreme weather events in recent years (Shepherd 2016, 2019; Shepherd et al. 2018; Stott et al. 2004, 2013, 2016; Trenberth et al. 2015). Two main approaches have been used in D&A of extreme weather events: the first is the risk-based approach, the best-known and

L. Miani (✉) · O. Levrini
Department of Physics and Astronomy "A. Righi", University of Bologna, Bologna, Italy
e-mail: lorenzo.miani2@unibo.it

O. Levrini
e-mail: olivia.levrini2@unibo.it

S. Faletič and J. Pavlin (eds.), *Teaching and Learning Physics Effectively in Challenging Times*, Challenges in Physics Education, https://doi.org/10.1007/978-3-031-72541-8_17

most widely used approach, and the second is the storyline approach. The introduction of this second approach has stimulated vivid debates not only among climate scientists, in that the two approaches are based on different assumptions and cohere with different conceptions about the nature of science and the role of probability in climate-related events. Because of its socio-political relevance, the debate has been also involving policymakers and philosophers of science (Lloyd and Oreskes 2018). This paper aims to enter the debate from an educational perspective and unpack its implications for science teaching and communication. Specifically, the peculiarities and differences between these two approaches are analysed in terms of their potential to develop competences for sustainability, as recommended by the European Union (Bianchi et al. 2022).

2 Climate Change Education

Science and Science Education can be characterized both as a cognitive-epistemic and socio-institutional system (Erduran and Dagher 2014). Scientific literacy is considered a major goal of Science Education to promote decision-making at a personal and a societal level (Yacoubian 2018) to face modern-day challenges. Science Education has the potential to play a primary role in enabling the development of a critical vision that is at the same time capable of adapting to and dealing with the challenges of tomorrow (Levrini et al. 2019). The search for effective ways to situate and address CC within school curricula is a more and more urgent goal since schools can play a crucial role in promoting change at a plurality of levels: knowledge, skills, and attitudes. A comprehensively different stance toward CC is necessary to make future citizens capable of holding positions of power with the possibility to change for the better world they will live in (Grayson 2020). More generally, Climate Change Education (CCE) "can be a strategic and meaningful entry point for promoting the principles and practice of sustainable development through education" (Mochizuki and Bryan 2015, p. 5). Over the past decades, since the first CC studies appeared in the scientific landscape, the number of CC courses in schools has increased exponentially. According to a literature review conducted for "Climate Change Education" research articles (Monroe et al. 2019), the number of articles went "from 12 articles published between 1990 and 1999, and 433 from 2000 to 2009, to 1489 from 2010 to 2015" (p. 1). The authors found 49 papers addressing effective CCE strategies in their review. Most of these courses focused on personally relevant and meaningful information (Bofferding and Kloser 2015; Wibeck 2014) and used active and engaging methods (Bybee et al. 2006; Jacobson et al. 2015; Walsh and McGowan 2017).

In the field of research on the societal implications of CC and the role of human influence on it, the main references for environmental policies are the reports of the IPCC. IPCC studies CC and its nature and produces reports aimed at both the scientific and the societal levels. If in the Second Assessment Report (SAR) the role of human influence in influencing climate (Anthropogenic Climate Change, ACC) was defined as "discernible" (IPCC 1995), in the Sixth Assessment Report (AR6) is stated that "it is unequivocal that human influence has warmed the atmosphere, ocean and land" (IPCC 2021, p. 4). The role that humankind has played in the changes over the past 150 years has now been defined as certain (IPCC 2021).

The role of the IPCC over the years has changed from trying to provide scientific proof of the anthropogenic origin of CC to providing governments with information and recommendations to prevent and mitigate the effects of CC (Hulme 2023). To do so, IPCC scientists need to use the language of uncertainty, in that full predictability of future climate events is not possible due to the intrinsic nature of climate. This necessity has been used by many CC contrarians to underscore the actual state of knowledge, as for them "action cannot be taken until knowledge is complete" (O'Reilly 2023, p. 160). The language of uncertainty has then been refined, formalised and reviewed to align all the Working Groups through different guidance for IPCC authors, to delineate and present a common approach to uncertainty (See, for example, Reisinger and colleagues (Reisinger et al. 2020). Different approaches and modalities of presenting and using uncertainty terms have been used in the reports (O'Reilly 2022). In the last reports in particular there has been a shift in the way uncertainties were considered and scenarios were created, leading to the construction of Shared Socioeconomic Pathways (SSP) where qualitative and quantitative elements are used to create alternative future worlds (Nikoleris et al. 2017). SSP narratives contain both qualitative descriptions of large world region trends and quantification of key variables to integrate different models for assessment and vulnerability (O'Neill et al. 2017) and "are intended as descriptions of plausible future conditions that can provide the basis for a range of scenarios to emerge" (Nikoleris et al. 2017, p. 309). These narratives are created to allow policymakers and stakeholders to make decisions by pondering different levels of sustainable behaviours for the future. This goal, therefore, dovetails with those that researchers in the field of Science Education have been trying to achieve for years, namely, to develop classroom practices and useful approaches to best educate toward CC and develop and promote sustainability competences.

Very recently, an official European sustainability competences framework, the GreenComp, has been developed by the Joint Research Center (JRC) of the European Union (Bianchi et al. 2022). GreenComp takes from a wide Science Education literature that addresses the role of science in shaping society (Bianchi et al. 2022; Irzik and Nola 2011; Lotz-Sisitka et al. 2015; Sjöström et al. 2017).

In GreenComp sustainability is defined as the intent of prioritising all life forms in everyday activities to not exceed planetary boundaries. Rockström and colleagues selected nine planetary boundaries whose limits "define the safe operating space for humanity" (2009, p. 472). These boundaries are biodiversity loss, atmospheric aerosol loading, chemical pollution, CC, ocean acidification, stratospheric ozone depletion, nitrogen and phosphorus cycle, global freshwater use and change in land use. Three of these limits have already been passed, i.e. biodiversity loss, nitrogen cycle, and especially CC (Rockström et al. 2009). There is therefore a need to promote sustainable behaviours in society to prevent the further passing of these boundaries. GreenComp identifies twelve sustainability competences "to feed into education programs to help learners develop knowledge, skills and attitudes that promote ways of think, plan and act with empathy, responsibility, and care for our planet and public health" (Bianchi et al. 2022, p. 2). GreenComp is supposed to be an important reference for designing activities and modules for CC Education, to make learners and educators competent concerning sustainability issues. The four areas and twelve competences described in the GreenComp are: Embodying sustainability values (Valuing sustainability; Supporting fairness; Promoting nature); Embracing complexity in sustainability (Systems thinking; Critical thinking; Problem framing); Envisioning sustainable futures (Futures literacy; Adaptability; Exploratory thinking); Acting for sustainability (Political agency; Collective agency; Individual initiative) (Bianchi et al. 2002). The competences present in GreenComp have been addressed already in Science Education literature (Duit et al. 1997; Erduran and Dagher 2014; Laherto et al. 2023a; Rasa and Laherto, 2022). Laherto et al. (2023b) showed this in detail. As with any political report of recommendations, its role is to provide some orientation. Decisions about educational approaches, contents, methods, and strategies to develop these competences are up to the implementer stakeholders (researchers, teachers, principals, national and local policy-makers, and so on). Being CC one of the nine planetary boundaries to preserve to achieve a sustainable way of living, in this study GreenComp is used as a reference framework to confront different methods usually used to describe, present, narrate and understand CC and compare their educational potential for sustainability training and developing of decision-making.

The approaches discussed in this study are the risk-based approach and the storyline, between which there has been a debate about the foundations of the scientific method and the role of research for society in detection and attribution (D&A) regarding extreme weather events with a big potential for educational purposes. The approaches are used for determining the impact of a specific factor over a multitude of possibilities, the first based on the concept of risk (Allen 2003; Stott et al. 2004, 2013, 2016) and the second that relies on storylines construction (Hoerling et al. 2013; Shepherd 2016, 2019; Shepherd et al. 2018; Trenberth 2011; Trenberth et al. 2015). These two approaches have been pitted against each other by various experts in the field because they are apparently divergent in the method used to provide answers in the case of detention and attribution of extreme weather events (Allen 2003; Lloyd and Oreskes, 2018; Trenberth 2011). Before getting to the heart of the debate and its educational potential, we present the two approaches and their characteristics in more detail.

3 Risk-Based Approach and Storyline Approach

The risk-based approach estimates the change in the likelihood of a certain effect arising from the presence of a certain factor. "Event attribution assessments seek to quantify to what extent anthropogenic or natural influences have altered the probability or magnitude of a particular type of event having occurred" (Stott et al. 2016,

p. 25). Three steps are fundamental for obtaining relevant data using this approach: to define the event, to construct a factual likelihood distribution for that event using a model, and then to construct a counterfactual distribution with which to compare the factual distribution (Shepherd 2016, pp. 30–31). CC effects are then calculated in the altered frequency of a class of events of a particular magnitude or the altered magnitude of a class of events having the same frequency in the factual world. This way of approaching the problem of event attribution by estimating the fraction of attributable risk (FAR) of climate extremes was first introduced in 2003 and used to address the European heat wave of 2003 (Allen 2003; Stott et al. 2004). Over the years it has developed as the main method for D&A of extreme weather events. In the FAR model, there is an estimated probability of a certain class of events related to the world as we see it today (p1), also called the factual world and a probability of the same class of events in a counterfactual world (p0) in which the estimated role of humankind in contributing to CC (ACC) is removed. The ratio of p1 to p0 (Risk Ratio, RR = p1/p0) says at what level that class of events is more or less likely interpreted as the result of human forcing. Because FAR is defined as 1–p0/p1, the threshold above which an actual human contribution to CC can be evidenced is 0.5 (Stott et al. 2016).

If this process can be reliable at a global scale when discussing thermodynamic aspects like temperature changes or sea surface temperatures (SST), the situation changes when dynamic aspects involve the atmospheric or oceanic circulation. At a regional scale, "the thermodynamic aspects are strongly modulated by the dynamic aspects so the latter must be taken into account" (Shepherd 2016, p. 32) and these are connected to high levels of uncertainties.

Storylines are defined as "physically self-consistent unfolding of past events, or plausible future events or pathways" (Shepherd 2016, p. 2). The storyline approach has been introduced and used for the first time by Trenberth and colleagues (2015), and one of the first examples can be seen in assessing the Texas drought of 2011 by Hoerling and colleagues (2013). Storylines "can raise risk awareness by framing risk in an event-oriented rather than a probabilistic manner" (Shepherd et al. 2018, p. 566). They provide a mechanism to improve decision-making by analyzing vulnerability points and partitioning probability factors, combining regional aspects and factors related to CC risk (Shepherd et al. 2018). A storyline-based approach identifies the causal chain of factors leading to a certain event, assessing their roles and their links. To make a so-called storyline one takes the data of a certain extreme event, and makes a kind of "autopsy" of the event (Lloyd and Oreskes 2018). The storyline approach makes it possible to qualitatively attribute the causes of a certain event when it is caused by not well-known dynamic factors and not only thermodynamic ones (Shepherd 2016). Through a Bayesian method, and thus the identification of causal links between events, it is possible to construct a causal network capable of highlighting areas where actions can be taken (Shepherd 2019). According to Trenberth and colleagues (2015) in those situations where a risk-based approach brings high-uncertainty results, it can be useful to physically investigate the unfolding of the events, trying to determine how each factor may have been affected by known thermodynamic aspects of CC (Shepherd 2016). Tracing the succession of certain events by determining their causes can help in predicting and preventing that event from occurring again with the same consequences and results. The storyline-based approach does not address the potential change in the probability of the dynamic situation that resulted in the event but can tie attribution directly to the observed event, rather than just being probabilistic in a frequentist sense (Shepherd 2016; Trenberth et al. 2015). "The direct confrontation with data as an essential component of the attribution is a very attractive feature of this approach, as is its emphasis on a physically based causal narrative" (Shepherd 2016, p. 34). Examples of this approach can be found in literature (Hazeleger et al. 2015; Hoerling et al. 2013; Trenberth et al. 2015).

4 Methodology

The aim of this paper is to understand the potential of developing sustainability competences of the two approaches from a Science Education perspective. To do so, we reviewed the risk-based and the storyline approach through the lenses of GreenComp, as we tried to evaluate if and how each competence was present, addressed and promoted.

To realize the study we focused on the papers that used the approaches directly (Allen 2003; Hoerling et al. 2013; Shepherd 2016; Shepherd et al. 2018; Shepherd 2019; Stott et al. 2004; Stott et al. 2013; Stott et al. 2016; Trenberth 2011; Trenberth et al. 2015) and on the papers that discussed the philosophical, societal and ethical implications of the approaches (Allen 2011; Curry 2011; Lloyd and Oreskes 2018; Mann et al. 2017). Each competence has been transformed into a specific guiding question (Table 1). The guiding questions have been used to navigate both the original papers and those that alimented the debate between the approaches. To conduct this focused reading, we looked for some particular keywords inside the papers mentioned above. For some of the competences, the keyword is the name of the competence itself, while for others we choose particular words that we consider representative of the competence. The keywords used for the analysis are also reported in Table 1 and are taken from the definitions given in the GreenComp. Both singular and plural versions were considered for the analysis. Keywords formed by two words have been searched in combination using the logic function 'AND' and not 'OR'. The analysis doesn't aim to provide a count of the number of times a certain keyword appears in a paper but rather to discuss if and how certain keywords are considered in the papers when discussing the approach.

Each approach has been analysed through the lens of these twelve questions and the related keywords. The results of this analysis are presented in the next section. These results are also used to present in detail the different points of view of the debate between the approaches.

Table 1 Competences, guiding questions and keywords based on GreenComp (Bianchi et al. 2022) competence definitions

Areas	Competences	Guiding questions	Keywords
Embodying sustainability values	Valuing sustainability	What kind of values are promoted? How do they align with sustainability values and ethics?	Values, sustainability, ethics
	Supporting fairness	Does the approach support equity and justice for all?	Fairness, equity, justice
	Promoting nature	Is the approach capable of promoting respect towards other species and nature itself to restore and regenerate healthy ecosystems?	Respect, nature
Embracing complexity in sustainability	Systems thinking	Does the approach help in tackling sustainability problems from all sides and focusing on how the elements interact within and between systems?	System thinking
	Critical thinking	Is the approach useful for promoting reflections about information, assumptions and the influence of social and personal backgrounds?	Critical thinking
	Problem framing	Does the approach help to formulate challenges as sustainability problems based on social, economic and territorial aspects to anticipate and prevent present and future challenges?	Problem framing
Envisioning sustainable futures	Futures literacy	Is the approach capable of generating alternative sustainable futures through imagination and the development of scenarios?	Futures literacy
	Adaptability	Does the approach help in managing transitions and challenges and in making future decisions in the face of uncertainty, risk and ambiguity?	Adaptability
	Exploratory thinking	Does the approach facilitate the adoption of a relational way of thinking by using creativity and experimentation?	Exploratory thinking, creativity
Acting for sustainability	Political agency	Does the approach help in identifying political responsibility and accountability?	Political agency, decision making
	Collective action	Is the approach useful for promoting action for change in collaboration with others?	Collaboration
	Individual initiative	Does the approach contribute to identifying one own potential to actively improve one's role in the community?	Initiative, attitude

5 The Study

We now discuss the results we obtained through the focused reading in terms of how the two approaches relate to each sustainability competence.

Regarding sustainability values and supporting fairness, the papers do not address directly the values promoted by each approach, but Lloyd and Oreskes (2018) showed how the two approaches differentiate themselves in terms of type I (i.e. false positive) or type II (i.e. false negative) errors, and therefore taking decisions of preferentially guarding against one or another is a question of values rather than a scientific one. Committing a type II error, i.e., a false negative, allows researchers to avoid exposure in the case of wrong predictions and to avoid unnecessary expenses and costs to prevent and prepare for extreme events that may never happen (Lloyd and Oreskes 2018). But at the same time, this choice puts at risk those exposed to extreme events who have not done adequate prevention and mitigation. According to (Lloyd and Oreskes 2018; Shepherd 2019; Trenberth et al. 2015), drawing parallels with the research that is carried out in the medical field for testing medicines and vaccines, it would be more appropriate from an ethical point of view to change the perspective and put the mantra of "do no harm" first. In the case of medical research, scientists need to verify both that the drug is effective, but also that it does not cause unwanted effects (Lloyd and Oreskes 2018). It then becomes necessary to understand which harm is most desired to be avoided in the case of D&A of extreme weather events. It has been shown that people are more forgiving when false alarms are made, rather than missed warning calls (Economou et al. 2016). Regarding promoting nature, no related keywords were found in the papers. Nonetheless, both approaches are aimed to improve the D&A process and to better describe extreme weather events to help society in dealing with them. The preservation of our ecosystem is a specific purpose of the detection and attribution process without distinction between approaches, as understanding the human role in influencing the climate is the necessary point to identify those activities that foster this influence.

Regarding "Systems thinking", the possibility of constructing causal networks to establish the role of each particular event as in the storyline approach can be useful for developing connections across systems. Being the climate a complex system, facing extreme weather events using a frequentist approach based on many factors could lead to a non-clear definition of the interaction between different elements. The frequentist approach is widespread in the scientific community but doesn't have deep roots (Mann et al. 2017), leading many scientists to re-examine the appropriateness of such an approach for hypotheses testing (Nuzzo 2014). "The result of bad and misleading statements about attribution, of which there have been many, is to grossly underestimate the role of humans in climate events of note in recent times to the detriment of perceptions about Climate Change and subsequent policy debates" (Trenberth 2011, p. 929). A Bayesian approach instead assumes, starting from many considerations on CC (Trenberth et al. 2015), that "Climate Change is likely to be impacting extreme weather events" (Mann et al. 2017, p. 133) as a prior. This approach proves much more useful in cases where there are aspects of the problem detectable with a certain high probability (e.g., thermodynamic aspects) and others much more uncertain (e.g., dynamic aspects) (Mann et al. 2017) but may lead to type I errors (Lloyd and Oreskes 2018). This could be also because the fundamental use of simulations and models requested by the risk-based approach can generate a black-box effect hiding important systems relations. Through the Bayesian approach and the assumption of the specific dynamics related to a certain event (past or future), the storyline approach helps in generating connections between factors (Shepherd 2019) and understanding the role of each one of those. Working with complexity has been shown to have a great potential to provide citizenship skills to navigate society (Barelli et al. 2018; Cilliers 1998; Morin 2003).

As for critical thinking and problem framing, there is a distinction between the two approaches that regards the ways of treating uncertainties. Unlike in the risk-based approach, in the storyline, a distinction is made between different types of uncertainty related to climate phenomena and extreme weather events. As explained in detail by Shepherd (2019) and Dessai and Hulme (2004), climate phenomena are marked by three different types of uncertainty: uncertainty in future climate forcing, which is the different scenarios that try to predict different types of futures, called reflexive uncertainty (Dessai and Hulme 2004); uncertainty in climate response to future climate forcing, which is known as epistemic uncertainty; and uncertainty in the actual realization of climate in a certain time window, called aleatoric uncertainty (Shepherd 2019). In the division of the factors that led to a certain event, the storyline approach makes explicit distinctions between the aspects connected to science, those connected to nature and those connected to humans. To promote action, an approach such as the storyline approach is more suitable than a risk-based one. In the storyline approach much distinction is made between epistemic and aleatoric uncertainty, whereas in the risk-based approach, these uncertainties are often considered together (Shepherd 2016). This distinction can help in addressing the different responsibilities that each one can have in preventing or facilitating a certain event, and therefore in framing extreme weather events under a different light based on the related type of uncertainty.

The competences presented in the "envisioning sustainable futures" area are also strictly related to the different ways of treating uncertainties in decision-making. The distinction between the two approaches stays in the difference between the concepts of forecast and foresight. Forecasting is the process of making predictions using past and present data, as in the risk-based approach. Foresight explores the range of plausible futures starting from a certain point to identify preferable scenarios to reach (Cuhls 2003), as in the storyline approach. Foresight is also tied with the process of backcasting, where the task is to figure out a way to a particular plausible or preferred future, as Laherto and colleagues showed (2023b). The possibility of imagining different future scenarios can therefore help in promoting creativity and experimentation for decision-making, two main aspects of "exploratory thinking". For example, a Future-Oriented Science Education approach that has been produced in recent years (Branchetti et al. 2018; Rasa and Laherto 2022) ties together complexity and future imagination with values through the development of different possible futures.

Regarding the area of "acting for sustainability", the distinction between the two approaches can be analysed through the lens of the three spheres of change model proposed by O'Brien and Sygna (2013). This model expresses the need to work at the same time on three distinct levels (practical, political, and personal) to promote a deep transformation for achieving the necessary sustainability goals for our society. When facing a wicked problem such as CC or others demanding socio-scientific issues, there is a tendency to extremize the dichotomy between the individual and the collective dimension of problems (Barelli et al. 2022). The possibility given by the storyline approach of distinguishing between the several factors that are tied to a particular extreme event, and therefore the possibility of generating a causal network to separate the different levels of intervention (Shepherd 2019) can be a strong incentive for promoting action both at the individual and collective level. Also, stakeholders or policymakers and scientists may respond to different types of information. This aspect is tied to the possibility of the storyline approach focusing more on local events and therefore facilitating the decision-making process. The storyline approach is case-specific, referring to well-defined events that have already had local consequences and thus it succeeds in being more effective at the time when decisions need to be made to prevent the event from recurring with worse consequences. As Shepherd noted, "It may be difficult to convince people to invest in defences against a hypothetical risk, but easier to do so if an event has previously occurred so clearly could occur again, but potentially with more impact" (Shepherd 2016, p. 33). In the case of risk-based, on the other hand, there is a focus on classes of events, which can be categorized as unlikely or very likely within ranges that are too wide to allow an effective decision to be made (e.g. unlikely corresponds to a 33% chance) (Shepherd 2019).

In light of these differences, several researchers inferred that the two approaches are incompatible and that the storyline approach challenges what the scientific community assumed was right in the past, i.e., preferring type II to type I errors (Allen 2011; Curry 2011; Lloyd and Oreskes, 2018). This debate can be seen in detail in Lloyd and Oreskes (2018), in which the authors analyse the debate showing how the two approaches are compatible and complementary as they answer two different research questions: if the FAR methodology's research question is "what are the odds or probabilities of a particular class of events, given changes in climate compared to historical pre-industrial or counterfactual climate?" (Lloyd and Oreskes 2018, p. 313), the research question the storyline approach tries to answer is "given the atmospheric dynamics about the event, how did climate change alter it?" (Lloyd and Oreskes 2018, p. 313). According to the authors, the choice of one approach over another lies in the preference and the decision that needs to be made about which risk is of greater concern. This discussion is based on several aspects, namely, what is the role of scientists in research, what role scientists have in society, what kind of information needs to be passed to the public, and what values are tied to this information passage. In cases where information needs to be passed to other researchers, local stakeholders, policymakers or laypeople, it is crucial to understand what kind of information to present and for what purpose. Likewise, from an educational perspective, it is necessary to understand which approach is most effective in enabling a more adequate understanding of the phenomenon of CC and the training competences necessary for sustainable development.

6 Storyline-Approach for Climate Change in Physics Education

As stated by Lotz-Sisitka and colleagues (2015) and UNESCO (2017), there is a request for a rethinking of values and purposes to develop transformation in individuals and society. Sjöström and colleagues (2017) expressed the need for science education to directly support value-based transformative agency. As discussed above, the two approaches offer different stances and positions about the four competence areas. On the value level, the two approaches are both useful in promoting awareness toward sustainability issues, with the storyline more focused on the safety of people while the risk-based focuses more on the accuracy of statements and assumptions. Regarding complexity, the storyline looks closer to the requests coming from the Science Education literature on developing skills for reading complexity in everyday life. The complexity of climate systems is fully addressed within the storyline approach, allowing the reconstruction of the links between the different factors that led to the occurrence of a certain event, while in the risk-based approach, this possibility is limited by the extensive use of models and simulations that can prevent a clear view of the mechanisms at play unless specific advanced knowledge is available. Regarding the future, the distinction between forecast and foresight helps in creating plausible scenarios rather than making predictions, leading to a possible enhanced capacity for making decisions and creating policy plans. To promote action, an approach such as the storyline approach is more suitable than a risk-based one, in that the clear division between the different types of uncertainty present within the storyline approach, along with the reconstruction of causal networks, allows for clear identification of which aspects can be acted upon. In this way, one goes to act on aspects of systemic thinking that are not touched upon in the risk-based approach instead.

Building on these methodological and epistemological differences, we infer that the storyline approach is better suited to the development of the sustainability competences presented in GreenComp. Starting from these results, we are now working to test if and how the epistemological assumptions underlying the storyline approach can be implemented and exploited for developing teaching courses and modules. CC touches on deep underlying epistemological and ontological issues about the concept of causality, temporality and prediction. "The widespread misunderstanding of climate change may arise from an error in people's ontological assumptions" (Chen 2011, p. 31). We are used, due to our cognitive development, to think about processes as objects, therefore preventing us to accept CC as a process (Chen 2011). We hope that the reflections presented above can be seen as a useful tool to differentiate the approaches for educational purposes and used to construct teaching sequences and programmes aimed to promote systemic thinking and generate action. An approach such as the storyline allows one to focus on process characteristics, the actual causes that lead to certain consequences, and possible actions that can be taken with a view to prevention and adaptation. Bringing a similar approach into the classroom and re-reading storylines from an educational perspective could be a way to foster action and enable the development of skills needed to promote sustainable development.

References

Allen M (2003) Liability for CC. Nature 421(6926):891–892. https://doi.org/10.1038/421891a

Allen M (2011) In defense of the traditional null hypothesis: remarks on the Trenberth and Curry WIREs opinion articles: Defending the traditional null hypothesis. Wiley Interdiscip Rev Clim Change 2(6):931–934. https://doi.org/10.1002/wcc.145

Barelli E, Tasquier G, Caramschi M, Satanassi S, Fantini P, Branchetti L, Levrini O (2022) Making sense of youth futures narratives: Recognition of emerging tensions in students' imagination of the future. Front Educ 7(911052):1–17. https://doi.org/10.3389/feduc.2022.911052

Barelli E, Branchetti L, Tasquier G, Albertazzi L, Levrini O (2018) Science of complex sytems and citizenship skills: A pilot study with adult citizens. EURASIA J Math Sci Tech Ed 14(4):1533–1545. https://doi.org/10.29333/ejmste/84841

Bianchi, G., Pisiotis, U., & Cabrera Giraldez, M. (2022). GreenComp – The European sustainability competence framework. In M. Bacigalupo & Y. Punie (Eds.), European Commission, Joint Research Centre. Luxembourg: Publications Office of the European Union. https://doi.org/10.2760/13286

Bofferding L, Kloser M (2015) Middle and high school students' conceptions of Climate Change mitigation and adaptation strategies. Environ Educ Res 21(2):275–294. https://doi.org/10.1080/13504622.2014.888401

Branchetti L, Cutler M, Laherto A, Levrini O, Palmgren EK, Tasquier G, Wilson C (2018) The I SEE project: An approach to futurize STEM education. Vis Sustain 9:10–26. https://doi.org/10.13135/2384-8677/2770

Bybee RW, Taylor JA, Gardner A, Van Scotter P, Powell JC, Westbrook A, Landes N (2006) The BSCS 5E instructional model: origins and Effectiveness. A report prepared for the Office of Science Education, National Institutes of Health

Chen X (2011) Why do people misunderstand CC? Heuristics, mental models and ontological assumptions. Clim Change 108(1–2):31–46. https://doi.org/10.1007/s10584-010-0013-5

Cilliers P (1998) Complexity and postmodernism: Understanding complex systems. Routledge, London

Cuhls K (2003) From forecasting to foresight processes? new participative foresight activities in Germany. J Forecast 22(2–3):93–111. https://doi.org/10.1002/for.848

Curry J (2011) Nullifying the climate null hypothesis. Wiley Interdiscip Rev Clim Change 2(6):919–924. https://doi.org/10.1002/wcc.141

DeBoer GE (1991) A history of ideas in science education: implications for practice. Teachers College Press

De Pryck K, Hulme M (Eds) (2023) A critical assessment of the intergovernmental panel on climate change. Cambridge University Press. https://doi.org/10.1017/9781009082099

Dessai S, Hulme M (2004) Does climate adaptation policy need probabilities? Clim Policy 4(2):107–128. https://doi.org/10.3763/cpol.2004.0411

Duit R, Komorek M, Wilbers J (1997) Studies on educational reconstruction of chaos theory. Res Sci Educ 27(3):339–357. https://doi.org/10.1007/bf02461758

Economou T, Stephenson DB, Rougier JC, Neal RA, Mylne KR (2016) On the use of Bayesian decision theory for issuing natural hazard warnings. Proc Math Phys Eng Sci 472(2194). https://doi.org/10.1098/rspa.2016.0295

Erduran S, Dagher Z (2014) Reconceptualizing the nature of science for science education: Scientific knowledge, practices and other family categories. Springer, New York, NY

Grayson DJ (2020) Physics education for 21st century graduates. J Phys Conf Ser 1512(1). https://doi.org/10.1088/1742-6596/1512/1/012043

Hazeleger W, van den Hurk BJJM, Min E, van Oldenborgh GJ, Petersen AC, Stainforth DA et al (2015) Tales of future weather. Nat Clim Chang 5(2):107–113. https://doi.org/10.1038/nclimate2450

Hoerling M, Kumar A, Dole R, Nielsen-Gammon JW, Eischeid J, Perlwitz J et al (2013) Anatomy of an extreme event. J Clim 26(9):2811–2832. https://doi.org/10.1175/jcli-d-12-00270.1

IPCC (1995) CC 1995: The science of CC: Contribution of working group I to the second assessment report of the intergovernmental panel on CC. Cambridge University Press, Houghton JT

IPCC (2014) Climate change 2014: Synthesis report. In: Pachauri RK and Meyer LA (eds) Contribution of working groups I, II and III to the fifth assessment report of the intergovernmental panel on climate change. Geneva, Switzerland

IPCC (2021) Summary for policymakers climate change 2021: The physical science basis. In: Masson-Delmotte VP, Zhai A, Pirani SL et al. (ed) Contribution of working group I to the sixth assessment report of the intergovernmental panel on climate change. Cambridge University Press, Cambridge, United Kingdom and New York, NY, USA

Irzik G, Nola R (2011) A family resemblance approach to the nature of science for science education. Sci Educ 20(7–8):591–607. https://doi.org/10.1007/s11191-010-9293-4

Jacobson SK, McDuff MD, Monroe MC (2015) Conservation education and outreach techniques, 2nd edn. Oxford University Press, London, England

Laherto A, Levrini O and Erduran S (2023a) Editorial: Future-oriented science education for agency and sustainable development. Front Educ 8(1155507). https://doi.org/10.3389/feduc.2023.1155507

Laherto A, Rasa T, Miani L, Levrini O, Erduran S (2023b) Future-oriented science education building sustainability competences: An approach to the European GreenComp framework. In: Fazio X (ed) Science curriculum for the anthropocene, vol 2. Palgrave Macmillan, Cham. https://doi.org/10.1007/978-3-031-37391-6_5

Levrini O, Tasquier G, Branchetti L, Barelli E (2019) Developing future-scaffolding skills through science education. Int J Sci Educ 41(18):2647–2674. https://doi.org/10.1080/09500693.2019.1693080

Lloyd EA, Oreskes N (2018) CC attribution: When is it appropriate to accept new methods? Earth's Future 6(3):311–325. https://doi.org/10.1002/2017ef000665

Lotz-Sisitka H, Wals AEJ, Kronlid D, McGarry D (2015) Transformative, transgressive social learning: rethinking higher education pedagogy in times of systemic global dysfunction. Curr Opin Environ Sustain 16:73–80. https://doi.org/10.1016/j.cosust.2015.07.018

Mann ME, Lloyd EA, Oreskes N (2017) Assessing CC impacts on extreme weather events: the case for an alternative (Bayesian) approach. Clim Change 144(2):131–142. https://doi.org/10.1007/s10584-017-2048-3

Mochizuki Y, Bryan A (2015) CC Education in the context of education for sustainable development: Rationale and principles. J Educ Sustain Dev 9(1):4–26. https://doi.org/10.1177/0973408215569109

Monroe MC, Plate RR, Oxarart A, Bowers A, Chaves WA (2019) Identifying effective Climate Change education strategies: a systematic review of the research. Environ Educ Res 25(6):791–812. https://doi.org/10.1080/13504622.2017.1360842

Morin E, UNESCO (2003) Seven complex lessons in education for the future. UNESCO

Nikoleris A, Stripple J, Tenngart P (2017) Narrating climate futures: shared socioeconomic pathways and literary fiction. Clim Change 143(3–4):307–319. http://link.springer.com/10.1007/s10584-017-2020-2

Nuzzo R (2014) Scientific method: statistical errors. Nature 506(7487):150–152. https://doi.org/10.1038/506150a

O'Brien K, Sygna L (2013) Responding to CC: The three spheres of transformation. In: Proc transformation changing climate int conf, Oslo, 19–21 June 2013, pp 16–23

O'Neill BC, Kriegler E, Ebi KL, Kemp-Benedict E, Riahi K, Rothman DS et al (2017) The roads ahead: Narratives for shared socioeconomic pathways describing world futures in the 21st century. Global Environ Change 42:169–180. https://linkinghub.elsevier.com/retrieve/pii/S0959378015000060

O'Reilly J (2022) Uncertainty A critical assessment of the intergovernmental panel on climate change. De Prick K and Hulme M, Cambridge University Press, Chapter 17:159–168

Rasa T, Laherto A (2022) Young people's technological images of the future: Implications for science and technology education. Eur J Futures Res 10(1). https://doi.org/10.1186/s40309-022-00190-x

Reisinger A, Howden M, Vera C et al (2020) The concept of risk in the IPCC sixth assessment report: A summary of cross-working group discussions. Intergovernmental Panel on Climate Change, Geneva, Switzerland. https://www.ipcc.ch/event/guidance-note-concept-of-risk-in-the-6ar-cross-wg-discussions/ (Accessed 6 November 2024)

Rockström J, Steffen W, Noone K, Persson Å, Chapin FS, Lambin EF et al (2009) A safe operating space for humanity. Nature 461(7263):472–475. http://www.nature.com/articles/461472a

Shepherd TG (2016) A common framework for approaches to extreme event attribution. Curr Clim Change Rep 2(1):28–38. https://doi.org/10.1007/s40641-016-0033-y

Shepherd TG (2019) Storyline approach to the construction of regional CC information. Proc Math Phys Eng Sci 475(2225):20190013. https://doi.org/10.1098/rspa.2019.0013

Shepherd TG, Boyd E, Calel RA, Chapman SC, Dessai S, Dima-West IM et al (2018) Storylines: an alternative approach to representing uncertainty in physical aspects of Climate Change. Clim Change 151(3):555–571. https://doi.org/10.1007/s10584-018-2317-9

Sjöström J, Frerichs N, Zuin VG, Eilks I (2017) Use of the concept of Bildung in the international science education literature, its potential, and implications for teaching and learning. Stud Sci Educ 53(2):165–192. https://doi.org/10.1080/03057267.2017.1384649

Stott PA, Stone DA, Allen MR (2004) Human contribution to the European heatwave of 2003. Nature 432(7017):610–614. https://doi.org/10.1038/nature03089

Stott PA, Allen M, Christidis N, Dole RM, Hoerling M, Huntingford C et al (2013) Attribution of weather and climate-related events. Climate Science for Serving Society. Springer, Netherlands, Dordrecht, pp 307–337

Stott PA, Christidis N, Otto FEL, Sun Y, Vanderlinden J-P, van Oldenborgh GJ et al (2016) Attribution of extreme weather and climate-related events. Wiley Interdiscip Rev Clim Change 7(1):23–41. https://doi.org/10.1002/wcc.380

Trenberth KE (2011) Attribution of climate variations and trends to human influences and natural variability: Attribution of the human influence. Wiley Interdiscip Rev Clim Change 2(6):925–930. https://doi.org/10.1002/wcc.142

Trenberth KE, Fasullo JT, Shepherd TG (2015) Attribution of climate extreme events. Nat Clim Chang 5(8):725–730. https://doi.org/10.1038/nclimate2657

UNESCO (2017) Education for sustainable development goals: Learning objectives. UNESCO, Paris. https://unesdoc.unesco.org/ark:/48223/pf0000247444

Walsh EM, McGowan VC (2017) 'Let your data tell a story:' Climate Change experts and students navigating disciplinary argumentation in the classroom. Int J Sci Educ 39(1):20–43. https://doi.org/10.1080/09500693.2016.1264033

Wibeck V (2014) Enhancing learning, communication and public engagement about Climate Change—some lessons from recent literature. Environ Educ Res 20(3):387–411. https://doi.org/10.1080/13504622.2013.812720

Yacoubian HA (2018) Scientific literacy for democratic decision-making. Int J Sci Educ 40(3):308–327. https://www.tandfonline.com/doi/full/10.1080/09500693.2017.1420266

Teacher Education

The Interplay of Curricular Knowledge and Perceived Agency of Pre-service Physics Teachers in Vienna and Tokyo

Michael M. Hull, Haruko Uematsu, and Andrew Elby

Abstract Teacher educators and education researchers have argued for the importance of responsive teaching. To teach responsively, however, requires more than creativity. It requires also a perception of agency, that one has the freedom to deviate from the lesson plan when needed. Effective responsive teaching requires also an understanding of why the textbook and other curricular materials were designed the way they were. In this paper, we investigate the interplay between this last factor ("curricular knowledge") and perceived agency in the context of an opportunity for pre-service teachers earning a MS degree to learn about and use Open Source Tutorials. Through qualitative analysis of written student work and transcripts from interviews and in-class recordings, we document growth in curricular knowledge that, at the end of the semester, seemed to be coupled to perceived agency.

1 Introduction

Teacher educators and education researchers have argued for the importance of responsive teaching (e.g., Debarger et al. 2017; Robertson 2018; Sawyer 2004). This comes in opposition to curriculum such as Success for All, which "provide word-for-word scripts that teachers are strongly encouraged to follow" in order to "teacherproof" instruction (Sawyer 2004). Sawyer, for example, writes "...the best teachers apply immense creativity and profound content knowledge to their jobs, both in advance preparation and from moment to moment while in the classroom" (Sawyer 2004). Effective responsive teaching requires also "curricular knowledge...

M. M. Hull (✉)
Physics Department, University of Alaska Fairbanks, Fairbanks, AK 99775, USA
e-mail: mmhull2@alaska.edu

H. Uematsu
Department of Physics, Tokyo Gakugei University, Koganeishi 184-0015, Tokyo, Japan

A. Elby
Department of Teaching and Learning, Policy and Leadership Division: STME,
University of Maryland, College Park, MD 20742, USA

S. Faletič and J. Pavlin (eds.), *Teaching and Learning Physics Effectively in Challenging Times*, Challenges in Physics Education, https://doi.org/10.1007/978-3-031-72541-8_18

knowledge of the purposes of particular questions or sequences of questions within the curricula [teachers] use" (Robertson 2018). Curricular knowledge helps teachers and pre-service teachers (PSTs) implement the materials in coherent ways (Goertzen 2010) and make modifications (if needed) consistent with the underlying approach (Harlow 2010).

Teachers do not, by default, perceive automatically the curricular knowledge embedded in the curriculum they implement. Rather, curricular knowledge is something that must be learned (Goertzen 2010; Robertson 2018). Just as education research has shown that physics students learn physics better when interactive engagement methods are used, it seems the same is true for teachers learning about curricular knowledge. Organized active learning about curricular knowledge that leads to positive results can take various forms, including a class where student teachers "dissect" curricular materials and discuss what curricular knowledge they think is contained (Robertson 2018) or in-service workshops where teachers are involved in co-creating new curriculum with education researchers (Debarger et al. 2017).

In addition to creativity and knowledge, we argue that teachers must also have a perception of agency to teach responsively. We do not mean "agency" in the sense of being an extension of an organization, like a secret agent. Rather, our usage is consistent with the definition of "the capacity to initiate purposeful action that implies will, autonomy, freedom, and choice" (Ball and Cohen 1996). We define "perceived (teacher or pre-service teacher) agency" to be "a feeling of being in control over what is taught and of how it is taught." A teacher or PST with a weak sense of agency, then, feels controlled in this regard, either by the education system, colleagues at the school, or by other factors including student expectations (Lipponen and Kumpulainen 2011). We find it intuitive that there is an interplay between curricular knowledge and perceived agency, and our research aims to investigate this intersection. We hypothesize that an experience of discussing what might at first seem like a foreign teaching approach, using that approach in an actual classroom, and then reflecting on what went well and what did not go well in that classroom can tremendously boost curricular knowledge. At the same time, especially if that curriculum is intended to be used in a responsive manner, this experience can also increase a teacher's perceived agency, as the curriculum calls on them to enact their agency moment by moment. In this paper, we will discuss two such opportunities for PSTs in a MS program, one at the University of Vienna (UV), and one at Tokyo Gakugei University (TGU).

2 Study Overview

For about a decade, Uematsu has been using Open Source Tutorials (OSTs) (Elby et al. 2007) in her physics recitation course ("Physics Exercises") for PSTs at TGU. OSTs are guided worksheets which students complete in groups of four. In addition to aiming to deepen the conceptual understanding of students around basic but tricky physics topics, OSTs prioritize the maturation of student "physics epistemology" (views about the nature of physics knowledge and learning (Hofer and Pintrich 1997). It accomplishes these goals by building productively upon student ideas in the learning process, both through the worksheets themselves, and through instructor moves recommended in the instructors' guides. Although some OSTs are based upon Tutorials in Introductory Physics (TIPs) (McDermott and Shaffer 2002), the epistemological emphasis is a novel feature of OSTs. Also novel is the underlying assumption that student reasoning about a given topic is potentially fluid and context-dependent. Whereas TIPs make frequent use of the "Elicit, Confront, Resolve" (ECR) strategy to remove "misconceptions" (McDermott 1991), OSTs generally do not do this, as they recognize that an incorrect student response does not necessarily imply that the student "has" a misconception to remove in the first place. Scherr (2007), building on the work of Hammer (2000), describes OSTs as representative of the "Pieces" model of student ideas (diSessa 1993, 2009; Hammer 2000). In contrast to the "misconceptions" model, which attributes (often only tacitly!) a stable, rigid, and context-independent character to student ideas, the pieces model explicitly treats student ideas as being context-dependent collections of smaller knowledge pieces.

A particularly well-known and effective OST is the Tutorial pertaining to Newton's Third Law (N3). In this OST, students are guided to "refine their intuitions" about the situation of a truck hitting a parked car. After a brief review of N3, the students are asked in their groups what their intuition says about the situation. Many students write down that, intuitively, they would suppose that the truck exerts a larger force on the car than the car on the truck. These same students are also able to recognize that this intuition contradicts N3, which says that the forces should be equal. The OST then demonstrates to students via force probes on carts gliding on a low-friction track that N3 does indeed hold and that their intuitions led to incorrect predictions. The OST does not stop there, however. Rather than thinking of the student idea (that the truck exerts a larger force) as being a misconception that must be eradicated via cognitive conflict, the underlying assumption rather is that this student idea arises from a combination of smaller pieces, including the idea of the car reacting more. The "car reacts more" idea is, in fact, correct, and the OST capitalizes on this fact to have students use that same "car reacts more" idea to reach an (intuitive!) conclusion that does agree with N3. Specifically, since students find it intuitive that the car reacts more, they can also find it intuitive that the car accelerates more. With the aid of Newton's Second Law ($F = ma$), students then arrive at the conclusion that the forces are indeed equal. Instead of rejecting their intuitive idea about the car reacting more, students are instead guided to "refine their intuitions" into a form consistent with the physics. Scherr's explanation of the OST is that "Students … recognize … that their intuitive idea that "bigger objects exert larger forces" (an application of "more effort gives more result") is incorrect regarding Newton's third law. However, a different refinement of "more effort gives more result" is useful: more massive objects cause more acceleration of the target object. Instead of rejecting intuitions, students are guided to find appropriate ways to incorporate them into their thinking." (p. 4 of Scherr 2007).

OSTs were developed by a team at the University of Maryland, where they were used in physics recitations with, ideally, two teaching assistants (TA's) and 24 students in each section. At TGU, the ratio of 1 TA to 12 students is more or less maintained, but in a large room with over 100 students. The TA's are often physics PSTs in a MS program, and the students are undergraduate PSTs of various disciplines. TA's meet weekly with Uematsu to prepare for the next week's Tutorial, and they meet again after the Tutorial to reflect on the experience.

In the winter semester of 2021–2022, Hull taught a seminar titled "Instructor's Guide to Open Source Tutorials" at UV. Like at TGU, PSTs at UV experienced a different Tutorial every week, but it was generally done in a mock lesson format. That is, one PST took on the role of "teacher" for the lesson while the other PSTs were "students". Feedback was provided to the "teacher" after the Tutorial, and the "teacher" then wrote a reflection about the mock lesson. In this seminar, PSTs were also assigned as homework reading about the theory underlying OSTs [including the epistemological emphasis and Scherr's paper (2007)], and the seminar culminated in each PST creating his or her own OST and arguing how it is consistent with the "spirit" of OSTs in that it includes an epistemological emphasis and "expects and utilizes the fact that student reasoning is often fluid (in accordance with the 'Pieces model')". Each PST taught his or her own OST in a high school, usually with about 20 students to instruct alone, wrote a reflection, and improved the OST.

The group-based instruction format of OSTs was perceived by PSTs at both TGU and UV as unusual, as was the epistemological focus. This provides an ideal environment for observing growth of curricular knowledge in the PSTs.

3 Methodology

Our qualitative study draws upon video and audio recordings from in-class discussions in courses for PSTs and from one-on-one semi-structured interviews with PSTs. Our focal participants come from two cohorts of PSTs, one instructed by Hull at UV and the other instructed by Uematsu at TGU. We present below a case study of Saki (all names are pseudonyms), a MS student at TGU who took the "Physics Exercises" course as an undergraduate PST in 2018 and then went on to become a TA for the next generation of PSTs in 2021. For her MS thesis advised by Uematsu, Saki made significant modifications to the Pressure OST and investigated the effects of these modifications. She was interviewed by Uematsu in May 2022. She was presented with a gift certificate worth $10 to the bookstore as compensation.

We also present case studies of three PSTs from UV (Brock, Arnold, and Sean), who were interviewed by Hull in February and March 2022 (after the seminar had concluded and final grades had been assigned). As compensation, Hull donated money to treenation.org for trees that the PSTs had chosen to be planted in their honor. Brock had been one of 16 PSTs to take an undergraduate version of the OSTs seminar (which did not include the creation of an OST) from Hull in the summer semester of 2018. In between the two seminars, Brock had experienced teaching with OSTs in his own classroom during a year-long internship. Arnold and Sean had previously taken a different seminar from Hull (on the topic of radioactivity) and had had no prior exposure to OSTs.

All interviews were video and audio recorded with interviewee consent. At UV, each class was also audio and video recorded, and PSTs were assured that Hull would not share this data with the other authors without consent, and that consent forms would only be checked after final grades had been assigned. Interview questions for Brock, Arnold, and Sean were based in part on things they had said and done in class, in order to further draw out their curricular knowledge about OSTs and their perceived agency. After the course ended, Hull made outlines for the interviews that he had conducted with Brock, Arnold, and Sean, highlighting and transcribing points that related to their curricular knowledge and perceived agency. Consistent with the constant comparative method (e.g., Kolb 2012), Hull drew upon the transcribed text as well as the written coursework and in-class utterances of the PSTs to draft an account that was shared with Uematsu and Elby. Claims about the PSTs were scrutinized by looking for consistency across each interview as a whole, and modifications were made until agreement between the three authors was reached. A similar process was carried out in creating Saki's case study.

Drawing on coursework, reflections, and interviews, this paper will tell the stories of PSTs at both institutions to consider how curricular knowledge and perceived agency co-developed. We will first discuss curricular knowledge development. Then we will discuss the data we have regarding perceived agency of these PSTs. Finally, we will discuss the interplay between these two aspects.

4 Curricular Knowledge Development

One of the key elements of the curricular knowledge necessary to successfully implement OSTs is the underlying assumption that student reasoning is often fluid and context-dependent. Although it is not necessary for OST facilitators to be knowledgeable about the nuanced aspects of the Pieces model, they do need to understand that student ideas, even ones that seem blatantly incorrect, can often be built upon in effective responsive teaching. One way in which OSTs accomplish this is with the "intuition refinement strategy" discussed above in the context of the N3 OST. In a one-on-one semi-structured interview between Uematsu and her MS student, Saki, the PST discussed how students are guided to refine their intuitions in a different OST, one about hydrostatic pressure. Saki had been focusing on this OST for her MS thesis, modifying and testing both the tutorial and the associated homework in response to her experiences of first learning with the materials as an undergraduate student in the course in 2018 and then implementing the materials as a TA in 2021. Saki explained:

> Like, a lot of students catch this rough idea that the magnitude of water pressure has something to do with height or displacement from the water surface... There is where that ambiguity comes in, but when you go through experiments and solve problems, you will find that it is

C. Imagine a container *wider* than those you used above, but with the holes A, B, and C still at the same depth. *Before you do the experiment,* consider the following exchange between two students.

Wendy: "The water above the hole pushes down on the water next to the hole, pressing it out. The strength of the squirting is determined by the weight of the water above the hole. There's more water above point A for this container, so the water will squirt out of that hole more strongly than it did with the other containers."

Doug: "I think the strength of the squirting is determined by the depth of the hole. The depth of point A is the same as for the other containers, so the water will squirt out the same."

1. With whom do you agree? Or do you have a different idea?

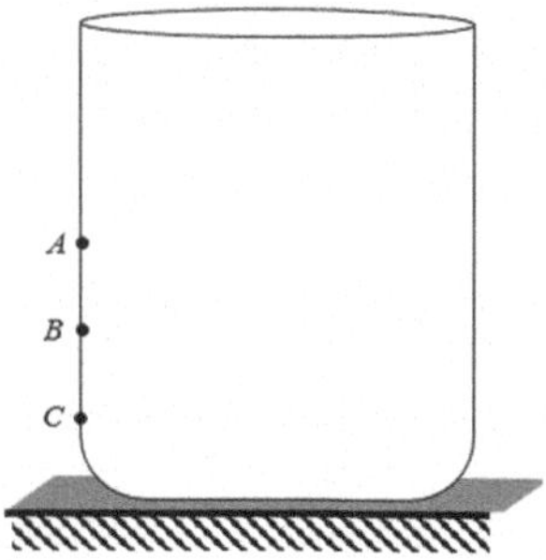

Fig. 1 An excerpt from the pressure OST in which students consider more carefully what causes pressure differences in a container of water

> not the amount [of water on top] … or distance to the ceiling, but the height from the water surface … that is close to refining intuitions.

Unlike the N3 OST, the Pressure OST does not explicitly talk about intuition refinement, either in the tutorial itself or in the teacher's guide. However, examination of the OST shows that Saki's description is accurate. At one point, students are asked to consider an exchange between two fictitious classmates, Doug and Wendy, about what determines pressure at a point (operationalized in the tutorial by how strong a stream of water will squirt out of a hole at that depth) (Fig. 1).

Whereas Wendy argues mechanistically (but incompletely) that the water pressure is determined by the weight of the water above a given depth, Dough answers correctly (but without explanation) that only depth is relevant for determining pressure. In the tutorial, many students agree with the mechanism proposed by Wendy and are then surprised to find experimentally that Doug is correct. As Saki pointed out, students have a "rough idea that the magnitude of water pressure has something to do with… the amount [of water on top]" but, through conducting experiments, their intuition becomes refined to gain conceptual insight into what, precisely, matters (the height to the water surface) and why. Although intuition refinement is not explicitly labeled in this OST, Saki was able to specify how the OST refines ideas. This is evidence that she has come to understand an element of the Pieces model and has gained curricular knowledge about this OST.

At UV, Hull interviewed three PSTs in his MS seminar about their experiences in learning about OSTs, and all three demonstrated the curricular knowledge that OSTs aim to work with student ideas. At one point of the interview, Sean articulated the process of refining intuitions in his own words:

> In OSTs, we kind of try to look for … the true part of the wrong idea. Like, there's a core in there somewhere that is true, and you kind of try to concentrate on that and, and, go from there to bend it towards the right direction, and … say "yeah, actually this part of your idea is correct, but maybe this is a little imprecise, so let's look at it from this angle … how can we formulate this idea that's already there in a way … that fits the real world?"

Sean further described trying to remove student ideas as a "really bad idea":

> [Student ideas] often conflict with what we should teach them in the classroom, because it is often that they have a simple idea, and we know the world is complex, so the simple idea is not going to explain everything ... I don't think you should get rid of them before you start teaching physics. So I think that would be a really bad idea to be like "forget everything that you know, we are starting from zero."

Brock demonstrated his curricular knowledge when he explained that "TIPs use the elicit, confront, resolve method explicitly, and OSTs don't use that, and it's more about epistemology as well, that's what I would say are the key elements, the key differences." This increased curricular knowledge was not just something the PSTs memorized like one might memorize dates for a history exam. Rather, it is something that connected with them on a personal level to the extent that they were able to articulate personal opinions about refining intuitions and the Pieces model. Brock valued thinking about student ideas as being something that can be "worked with" instead of "removed":

I think these intuitive ideas... will remain with you almost forever, because they are intuitive... it kind of relates to the Pieces model... more about restructuring them maybe, or adding new pieces to that piece, to previous information or previous intuitions. ... I think it is not worth the time, even if it is, even if it is possible, to remove all these ideas, but it is better to add additional ideas and to teach students how to, how to work with their intuitions and new physics ... (emphasis his).

To be clear, the opinions of our PSTs about the Pieces model were not always unequivocally positive. Arnold, for example, expressed a tradeoff on an emotional level. He showed value for the Pieces model when he said that "The pieces model... is nice that it reconciles your initial ideas... it makes you feel better in the sense that, uh... my own experiences are not completely disconnected from the physics class and they have some value." At the same time, however, he freely shared his partiality to the misconceptions model:

> I'm going to be real, I like ... the misconceptions model ... it's just a feeling of instant gratification if you see something that you thought about wrongly and then you know what the thing was that you were missing or that was not right in your head and then suddenly you understand the concept.

Arnold's admission that he was "going to be real" indicates that he was aware that the interviewer was a proponent of the pieces model that undergirded the OSTs that had been taught that past semester. That awareness didn't stop Arnold from "being real". The authenticity with which Arnold shared his preference for the misconceptions model demonstrates a more personal understanding about the pieces and misconceptions models. Arnold is not just reciting some answer he had memorized during the seminar: his curricular knowledge had made a deep enough impression that he could articulate a preference, a preference strong enough that he was willing to voice it to an interviewer who, he anticipated, might not share that preference.

Furthermore, PSTs at UV were able to make their own tutorials that demonstrated their curricular knowledge about OSTs. All five PSTs in the MS seminar succeeded in this task. An excerpt from an English translation of Brock's OST on heat and temperature is in Fig. 2. By this point in the tutorial, most students will have written down the well-known incorrect idea that metals are colder in the winter time and hotter in the summer, having a more extreme temperature than their surroundings. Rather than confronting this idea to have students reject it in favor of the normative idea that metals are actually the same temperature as their surroundings, Brock's approach is to explicitly point out to students that their idea of metals having a more extreme temperature (lower left box) has a valid origin: metals do indeed feel hotter in the summer and colder in the winter (top box)! However, that same underlying experience-based intuition can also be refined to be consistent with the normative idea: metals feel colder not because they are at a lower temperature, but because they are better at drawing heat away from warm hands (lower right box). This diagram follows the format of the intuition refinement diagram in the OST on Newton's Second Law, where the boxes read "A force is needed for motion" (top box), "A net force is needed to maintain an object's motion (velocity)" (bottom-left) and "A net force is needed to initiate or change an object's motion (velocity)" (bottom-right). The fact that Brock was able to apply this pattern to a new physics topic indicates his curricular understanding about intuition refinement diagrams and, by extension, OSTs. Brock's use of the intuition refinement diagram in his tutorial is something he remembered in the interview and he mentioned without prompting from the interviewer.

Finally, we observed that our TA's demonstrated their curricular knowledge about OSTs and the Pieces model in the teaching that they carried out in mock lessons (at UV) and as a TA (at TGU). After the seminar, Sean expressed how this curricular knowledge was informing his ongoing teaching:

> [I am using] the strategy of not … being "no, your idea is wrong. Look, here's the correct one." …[In my math class] there was text about some loan and interest rates and you had to put those relations into an equation… the interest rate was 3.5% and … x, that's the amount of this one loan, and then [some students] said "the interest that I need to pay is 3.5 times x." … it's the correct idea that you have to multiply something, but … there's some detail missing… [furthermore] every student understood that they don't have to pay 350 Euros for the one hundred Euro loan.

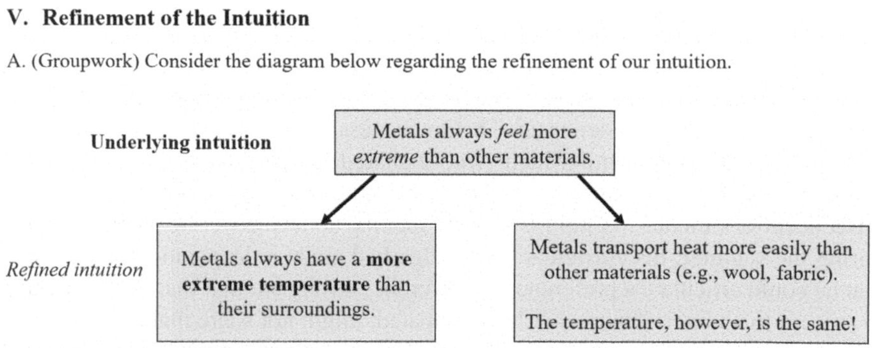

Fig. 2 An excerpt from Brock's tutorial demonstrating his understanding of the refining intuitions process utilized in OSTs

In this quote, Sean demonstrated that he had learned that OSTs have students check for consistency and draw upon student ideas. He recognized the value of doing this in his math class and recognized student ideas that could be productive to draw upon.

So far we have shown that, at the end of the semester, PSTs could articulate and appreciate the pieces model, even if they were not completely committed to it (Arnold). In the next section, we will share data regarding the perceived agency of our PSTs.

5 Perceived Agency of the PSTs

We hypothesize that learning to facilitate student-centered curriculum like OSTs not only increases a teacher's or PST's curricular knowledge about that curriculum, but his or her perceived agency as well. Our argument is that curriculum which calls on the facilitator to be responsive to learners calls on them to make moment-by-moment decisions. Such decisions can come at the expense of adhering to pre-determined schedules and, by extension, national standards. Such decisions can even come at the expense of adhering to the curriculum itself, when the facilitator deems it most productive to deviate (Harlow 2010). When educators make such decisions, they enact their agency. Curriculum that emphasizes to instructors the importance of enacting their agency, we suspect, increases their perceived agency. Unfortunately, due to logistical issues, we conducted our interviews exclusively at the end of the semester. Although it is likely that our PSTs had very limited curricular knowledge about OSTs at the start of the seminar (and a reflection from Saki and in-class recordings of Brock at the start of the semester corroborates this claim), we can do little more than speculate about their perceived agency at the start of the semester. Generally speaking, our prior survey-based research on perceived agency has indicated that PSTs at TGU have lower perceived agency that counterparts at UV (Hull et al. 2019). Saki thus voices comparably strong perceived agency when she told Uematsu in her interview that she does not feel confined to fit with student expectations that instruction should follow the status quo:

> A lot of students ask me if something is correct or not, but… [it should not be] "the teacher says it is correct so it is correct", but rather, I think it is more important whether that will convince someone or not.

Furthermore, she does say in the interview that this is a view she developed since working with OSTs. However, we acknowledge the possibility that she may be a relatively agentive teacher who would have voiced such an opinion if interviewed before working as a TA as well.

In the case of UV PSTs, it is even less clear whether they developed a stronger perceived agency from the MS seminar. In fact, Hull administered the Perceived Agency Survey (Hull et al. 2019) to the PSTs both at the start and at the end of the seminar, and no clear differences emerged, both due to the small sample size, and also due to a ceiling effect (the PSTs indicated a high degree of perceived agency on the pretest as well). What is clear, however, is that the PSTs at UV ended the semester with a strong perceived agency. In his interview with Hull, Brock made it clear that he sees himself as being responsible for making the decision of what kind of approach to take in the classroom when he said that "…the difficult part is thinking about, 'ooh, … how do I deal with misconceptions in general? Is that such a strong relevant misconception that I have to confront it using 'confront, resolve'?" In contrast, Brock discussed the alternative approach of "refining intuitions" that is more appropriate if the student idea (Brock uses the word "misconception") is less "strong". Similarly, Sean articulated his perceived agency to make decisions based upon what he thinks will be effective for his students:

> …you really have to look at it from the student's perspective of what can they bring to the table… what are some difficulties, what are the important things that they should learn … it's easy to see teaching as a teacher's activity… it's really easy to fall into that "OK, what do I have to do to get through the, through the class?", … [but] the more you think about [the student's perspective] the better … be it in a … traditional physics class, or be it a format like OSTs.

Rather than having to just adhere to curriculum that is already there to "get through the class", it is important to exercise perceived agency in "looking from the student's perspective".

Although the PSTs at UV began and ended with much perceived agency, it is noteworthy that this perceived agency appeared to have coupled to their curricular knowledge about OSTs at the end of the semester. We will end with qualitative examples of this coupling.

6 Discussion and Conclusion: Interplay Between Curricular Knowledge and Perceived Agency

The opportunities that Saki, Brock, Arnold, and Sean experienced as a TA (Saki) and in a MS seminar (Brock, Arnold, and Sean) called on them to enact their agency as they deepened their curricular knowledge about OSTs. For her MS thesis, Saki improved the Pressure OST based upon student responses to it. In so doing, she enacted her agency driven by curricular knowledge about that OST. The PSTs at UV did likewise when they created their own tutorials. When these PSTs facilitated the OSTs, they were essentially instructed to enact their agency, making their own moment-by-moment decisions that they felt were called for in response to student ideas. It is thus plausible that perceived agency and curricular knowledge about OSTs co-developed, either growing side by side (as may have been the case for Saki) or with an already strong perceived agency providing a support around which curricular knowledge could grow like a vine (which more likely was the case for the UV PSTs). It is also not altogether surprising that there were statements made by PSTs in the interviews that clearly showed an intersection between curricular knowledge about OSTs and PST perceived agency. One example of such a statement, already discussed above, is Saki's statement about how she tells students to discuss on their own to find an explanation that will "convince someone" (essentially, what takes place during OSTs) instead of being stuck in the mindset that "the teacher said it is correct, so it is correct". An additional example of a statement indicating an intersection between curricular knowledge about OSTs and PST perceived agency comes from Arnold:

> I started thinking through the OST seminar that actually, maybe, it is important to teach [my students] about the nature of physics itself … I did not use [the OSTs]… I am mainly teaching 12 year olds, you can forget that, they are not going to be able to work with that… [But] what I did do … it's the same thing as the torque OST, basically… but … I did not have enough of those torque sticks. So I pulled up a simulation… they had to balance stuff and … come up with a formula… which did not work, I had to give them the formula, but, I tried… basically the same thing as the OST…

If Arnold had had no curricular knowledge about the torque OST, then he would have been less likely to create his own improvised version of it in his classroom. He would have also been less likely to then deviate from that improvised version, giving students the answer of what the equation is, when he saw that they were not going to get it on their own. We see his curricular knowledge about this OST—not only that it uses "torque sticks" as apparatus, but also that it emphasizes epistemology—as enabling him to act out on his perceived agency in the classroom.

In this paper, we have limited our discussion to a qualitative discussion about four PSTs who we were able to interview out of convenience. We are not at a point where we can make any broad claims about the effects of learning to facilitate student-centered curriculum like OSTs on perceived agency. However, there is some indication that these four PSTs are not altogether idiosyncratic. Uematsu administered a survey to her TA's during the debriefing after the last OST of 2021. One of these TAs, who had facilitated OSTs also in 2020, wrote the following reflection:

> I thought I was used to [teaching with OSTs], but there are many times where I wonder "what should I do in this case?" There were many situations where I taught rather than stimulated the discussion… I still have a lot to think about, because I don't know how to ask students key questions without interrupting their discussions.

Here, the PST demonstrates curricular knowledge in recognizing that student discussions are key in OSTs. This curricular knowledge intersects perceived agency, in that the PST recognizes the decision to choose between myriad options available ("teaching" vs. "stimulating") for supporting those discussions.

Our research concerns investigation of the interplay between curricular knowledge and perceived agency. Our work expands on previous literature that has studied teacher perceived agency (e.g., Lipponen and Kumpulainen 2011) and curricular knowledge (e.g., Goertzen 2010; Harlow 2010; Robertson 2018) by conceptualizing these two aspects as things that can co-develop and by looking for mechanisms by which this intertwining can occur. At both TGU and at UV, PSTs who learned about and facilitated OSTs grew in their curricular knowledge about the curriculum. Their perceived agency at the end of the semester seemed to be coupled to this curricular knowledge. Future work should attend to describing in detail pathways by which curricular knowledge and perceived agency of these PSTs may have interacted with each other as they co-developed (Hull et al. 2022).

References

Ball DL, Cohen DK (1996) Reform by the book: what is—or might be—the role of curriculum materials in teacher learning and instructional reform? Educ Res 25(9):6–14. https://doi.org/10.3102/0013189X025009006

Debarger AH, Penuel WR, Moorthy S, Beauvineau Y, Kennedy CA, Boscardin CK (2017) Investigating purposeful science curriculum adaptation as a strategy to improve teaching and learning. Sci Educ 101(1):66–98. https://doi.org/10.1002/sce.21249

diSessa AA (1993) Toward an epistemology of physics. Cogn Instr 10(2–3):105–225. https://doi.org/10.1080/07370008.1985.9649008

diSessa AA (2009) A bird's-eye view of the "pieces" vs. "coherence" controversy (from the "pieces" side of the fence). In: Vosniadou S (ed) International handbook of research on conceptual change, p 35

Elby A, Scherr R, McCaskey T, Hodges R, Redish EF, Hammer D, Bing T (2007) Open source tutorials in physics sense-making. DVD, funded by NSF DUE-0341447

Goertzen RM (2010) Investigating and accounting for physics graduate students' tutorial classroom practice [PhD]. University of Maryland, College Park

Hammer D (2000) Student resources for learning introductory physics. Am J Phys 68(S1):S52–S59. https://doi.org/10.1119/1.19520

Harlow DB (2010) Uncovering the hidden decisions that shape curricula. AIP Conf Proc 21–24.

Hofer BK, Pintrich PR (1997) The development of epistemological theories: beliefs about knowledge and knowing and their relation to learning. Rev Educ Res 67(88)

Hull MM, Vormayr K, Uematsu H (2019) Validation of a survey to measure pre-service teachers' sense of agency. GIREP

Hull MM, Uematsu H, Elby A (2022) A progression of pre-service teachers towards deep curricular knowledge (the pieces model in open source tutorials). In: Jones (ed) 2022 PERC proceedings. https://doi.org/10.1119/perc.2022.pr.Hull

Kolb SM (2012) Grounded theory and the constant comparative method: valid research strategies for educators. J Emerg Trends Educ Res Policy Stud 3(1):83–86

Lipponen L, Kumpulainen K (2011) Acting as accountable authors: creating interactional spaces for agency work in teacher education. Teach Teach Educ 27(5):812–819. https://www.sciencedirect.com/science/article/pii/S0742051X11000023

McDermott LC (1991) Millikan Lecture 1990: what we teach and what is learned—closing the gap. Am J Phys 59(1):301

McDermott LC, Shaffer PS (2002) Tutorials in introductory physics, 1st edn. Prentice Hall

Robertson AD (2018) Supporting the development of curricular knowledge among novice physics instructors. Am J Phys 86(4):305–315

Sawyer RK (2004) Creative teaching: collaborative discussion as disciplined improvisation. Educ Res 33(2):12–20. https://doi.org/10.3102/0013189X033002012

Scherr RE (2007) Modeling student thinking: an example from special relativity. Am J Phys 75(3):272–280

Examining Differences in Teacher Self-efficacy for In-field Versus Out-of-Field Teachers of Physics

Kyla M. Smith, Judith Hillier, and Sibel Erduran

Abstract Out-of-field teaching has been shown to have negative effects on students' decisions around their future studies and on the types of instructional strategies used to support their learning. Behind these challenges for students are teachers working to overcome challenges of their own: teaching out-of-field negatively impacts teachers' wellbeing and retention in the profession. This paper investigates differences between the teacher self-efficacy of in-field versus out-of-field teachers of secondary school physics. A theoretical framework of social cognitive theory places teachers' perspectives at the centre of the study. Using data from a cross-sectional survey of teachers of grades 9 through 12 science and mathematics in the four western provinces of Canada (British Columbia, Alberta, Saskatchewan, and Manitoba), a significant difference in overall teacher self-efficacy is found between in-field and out-of-field teachers of physics with a medium effect size. On average, out-of-field teachers of physics had an overall teacher self-efficacy of approximately half a standard deviation lower than their in-field counterparts. Teacher self-efficacy with instructional strategies is where the largest differences are found between the in-field and out-of-field groups. These findings have implications for continuing professional development needed to support out-of-field teachers of physics.

1 Introduction

1.1 Out-of-Field Teaching

Out-of-field teaching occurs when the courses a teacher is assigned to teach do not match with their specialised knowledge, and this mismatch can occur in either subject area or grade level (Hobbs and Törner 2019a). As teachers' degree backgrounds and certifications are often used to determine who is in- versus out-of-field in a particular jurisdiction (Shah et al. 2019), operationalisations of this concept vary from place to

K. M. Smith (✉) · J. Hillier · S. Erduran
University of Oxford, Oxford, UK
e-mail: kyla.smith@education.ox.ac.uk

S. Faletič and J. Pavlin (eds.), *Teaching and Learning Physics Effectively in Challenging Times*, Challenges in Physics Education, https://doi.org/10.1007/978-3-031-72541-8_19

place (Ingersoll 2019; Price et al. 2019; Vale and Drake 2019). Out-of-field teaching has been described as "education's 'dirty little secret'" (Shanker 1985, p. E9) and a "misuse of teachers" (Conant 1963, p. 67). Ingersoll characterises out-of-field teaching as an "an important problem because highly qualified teachers may actually become highly unqualified if they are assigned to teach subjects for which they have little training or education" (2019, p. 22).

The issues with out-of-field teaching lie in its negative impacts on both students, in terms of their achievement (Goldhaber and Brewer 1996; Rowan et al. 1997; Hill and Dalton 2013; Sheppard et al. 2020) and the instructional strategies used to support their learning (Carlsen 1993; Arzi and White 2008), and teachers, in terms of their wellbeing and retention in the profession (Du Plessis et al. 2014). Despite policy changes aimed at reducing out-of-field teaching in some countries, for example, the highly qualified teachers requirement of the No Child Left Behind Act in the USA (2002), the phenomenon of out-of-field teaching is not going away (Ingersoll 2019).

1.2 Out-of-Field Teaching in Physics

With long-standing shortages of physics-trained teachers in many countries (OECD 2005), out-of-field teaching of physics is commonplace in many education systems (Hobbs and Törner 2019b). Within physics, this shortage and its impact have been well-researched in the United Kingdom and the United States of America. Whereas much is known in these contexts, there is a lack of information and research on out-of-field teaching in Canada, not just within physics but all subject areas.

In the USA, Monk showed that teachers' undergraduate training in physical sciences has a significant positive effect on students' achievement in general science courses at the sophomore and junior levels (ages 15–17), more so than teachers' undergraduate training in maths or life sciences (Monk 1994). More specific to physics courses, it has also been shown that, after student ability, teacher expertise has the greatest impact on student achievement in physics at both the GCSE (ages 14–16) and A-level (ages 16–18) stages of secondary school in England (Smithers and Robinson 2005). While student achievement is an important outcome and a common indicator of education quality, given the need for physics trained people in many sectors of society (Institute of Physics 2010a; UK Commission for Employment and Skills 2015), students continuing on in physics is perhaps just as important. Here, too, there exists a strong link with teachers' background in physics, as students are more likely to choose A-level physics if they are taught physics at GCSE by a physics trained, rather than biology or chemistry trained, teacher (Smithers and Robinson 2005).

In the USA, there has been significant concern about out-of-field teaching (Ingersoll 1999), which led to policy changes that attempted to address the issue (2002). All teachers were to become "highly qualified" in each subject they teach by 2005–06 (Escalada and Moeller 2006). Prior to this standard, a 2001 nationwide survey of physics teachers found that only 32% of physics teachers had a degree-level qualification of a major or minor in physics (Neuschatz and McFarling 2003). By 2015, this had risen to almost 40% of physics teachers holding a degree-level qualification in physics (Shah et al. 2019). Compounding difficulties with a lack of subject background, it was also reported that 82% of physics teachers were the sole teacher of physics at their school, with a further 13% having only one physics teacher colleague (Neuschatz and McFarling 2003).

In the UK, there is a concerning shortage of physics teachers (Institute of Physics 2010b), and the recruitment and retention of physics teachers has long been a problem (Physics Education Committee 1981). In 2014, the percentage of physics courses taught by those without degree-level qualifications in physics was 28%. By 2018, nearly 50% of schools in England were found to use out-of-field teachers to teach physics at GCSE level (Lauchlan 2018), and in 2021 initial teacher training programmes in England reached only 22% of their target for physics teacher recruitment (Gov.uk 2021). Interventions to recruit physics teachers have been varied in their uptake and success (Smithers and Robinson 2008; Campbell 2011; Hillier et al. 2013; Straw et al. 2017).

1.3 Social Cognitive Theory

Teachers' subject knowledge is central to what and how teachers teach (Shulman 1986; Loewenberg Ball et al. 2008), which poses difficulties for both out-of-field teachers and their students (Crisan and Hobbs 2019; Porsch and Whannell 2019). One of the causes of out-of-field teaching is the lack of control teachers have over their teaching assignments, i.e. the classes they are assigned to teach (Du Plessis et al. 2017). Therefore, this study embraces a theoretical framework of social cognitive theory (Bandura 1986), which fosters teacher agency by placing their experiences and voices at the centre of the study. Teachers in this study are assumed to be the experts when it comes to their own training, knowledge, and capabilities and are, therefore, the only source of data.

1.4 Teacher Self-efficacy

Self-efficacy is a primary concept of social cognitive theory and allows teachers to assess what they believe their capabilities are (Bandura 1977, 1997). The concept of self-efficacy originated in the field of psychology and was established by Bandura in his seminal paper, "Self-efficacy: toward a unifying theory of behavioural change" (1977). Bandura defines self-efficacy as "beliefs in one's capabilities to organize and execute the courses of action required to produce given attainments" (Bandura 1997, p. 3).

One important distinction between self-efficacy and other measures of competence or ability is that "self-efficacy has to do with self-perception of competence rather than actual level of competence" (Tschannen-Moran et al. 1998, p. 211). Additionally, self-efficacy is distinct from other confidence-related concepts, such as self-esteem, which is related to self-worth rather than capabilities (Bandura 2006). Self-efficacy is also distinct from self-concept, which is a more general sense of identity and includes the traits and attributes you associate with yourself, as well as which social groups and identities you feel you belong to (Oyserman et al. 2012). Self-efficacy is not a general self-belief or a "global trait," meaning that one does not have an overall sense of self-efficacy (Bandura 2006, p. 307); rather, it is "a differentiated set of self-beliefs linked to distinct realms of functioning" (Bandura 2006, p. 307). Therefore, self-efficacy is dependent upon the task or action one is evaluating their self-efficacy in relation to.

Teacher self-efficacy, as a domain specific version of self-efficacy, is a "teacher's belief in [their] capability to organize and execute courses of action required to successfully accomplish a specific teaching task in a particular context" (Tschannen-Moran et al. 1998, p. 233). Research has revealed positive relationships between teacher self-efficacy and a number of desirable factors, such as support for learning and teacher retention and wellbeing (Zee and Koomen 2016).

Research has found that teachers with high teacher self-efficacy engage in professional development more frequently (Geijsel et al. 2009) and are more likely to implement new instructional strategies and skills from professional development programs (Eun and Heining-Boynton 2007). Lakshmanan found that participation in a professional development program for elementary and middle school science teachers was able to increase both teacher self-efficacy and teachers' implementation of new science inquiry instructional strategies (Lakshmanan et al. 2011). Additionally, higher teacher self-efficacy has been associated with a less controlling approach to instructional management (Martin and Sass 2010), which supports findings that efficacious teachers are more likely to use more learner-centered, constructivist instructional methods (Nie et al. 2013).

Martin, Sass, and Schmitt examined the complex interplay between teacher self-efficacy in student engagement, instructional management, student stressors, burnout, and intent-to-leave, finding that teachers with lower teacher self-efficacy in student engagement were found to have higher intents-to-leave (Martin et al. 2012). Egyed and Short found a significant negative correlation between teacher self-efficacy and burnout in elementary teachers, suggesting that teachers with higher levels of teacher self-efficacy experience lower levels of burnout (Egyed and Short 2006). Additionally, teachers with higher teacher self-efficacy have been found to have higher levels of job satisfaction (Caprara et al. 2006).

2 Research Question

Given the marked importance of teacher self-efficacy and the prevalence of out-of-field teaching in physics, this study was designed to explore the relationship between teacher self-efficacy and teaching out-of-field in physics. Therefore, this paper addresses the following research question: in what ways, if at all, does teacher self-efficacy differ between in-field and out-of-field teachers of physics?

3 Methodology

This paper uses data collected in the questionnaire phase of a larger study. The cross-sectional questionnaire received 220 responses from teacher participants in the four western provinces of Canada: British Columbia, Alberta, Saskatchewan, and Manitoba. Of the 220 secondary science and mathematics teachers who responded, 72 reported to be teaching physics.

The questionnaire collected demographic information, teachers' full teaching assignments, and educational background, and the main component of the questionnaire was the instrument used to measure teacher self-efficacy. For this, the Teacher Sense of Efficacy Scale (TSES) (Tschannen-Moran and Woolfolk Hoy 2001), a validated and widely-used instrument (Morris et al. 2017), was selected. The TSES includes 24 items, each of which is rated on a Likert-type rating scale. The 9-point scale is numbered 1 to 9 and includes worded anchors at 1, 3, 5, 7, and 9 of "nothing/ not at all", "very little", "some degree", "quite a bit", and "a great deal", respectively. The scale has three factors: instructional strategies, classroom management, and student engagement. Eight items load onto each of these three factors. As has been done previously for use with teachers of English as a foreign language (Chacón 2005), subject specific versions of the TSES were created for this study. The wording of items was adjusted as little as possible, such that they refer specifically to each science discipline, mathematics, and other subject areas. Participants completed a subject specific teacher self-efficacy instrument for each subject area they taught within the 2020–2021 school year. The data utilised in this paper is from the physics-specific version of the TSES, which was completed by the 72 teachers who reported teaching physics.

Prior research often categorises teachers as in-field or out-of-field based on their degree background. This does not allow for considerations of how teachers may grow over their careers, or for informal sources of learning. The last decade has seen a shift towards more nuanced, non-binary classifications of out-of-field teaching (Hobbs 2012, 2013; Du Plessis et al. 2014). Therefore, given the framework of social cognitive theory at play in this study, teachers were asked to rate the extent to which they felt out-of-field in each of the subject areas they teach. For this, a 4-point rating scale was introduced with options of "not at all, I feel in-field", "slightly out-of-field", "moderately out-of-field", and "very out-of-field". In the analysis for this paper, the

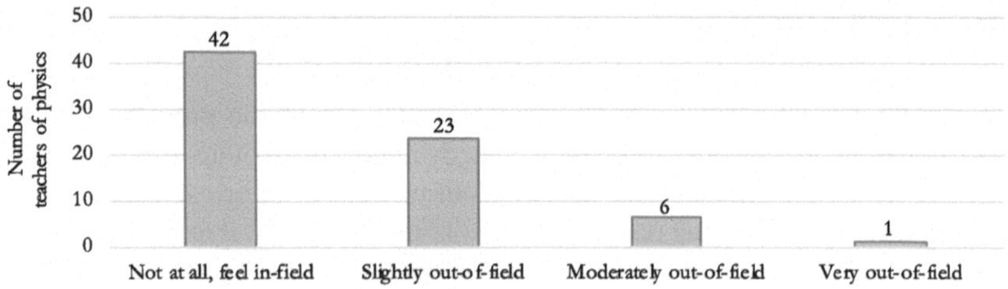

Fig. 1 Distribution of teachers of physics by extent out-of-field

three out-of-field categories are combined, such that teachers who feel any extent out-of-field are grouped together. Of the 72 teachers of physics, 42 felt in-field in physics and 30 felt some extent out-of-field. Figure 1 shows the distribution across categories of out-of-field-ness.

4 Findings

In order to investigate differences in teacher self-efficacy between in-field and out-of-field teachers of physics, independent *t*-tests were run on each of the 24 items of the Teacher Sense of Efficacy Scale (TSES). Additionally, independent t-tests were run between the in-field and out-of-field groups on the three TSES factors of instructional strategies, classroom management, and student engagement, as well as on overall teacher self-efficacy. The teacher self-efficacy relating to each factor is a mean of the ratings from the eight items that load onto that factor, and only teachers who rated at least four of the eight items are included in the factor means. Similarly, only teachers who rated at least 12 of the 24 items are included in the overall teacher self-efficacy means. Table 1 shows the results of these *t*-tests, including two-tailed *p*-values and Cohen's *d*. Also included in the table are the means for the in-field and out-of-field groups on each item, factor, and overall: means are on the original 9-point rating scale.

There was a significant difference ($p \leq .05$) between the in-field and out-of-field groups on five of the individual teacher self-efficacy items, with out-of-field teachers reporting lower self-efficacy on these items than their in-field counterparts. The items are:

- How well can you respond to difficult questions about physics concepts from your students?
- To what extent can you craft good physics questions for your students?
- To what extent can you use a variety of assessment strategies in your physics teaching?

Table 1 Independent *t*-tests of teacher self-efficacy between in-field and out-of-field teachers of physics

Teacher self-efficacy items, or mean factor ratings (bold)	In-field mean	Out-of-field mean	Mean difference	*t*-statistic	*p*-value	Cohen's *d*
Instructional strategies teacher self-efficacy in physics	**7.62**	**6.76**	**−0.86**	**−3.58****	**<0.001**	**−0.86**
How well can you respond to difficult questions about physics concepts from your students?	8.31	6.76	−1.55	$-4.64^{\dagger}$**	<0.001	−1.24
To what extent can you gauge student comprehension of the physics concepts you have taught?	7.76	7.28	−0.48	$-1.82^{\dagger}$	0.075	−0.48
To what extent can you craft good physics questions for your students?	8.12	6.55	−1.57	$-4.89^{\dagger}$**	<0.001	−1.31
How much can you do to adjust your physics lessons to the proper level for individual students?	6.67	6.29	−0.38	−0.91	0.365	−0.22
To what extent can you use a variety of assessment strategies in your physics teaching?	7.31	6.59	−0.72	−2.10*	0.039	−0.51
To what extent can you provide an alternative explanation or example when students are confused about a physics concept?	8.02	7.18	−0.85	−3.10**	0.003	−0.76
How well can you implement alternative instructional strategies in your physics classroom?	6.98	6.43	−0.55	−1.44	0.155	−0.35
How well can you provide appropriate challenges for very capable students in physics?	7.76	7.14	−0.62	−1.81	0.075	−0.44
Classroom management teacher self-efficacy in physics	**7.91**	**8.01**	**+0.10**	**+0.49**	**0.629**	**+0.12**
How much can you do to control disruptive behaviour in your physics classroom?	7.79	8.21	+0.43	+1.47	0.148	+0.36
To what extent can you make your expectations clear about student behaviour in your physics classroom?	8.34	8.11	−0.23	−1.01	0.316	−0.25
How well can you establish routines to keep activities running smoothly in your physics classroom?	8.19	7.97	−0.23	−0.97	0.335	−0.24
How much can you do to get students to follow classroom rules in your physics classroom?	7.95	8.32	+0.37	+1.65	0.103	+0.41

(continued)

Table 1 (continued)

Teacher self-efficacy items, or mean factor ratings (bold)	In-field mean	Out-of-field mean	Mean difference	*t*-statistic	*p*-value	Cohen's *d*
How much can you do to calm a student who is disruptive or noisy during your physics lessons?	7.50	7.75	+0.25	+0.86	0.396	+0.21
How well can you establish a classroom management system with each group of students in physics?	8.07	7.86	−0.21	−0.83	0.410	−0.20
How well can you keep a few problem students from ruining an entire lesson in physics?	7.88	8.04	+0.16	+0.55	0.582	+0.14
How well can you respond to defiant students during your physics lessons?	7.64	7.79	+0.14	+0.46	0.650	+0.11
Student engagement teacher self-efficacy in physics	**6.76**	**6.41**	**−0.35**	**−1.33**	**0.189**	**−0.32**
How much can you do to get through to the most difficult students in physics?	6.62	6.03	−0.59	−1.44	0.154	−0.35
How much can you do to help your students think critically about physics?	7.48	6.90	−0.58	−1.76	0.084	−0.42
How much can you do to motivate students who show low interest in their physics work?	6.02	6.10	+0.08	+0.21	0.832	+0.05
How much can you do to get students to believe they can do well in their physics work?	7.17	6.55	−0.62	−2.13*	0.037	−0.51
How much can you do to help your students value learning about physics?	7.51	7.07	−0.44	−1.39	0.170	−0.34
How much can you do to foster student creativity in physics?	7.05	6.48	−0.57	−1.55†	0.129	−0.40
How much can you do to improve the physics understanding of a student who is failing?	6.36	6.14	−0.22	−0.63	0.533	−0.15
How much can you assist families in helping their children to do well in physics?	5.90	6.00	+0.10	+0.22	0.830	+0.05
Overall teacher self-efficacy in physics	**7.43**	**7.04**	**−0.39**	**−2.02***	**0.047**	**−0.49**

* $p < 0.05$; ** $p < 0.01$

† equal variances not assumed

- To what extent can you provide an alternative explanation or example when students are confused about a physics concept?
- How much can you do to get students to believe they can do well in their physics work?

Notably, four of these are within the instructional strategies factor. Therefore, there was also a significant difference between the in-field and out-of-field groups on instructional strategies teacher self-efficacy ($t(69) = -3.58$, $p < 0.001$) with a large effect size ($d = -0.86$). This demonstrates the relative lack of confidence that out-of-field teachers of physics experience on average with instruction-related tasks such as questioning, providing explanations, and assessment.

The highest rated factor for both the in-field and out-of-field groups is classroom management teacher self-efficacy, with mean ratings of 7.91 and 8.01, respectively, and no significant difference between the groups ($t(68) = 0.49$, $p = 0.629$). This indicates that both in-field and out-of-field teachers are typically confident with classroom management when teaching physics. Some teachers may even have seen the classroom management items as less related to their physics teaching specifically, given that classroom management is thought of as a more transferable skill between subjects. Additionally, the single items with the highest mean rating for each group are both within the classroom management factor. However, there is a subtle, but interesting, difference between the item with the highest mean rating for in-field versus for out-of-field teachers. In-field teachers' highest rated item, with a mean rating of 8.34, was "To what extent can you make your expectations clear about student behaviour in your physics classroom?" Out-of-field teachers' highest rated item, with a mean rating of 8.32, was "How much can you do to get students to follow classroom rules in your physics classroom?" Therefore, out-of-field teachers are, on average, most confident in enforcing rules in their physics classrooms, whereas in-field teachers are most confident communicating their expectations for behaviour in their physics classrooms.

The lowest rated factor for each of the in-field and the out-of-field groups is student engagement teacher self-efficacy, with mean ratings of 6.76 and 6.41, respectively. There was not a significant difference between in-field and out-of-field teachers of physics on student engagement teacher self-efficacy ($t(69) = -1.33, p = 0.189$). The lowest rated item for in-field teachers was the same as that for out-of-field teachers, and it falls within the student engagement factor. It was: "How much can you assist families in helping their children to do well in physics?"

There was a significant difference between the in-field and out-of-field teachers on overall teacher self-efficacy in physics ($t(69) = -2.02, p = 0.047$) with a medium effect size of $d = -0.49$. This signifies that on average the overall teacher self-efficacy of an out-of-field teacher of physics is approximately half a standard deviation lower than that of an in-field teacher of physics. This difference in teacher self-efficacy has implications for how teachers feel about their experiences teaching out-of-field, teacher burnout, and the uptake of new instructional strategies that may benefit students' learning in physics.

5 Discussion

Significant differences were found between in-field and out-of-field teachers of physics on five of the individual teacher self-efficacy items. This section will explore possible reasons for differences in these specific areas.

5.1 *How Well Can You Respond to Difficult Questions About Physics Concepts from Your Students?*

Out-of-field teachers of physics rated their capability in responding to students' difficult questions significantly lower than their in-field counterparts ($t(69) = -4.64$, $p < 0.001$, $d = -1.24$). This difference has a large effect size, and, given the centrality of dialogic approaches in modern classrooms, a lack of teacher confidence in responding to questions can lead to less interactive teaching (Nie et al. 2013). According to Aguiar, Mortimer, and Scott, categories of student 'wonderment' questions in science, defined according to their impact on the explanatory structure, include continuation, extrapolation, and contestation (Aguiar et al. 2010). These questions can present applications or extensions of a concept that out-of-field teachers may not have prepared for or had the time to reason through in advance. Contestation questions use "alternative views to approach a problem, thus challenging the teaching explanatory structure" (Aguiar et al. 2010, p. 180). Out-of-field teachers may not be able to shift their own understanding or take a new perspective so quickly, leaving them in a difficult position when it comes to the dialogic-authoritative continuum (Mortimer and Scott 2003): they may want to present one "correct" view because opening up to further dialogue is uncomfortable and risks more difficult questions, but they may not be sure of the "correct" answer, or whether one exists.

5.2 *To What Extent Can You Craft Good Physics Questions for Your Students?*

The reverse of being able to answer students' questions is being able to design and ask good questions of your students. Out-of-field teachers rated their capability in crafting good questions to be, on average, 1.31 standard deviations below that of in-field teachers of physics ($t(69) = -4.89$, $p < 0.001$, $d = -1.31$). Questioning may occur interactively as part of classroom discourse, as well as on student tasks and assessments. In order to craft good questions, teachers must first understand: What are the key ideas students really need to understand? What questions will help them develop their reasoning or understanding? These require both content knowledge and pedagogical content knowledge, which out-of-field teachers may struggle with.

5.3 To What Extent Can You Use a Variety of Assessment Strategies in Your Physics Teaching?

Effective and ongoing formative assessment is the key to understanding how students are progressing and being able to adapt instruction accordingly (Black and Wiliam 1998), but formative assessment, more so than summative assessment, often requires the use of a range of assessment strategies. Out-of-field teachers of physics have significantly lower ratings of their use of a variety of assessment strategies ($t(69) = -2.10, p = 0.039, d = -0.51$). As Gioka points out, teachers' content knowledge and pedagogical content knowledge inform their use of formative assessment strategies in physics (Gioka 2006), and traditional physics teaching often focusses on "plug and chug" problem solving procedures, which may be reflected in teachers' assessments (Kortemeyer 2016). Out-of-field teachers may, therefore, be less comfortable and flexible in designing assessments that move beyond numerical answers, such as asking students to explain concepts or to reason conceptually.

5.4 To What Extent Can You Provide an Alternative Explanation or Example When Students Are Confused About a Physics Concept?

Out-of-field teachers of physics report significantly lower ratings of their capability in providing alternative explanations or examples than their in-field counterparts ($t(68) = -3.10$, $p = 0.003$, $d = -0.76$) with a large effect size. Explanations and explanatory stories are fundamental to science learning (Millar and Osborne 1998). A coherent internal account of a concept or phenomena strengthens science teachers' ability to provide an explanation (Hillier 2013), and out-of-field teachers of physics may have yet to develop their own coherent internal accounts. Without this, it can be difficult to balance between simplicity, coherence, correctness, and completeness of a physics explanation (Viennot 2021), particularly when deviating from a single, prepared explanation. Additionally, out-of-field teachers report challenges in generating examples on the spot.

5.5 How Much Can You Do to Get Students to Believe They Can Do Well in Their Physics Work?

Out-of-field teachers rate their ability to get students to believe they can do well in physics lower than in-field teachers with a medium effect size ($t(69) = -2.13$, $p = 0.037$, $d = -0.51$). This is the only non-instructional strategies item to have a significant difference between the two groups. Some studies have reported difficulties with students' respect of out-of-field teachers due to challenges with content knowledge and subsequent struggles with behaviour management in an attempt to maintain control (Mizzi 2013). "Teachers who doubt their abilities to effectively engage students are likely to increase efforts to control instruction, which ultimately leads to greater stress from student behaviors and a lessened sense of personal accomplishment." (Martin et al. 2012, p. 555). There is also a sense here that to instil confidence in another, one must have confidence themselves.

6 Conclusion

By trusting teachers as experts on their own knowledge and skills, we can understand which elements of teaching physics out-of-field teachers feel they are struggling with and how these differ from tasks that in-field teachers may also find challenging, such as assisting families to help their children learn physics. Professional development programmes targeting these tasks are needed to assist out-of-field teachers in gaining confidence in their physics teaching and in providing effective learning opportunities for students. Techniques such as the development of coherent internal accounts (Hillier 2013) to solidify physics subject knowledge may help teachers in providing alternative explanations and answering questions. Additionally, support with designing a variety of physics questions and tasks may help teachers to craft better questions and utilise a broader range of assessment strategies, the use of which ultimately aims to build students' physics understanding and beliefs that they can succeed in physics.

Acknowledgements This work was supported in part by funding from the Social Sciences and Humanities Research Council of Canada and the Canadian Centennial Scholarship Fund.

References

Aguiar OG, Mortimer EF, Scott P (2010) Learning from and responding to students' questions: the authoritative and dialogic tension. J Res Sci Teach 47:174–193. https://doi.org/10.1002/tea.20315

Arzi HJ, White RT (2008) Change in teachers' knowledge of subject matter: a 17-year longitudinal study. Sci Educ 92:221–251. https://doi.org/10.1002/sce.20239

Bandura A (1977) Self-efficacy: toward a unifying theory of behavioral change. Psychol Rev 84:191–215. https://doi.org/10.1037/0033-295X.84.2.191

Bandura A (1986) Social foundations of thought and action: a social cognitive theory. Prentice-Hall, Englewood Cliffs, NJ

Bandura A (1997) Self-efficacy: the exercise of control. W. H. Freeman, New York

Bandura A (2006) Guide for constructing self-efficacy scales. In: Pajares F, Urdan T (eds) Self-efficacy beliefs of adolescents. Information Age Publishing, Greenwich, Connecticut, pp 307–337

Black P, Wiliam D (1998) Inside the black box: raising standards through classroom assessment. Phi Delta Kappan 80:139–148. https://doi.org/10.1177/003172171009200119

Campbell P (2011) Enabling non-specialists to teach school physics effectively. Phys Educ 46:152–158. https://doi.org/10.1088/0031-9120/46/2/002

Caprara GV, Barbaranelli C, Steca P, Malone PS (2006) Teachers' self-efficacy beliefs as determinants of job satisfaction and students' academic achievement: a study at the school level. J Sch Psychol 44:473–490. https://doi.org/10.1016/j.jsp.2006.09.001

Carlsen WS (1993) Teacher knowledge and discourse control: quantitative evidence from novice biology teachers' classrooms. J Res Sci Teach 30:471–481. https://doi.org/10.1002/tea.3660300506

Chacón CT (2005) Teachers' perceived efficacy among English as a foreign language teachers in middle schools in Venezuela. Teach Teach Educ 21:257–272. https://doi.org/10.1016/j.tate.2005.01.001

Conant J (1963) The education of American teachers. McGraw-Hill, New York

Crisan C, Hobbs L (2019) Subject-specific demands of teaching: implications for out-of-field teachers. In: Hobbs L, Törner G (eds) Examining the phenomenon of "teaching out-of-field." Springer Nature Singapore, Singapore, pp 151–178

Du Plessis AE, Gillies RM, Carroll A (2014) Out-of-field teaching and professional development: a transnational investigation across Australia and South Africa. Int J Educ Res 66:90–102. https://doi.org/10.1016/j.ijer.2014.03.002

Du Plessis AE, Carroll A, Gillies RM (2017) The meaning of out-of-field teaching for educational leadership. Int J Leadersh Educ 20:87–112. https://doi.org/10.1080/13603124.2014.962618

Egyed CJ, Short RJ (2006) Teacher self-efficacy, burnout, experience and decision to refer a disruptive student. Sch Psychol Int 27:462–474. https://doi.org/10.1177/0143034306070432

Escalada LT, Moeller JK (2006) The challenges of designing and implementing effective professional development for out-of-field high school physics teachers. In: AIP conference proceedings, pp 11–14

Eun B, Heining-Boynton AL (2007) Impact of an English-as-a-second-language professional development program. J Educ Res 101:36–49. https://doi.org/10.3200/JOER.101.1.36-49

Geijsel FP, Sleegers PJC, Stoel RD, Krüger ML (2009) The effect of teacher psychological and school organizational and leadership factors on teachers' professional learning in Dutch schools. Elem Sch J 109:406–427. https://doi.org/10.1086/593940

Gioka O (2006) Assessment for learning in physics investigations: assessment criteria, questions and feedback in marking. Phys Educ 41:342–346. https://doi.org/10.1088/0031-9120/41/4/009

Goldhaber DD, Brewer DJ (1996) Evaluating the effect of teacher degree level on educational performance. Dev Sch Financ 197–210. https://doi.org/10.1177/004057368303900411

Gov.uk (2021) Initial teacher training census: academic year 2021/22. https://explore-education-statistics.service.gov.uk/find-statistics/initial-teacher-training-census/2021-22. Accessed 4 Feb 2023

Hill JG, Dalton B (2013) Student math achievement and out-of-field teaching. Educ Res 42:403–405. https://doi.org/10.3102/0013189X13505682

Hillier J (2013) How does that work? Developing pedagogical content knowledge from subject knowledge. Teach Educ Pract 26:321–338

Hillier J, de Winter J, Twidle J (2013) I could enjoy teaching: the case of physics. Can J Sci Math Technol Educ 13:287–302. https://doi.org/10.1080/14926156.2013.816392

Hobbs L (2012) Teaching out-of-field: factors shaping identities of secondary science and mathematics. Teach Sci 58:21–29

Hobbs L (2013) Teaching "out-of-field" as a boundary-crossing event: factors shaping teacher identity. Int J Sci Math Educ 11:271–297. https://doi.org/10.1007/s10763-012-9333-4

Hobbs L, Törner G (2019a) Teaching out-of-field as a phenomenon and research problem. In: Hobbs L, Törner G (eds) Examining the phenomenon of "teaching out-of-field." Springer Nature Singapore, Singapore, pp 3–20

Hobbs L, Törner G (eds) (2019b) Examining the phenomenon of "teaching out-of-field." Springer Nature Singapore, Singapore
Ingersoll RM (1999) The problem of underqualified teachers in American secondary schools. Educ Res 28:26–37. https://doi.org/10.3102/0013189X028002026
Ingersoll RM (2019) Measuring out-of-field teaching. In: Hobbs L, Törner G (eds) Examining the phenomenon of "teaching out-of-field." Springer Nature Singapore, Singapore, pp 21–52
Institute of Physics (2010a) Physics an investment for the future
Institute of Physics (2010b) Physics and: teacher numbers
Kortemeyer G (2016) The losing battle against plug-and-chug. Phys Teach 54:14–17. https://doi.org/10.1119/1.4937964
Lakshmanan A, Heath BP, Perlmutter A, Elder M (2011) The impact of science content and professional learning communities on science teaching efficacy and standards-based instruction. J Res Sci Teach 48:534–551. https://doi.org/10.1002/tea.20404
Lauchlan E (2018) Science timetable models research
Loewenberg Ball D, Thames MH, Phelps G (2008) Content knowledge for teaching: what makes it special? J Teach Educ 59:389–407. https://doi.org/10.1177/0022487108324554
Martin NK, Sass DA (2010) Construct validation of the behavior and instructional management scale. Teach Teach Educ 26:1124–1135. https://doi.org/10.1016/j.tate.2009.12.001
Martin NK, Sass DA, Schmitt TA (2012) Teacher efficacy in student engagement, instructional management, student stressors, and burnout: a theoretical model using in-class variables to predict teachers' intent-to-leave. Teach Teach Educ 28:546–559. https://doi.org/10.1016/j.tate.2011.12.003
Millar R, Osborne J (1998) Beyond 2000: science education for the future—a report with ten recommendations. King's College London, London
Mizzi D (2013) The challenges faced by science teachers when teaching outside their specific science specialism. Acta Didact Napocensia 6:1–6
Monk DH (1994) Subject area preparation of secondary mathematics and science teachers and student achievement. Econ Educ Rev 13:125–145. https://doi.org/10.1016/0272-7757(94)90003-5
Morris DB, Usher EL, Chen JA (2017) Reconceptualizing the sources of teaching self-efficacy: a critical review of emerging literature. Educ Psychol Rev 29:795–833. https://doi.org/10.1007/s10648-016-9378-y
Mortimer EF, Scott PH (2003) Meaning making in secondary science classrooms. Open University Press, Maidenhead
Neuschatz M, McFarling M (2003) Broadening the base: high school physics education at the turn of a new century. College Park, Maryland
Nie Y, Tan GH, Liau AK et al (2013) The roles of teacher efficacy in instructional innovation: its predictive relations to constructivist and didactic instruction. Educ Res Policy Pract 12:67–77. https://doi.org/10.1007/s10671-012-9128-y
OECD (2005) Teachers matter: attracting, developing and retaining effective teachers. OECD Publishing
Oyserman D, Elmore K, Smith G (2012) Self, self-concept, and identity. In: Leary MR, Tangney JP (eds) Handbook of self and identity. The Guilford Press, pp 69–104
Physics Education Committee (1981) Shortage of physics teachers. Royal Society and Institute of Physics, Luton
Porsch R, Whannell R (2019) Out-of-field teaching affecting students and learning: whats known and unknown. In: Hobbs L, Törner G (eds) Examining the phenomenon of "teaching out-of-field." Springer Nature Singapore, Singapore, pp 179–191
Price A, Vale C, Porsch R, et al (2019) Teaching out-of-field internationally. In: Hobbs L, Törner G (eds) Examining the phenomenon of "teaching out-of-field." Springer Nature Singapore, Singapore, pp 53–83

Rowan B, Chiang F-S, Miller RJ (1997) Using research on employees' performance to study the effects of teachers on students' achievement. Sociol Educ 70:256–284. https://doi.org/10.2307/2673267

Shah L, Jannuzzo C, Hassan T et al (2019) Diagnosing the current state of out-of-field teaching in high school science and mathematics. PLoS ONE 14:e0223186. https://doi.org/10.1371/journal.pone.0223186

Shanker A (1985) Education's "dirty little secret." New York times Section 4:E9

Sheppard K, Padwa L, Kelly AM, Krakehl R (2020) Out-of-field teaching in chemistry and physics: an empirical census study. J Sci Teacher Educ 00:1–22. https://doi.org/10.1080/1046560X.2019.1702268

Shulman LS (1986) Those who understand: knowledge growth in teaching. Educ Res 15:4–14. https://doi.org/10.3102/0013189X015002004

Smithers A, Robinson P (2005) Physics in schools and colleges: teacher deployment and student outcomes. Carmichael Press, Buckingham, England

Smithers A, Robinson P (2008) Physics in schools IV: supply and retention of teachers. Carmichael Press, Buckingham

Straw S, Poet H, Worth J, et al (2017) Maths and physics teacher supply package. Department for Education; National College for Teaching & Leadership

Tschannen-Moran M, Hoy AW, Hoy WK (1998) Teacher efficacy: its meaning and measure. Rev Educ Res 68:202–248. https://doi.org/10.3102/00346543068002202

Tschannen-Moran M, Woolfolk Hoy A (2001) Teacher efficacy: capturing an elusive construct. Teach Teach Educ 17:783–805. https://doi.org/10.1016/S0742-051X(01)00036-1

UK Commission for Employment and Skills (2015) Reviewing the requirement for high level STEM skills

Vale C, Drake P (2019) Attending to out-of-field teaching: implications of and for educational policy. In: Hobbs L, Törner G (eds) Examining the phenomenon of "teaching out-of-field." Springer Nature Singapore, Singapore, pp 195–215

Viennot L (2021) Incomplete explanations in physics teaching: discussing the rainbow with student teachers. Eur J Phys 42. https://doi.org/10.1088/1361-6404/ac1500

Zee M, Koomen HMY (2016) Teacher self-efficacy and its effects on classroom processes, student academic adjustment, and teacher well-being: a synthesis of 40 years of research. Rev Educ Res 86:981–1015. https://doi.org/10.3102/0034654315626801

(2002) The no child left behind act. USA Congress

How Do Teachers Interpret and Respond to Students' Explanations?

Danijela Dodlek, Eugenia Etkina, and Gorazd Planinsic

Abstract We present the findings of the study of in-service and pre-service teachers' ability to interpret student explanations and respond to them productively. We used the Tasks of Teaching from the content knowledge for teaching energy (CKT-E) framework to analyse the participants' responses to a survey that asked them to comment on students' explanations for a non-traditional problem involving momentum and energy. The survey contained the text of the problem and four different students' explanations of their answer choice. The participants were asked to comment on the student's strengths/weaknesses, and describe how they would respond to the student.

1 Introduction: Model of Teacher Knowledge

To help students learn, teachers need to be able to recognize students' strengths and weaknesses. They have to recognize productive ideas in student responses even when those ideas are embedded in an incorrect answer or if the student's language is sloppy but the student is on the right track. This recognition is needed in order to give that student feedback in real time. The feedback needs to be meaningful and needs to address the students' strengths as well as weaknesses and it has to build on the students' existing knowledge (Shepard 2000).

The above-mentioned skills are a part of teachers' content knowledge for teaching (CKT). The construct of CKT fuses teaching knowledge and skill and specific subject matter. This means that teachers have to understand subject matter in ways that are unique to teaching (e.g. teachers need to know what challenges a specific topic might

---

D. Dodlek (✉)
Department of Physics, Josip Juraj Strossmayer University of Osijek, Trg Ljudevita Gaja 6, 31000 Osijek, Croatia
e-mail: ddodlek@fizika.unios.hr

D. Dodlek · G. Planinsic
Faculty of Mathematics and Physics, University of Ljubljana, Jadranska ulica 19, 1000 Ljubljana, Slovenia

E. Etkina
Rutgers University, 10 Seminary Place, New Brunswick, NJ 08901, USA

S. Faletič and J. Pavlin (eds.), *Teaching and Learning Physics Effectively in Challenging Times*, Challenges in Physics Education, https://doi.org/10.1007/978-3-031-72541-8_20

present to students, they need to know how to ask productive questions, how to build on student ideas or how to provide explanations that can move students understanding forward (Ball and Bass 2003).

The idea of CKT was developed by Ball and colleagues (2008). It arose from the construct of pedagogical content knowledge (PCK) that was introduced by Shulman (1986). While CKT assessments for teaching mathematics and reading have been created a few years ago (Phelps and Schilling 2004; Ball et al. 2008), the CKT assessment for physics teachers has been developed only recently (Etkina et al. 2018). The authors developed an assessment of physics teachers' content knowledge for teaching in a specific content area—the teaching of energy in the context of mechanics in high schools (CKT-E).

The model of CKT-E includes two components. The first component is the Tasks of Teaching (ToTs) and the second component is the Student Energy Targets (SETs). ToTs represent activities which teachers do while helping students learn any content, and SETs represent important disciplinary ideas, science practices and cross-cutting concepts that are significant for student learning of energy in the context of mechanics (Etkina et al. 2018). The goal of the study reported here is to use the framework of ToTs to analyse the participants responses to student ideas and to answer the following research questions:

- How do in-service and pre-service teachers interpret productive and problematic aspects of student explanations?
- How do in-service and pre-service teachers' interpretations compare to the interpretations of physics education experts?
- What is the difference between answers to Questions 1–2 for in-service and pre-service teachers?

2 Models of Student Ideas

Even though many science teachers have a dichotomous perspective on student ideas (Otero and Nathan 2008), research showed that student thinking is complex, context-dependent, and progressive (Smith et al. 1993). This throws us back in time when researchers investigated how to treat student ideas—the idea as old as the one that students are "tabula rasa" when they come to classrooms. The starting point in helping students learn is acknowledging that they come to classrooms with prior knowledge (Zull 2002; Dehaene 2021). Exploring students' existing knowledge is the next essential step.

There are two opposing approaches which researchers take in exploring student ideas. One approach classifies student ideas as "misconceptions" (Clement 1982; McCloskey 1983; Halloun and Hestenes 1985; Kaltakci Gurel et al. 2016), and the other classifies them as "p-prims" or "resources" (diSessa 1993; Hammer 2000).

When researchers talk about student misconceptions they want to say that students have a robust conception which is incorrect and needs to be eliminated from their minds so that the correct conception can take its place (Posner et al. 1982). If we remove the incorrect conceptions from the students' minds we cannot help them build new ones because, as brain research tells us (Dehaene 2021), new ideas can only be built if they are connected to previously held beliefs. The "misconceptions" view focuses only on deficiencies of the students' ideas and does not consider any productive ideas that may be used as learning resources.

An alternative to the "conceptual change" perspective shortly described above is "knowledge-in-pieces (KiP)" perspective (Hammer and Elby 2003). The beginning of the development of this perspective can be traced to the work of Andrea diSessa. In 1993 diSessa (diSessa 1993) introduced the concept of "phenomenological primitives", or "p-prims". The name itself illustrates that p-prims are connected to how events in the real world (phenomena) are interpreted and are basic elements of memory (Harlow and Bianchini 2020). Basically, p-prims are tiny and context-dependent student ideas that are neither correct nor incorrect in and of themselves. Sometimes they are appropriate for the situation and sometimes they are not, depending on the students' prior experiences and the language that we use. We can use the existing p-prims as productive resources on which to build students' new knowledge (Smith et al. 1993).

The concept of resources was further refined by David Hammer and his associates (Hammer 2000; Hammer and Elby 2003; Hammer et al. 2005). They suggested that resources are fragments of existing knowledge that can be activated independently or together with other resources while student reasons about a physics topic. Hammer and colleagues differentiate between conceptual and epistemological resources. Conceptual resources differ from p-prims in size and breadth and many of them are physics related. Epistemological resources explain how students and teachers approach learning and teaching, respectively (Hammer and Elby 2003). The work of Hammer and Elby led to the articulation of the knowledge-in-pieces model of students' ideas or KiP.

The KiP model of student ideas is more consistent with our knowledge of the brain (Zull 2002; Dehaene 2021). Being able to recognize productive ideas and building on them to help students learn is at the heart of this model.

In our study we investigated how teachers interpret student ideas, more specifically, we looked to what extent teachers recognise productive and problematic reasoning from the student's written work. We were also interested in how teachers respond to student ideas. We examined teachers' responses to see whether there are signs of building on productive elements that they identified as student's strengths, as well as whether the teachers addressed problematic elements that they identified as apparent weaknesses in the student's explanations. With all this in mind we reviewed what the research community had to say about interpreting and responding to student ideas.

Knowledge-in-pieces perspective suggests that teachers should build on instead of discarding or ignoring student ideas. Therefore, using this model as a foundation of our understanding of student ideas, we emphasise the importance of recognising the strengths in students' explanations and building on them.

3 The Study

In this section we describe the problem and provide the solution of the problem, and describe our participants.

3.1 *The Problem and the Solution*

In this research we studied how teachers interpret and respond to students' explanations to a non-traditional problem (see Fig. 1). The problem is non-traditional as it requires students to analyse limiting cases in order to assess the answer and to justify their conclusions.

The correct answer to this problem is graph (b). We assume that the spring is ideal and that the friction between the surface and the cars can be ignored. The spring and both of the cars are in our system. The spring transfers elastic potential energy to the cars by doing internal work on the cars.

The forces that the spring exerts on the cars are equal in magnitude at every moment during the recoil. These forces are exerted on the cars for the same amount of time and during the recoil the heavier car moves with smaller acceleration than

You place two toy cars on a horizontal table and connect them with a light compressed spring as shown in figure. The spring tries to push the cars apart, but they are tied together by a thread. When the thread is burned, the spring pushes the cars apart. You decide to investigate how the final speed of car 2 depends on the mass of car 1. You run several experiments changing m_1 and measuring v_2 while keeping the compression of the spring and the mass of car 2 constant. Which of the v_2-versus-m_1 graphs do you expect to obtain? Evaluate the graphs by analysing limiting cases. Explain your reasoning. Solve the problem without deriving the equations.

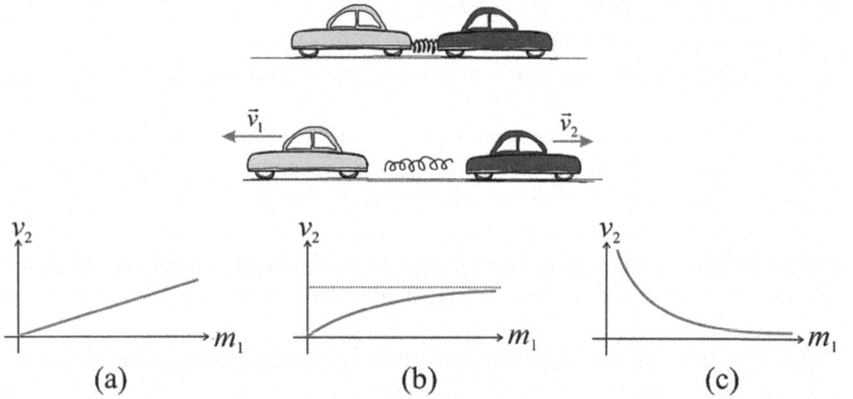

Fig. 1 The problem students were solving [published in Etkina et al. (2019)]. Figure from Dodlek et al. (2024). Published by the American Physical Society under the terms of the Creative Commons Attribution 4.0 International license

the lighter car. The displacement of the heavier car is greater than that of the lighter car due to the smaller acceleration of the heavier car and the equal amounts of time that the forces are exerted on the cars. As a result, the internal work done by the spring on the heavier car is smaller than that for the lighter car. Given that both cars start from rest and due to the smaller internal work done on the heavier car, the final kinetic energy of the heavier car will be smaller than the final kinetic energy of the lighter car.

Therefore, if car 1 has very small mass, it will have greater kinetic energy than car 2, and car 2 will move at lower speed. This rules out answer (c). On the other hand, if car 1 has very large mass, car 2 will have larger kinetic energy than car 1, and car 2 will move at larger speed. Since all energy is initially stored in the spring and we keep the compression of the spring the same in all experiments, no matter how much we increase the mass of car 1, the speed of car 2 will reach some asymptotic value. This rules out answer (a).

3.2 The Participants

The participants of the study were in-service (42) and pre-service physics teachers (24) who voluntarily took an anonymous survey. All together there were six groups; three groups of in-service teachers and three groups of pre-service teachers. Five out of six groups were from two small European countries—A and B, and one group of in-service teachers was international—they were from Europe, Asia and North America. All of the groups are convenience samples.

Women made up 43% and men made up 52% of the in-service physics teachers. Five percent of the in-service teachers did not identify as male or female. Regarding the number of years of experience in teaching physics, 21% of the in-service teachers have up to nine years of experience, 55% have between 10 and 24 years, and 24% have 25 or more years of experience in teaching physics.

Women made up 38% and men made up 58% of the pre-service physics teachers. One pre-service teacher opted not to answer the gender-related question. Undergraduate students accounted for 62.5% of the pre-service teachers, while graduate students accounted for 37.5%.

4 Methods

In this chapter we describe how we collected the data for our study and describe how we developed coding schemes to analyse teachers' responses.

4.1 Data Collection

We reached teachers A by emailing invitations to physics consultants across the country who sent our request to physics teachers at high schools. Additionally, we emailed our colleagues from major universities in countries A and B who then helped to forward the survey to the relevant in-service and pre-service teachers. Also, to reach in-service teachers outside of countries A and B, we used the Facebook group "Exploring and Applying Physics". We called this group of teachers "international teachers" as they participated in an international on-line community of teachers using or learning about the Investigative Science Learning Environment (ISLE) approach to teaching physics (Etkina 2015). We used a six-item survey made with Google Forms to collect the data.

4.2 Data Analysis

In order to answer research questions, we analysed teachers' responses to the explanations written by the students. The CKT-E framework served as the foundation for the analysis. We used two Tasks of Teaching (Etkina et al. 2018). One was Task II. Designing, selecting, and sequencing learning experiences and activities, and the other was Task III. Monitoring, interpreting, and acting on student thinking. Four subtasks were relevant to our work, and we further classified these subtasks into the following six categories:

- Ia—interpreting productive reasoning,
- Ib—interpreting problematic reasoning,
- II—providing descriptive feedback,
- III—engaging in metacognition,
- IVa—building on productive elements, and
- IVb—addressing problematic elements.

Three members of the research team (co-authors of this paper) and an independent expert in physics education responded to all of the questions for each student's explanation. Using the aforementioned categories, which we used to name our codes, and the analysis of the expert responses we developed coding schemes to analyse participants' responses. We took experts' and teachers' responses and compared what they wrote for each of the codes. In order to quantify the differences between experts' and teachers' responses we developed numerical scores for each of the codes (0, 1 or 2), together with the description of the criteria for assigning these scores.

5 Findings

In this chapter we present the results on teachers' interpretation of the productive and problematic reasoning and to what extent do teachers build on productive elements and address problematic ones. Presented findings are for responses to all three incorrect explanations combined.

5.1 Interpreting Productive and Problematic Reasoning

Figures 2 and 3 show that more than 70% of responses from all of the groups of participants, with the exception of the less experienced pre-service teachers B, received a score of 1 for interpreting productive and problematic reasoning in all incorrect students' explanations. This means that the participants mentioned some aspects of the productive and problematic reasoning but not all of the aspects found by experts.

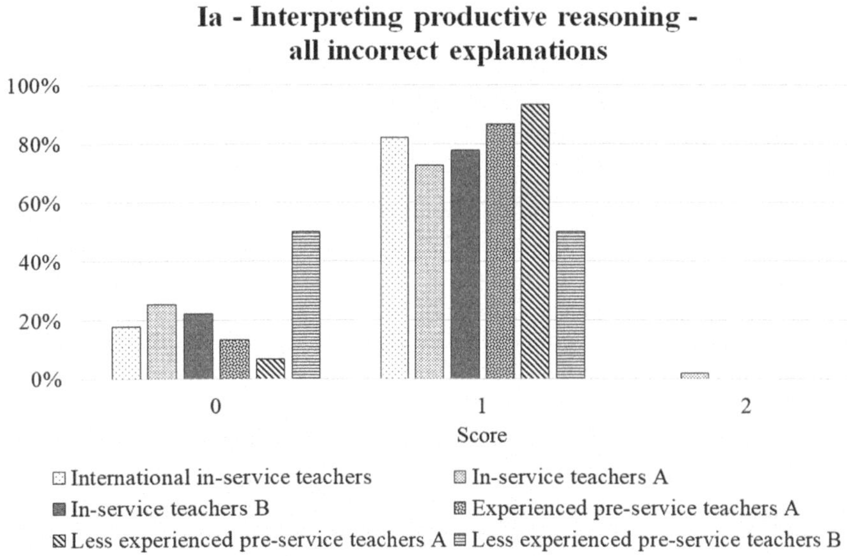

Fig. 2 The scores for interpreting productive reasoning (code Ia)

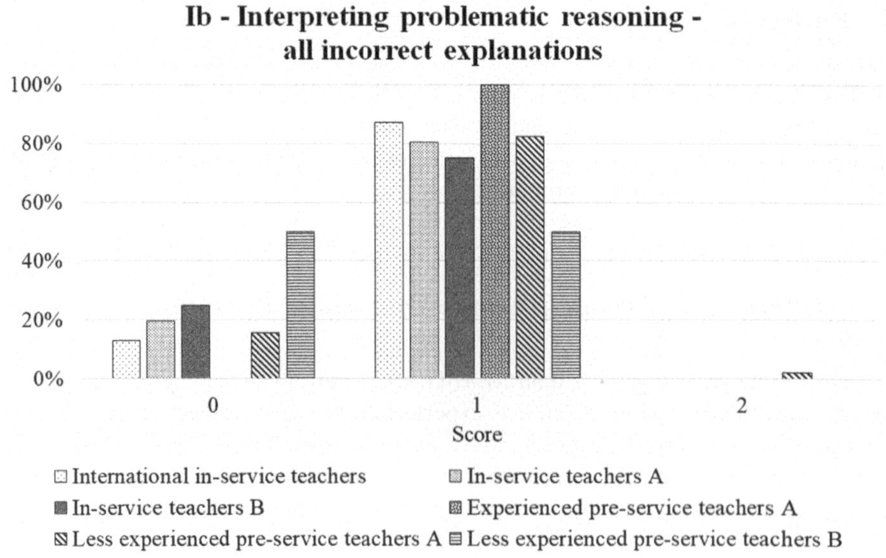

Fig. 3 The scores for interpreting problematic reasoning (code Ib)

5.2 *Building on Productive Elements and Addressing Problematic Elements*

Figures 4 and 5 show that responses from all of the groups of participants mostly received a score of 0 for building on productive elements, and that responses from all of the groups of participants, except for the in-service teachers B and the less experienced pre-service teachers B, mostly received a score of 2 for addressing problematic elements they mentioned in their response. This means that the participants usually did not build on the students' productive ideas but they did address all of the problematic elements they mentioned in their response.

6 Discussion

We see that the majority of participants were able to identify and interpret at least some aspects of the productive and problematic reasoning in incorrect students' explanations. We could not say that the participants were not able to identify students' productive and problematic reasoning as we only know how they did on our survey. Therefore, we can only state what they wrote in their responses. One of the possible explanations for this pattern was that participants simply did not mention all of the productive and problematic elements they found in the students' explanations.

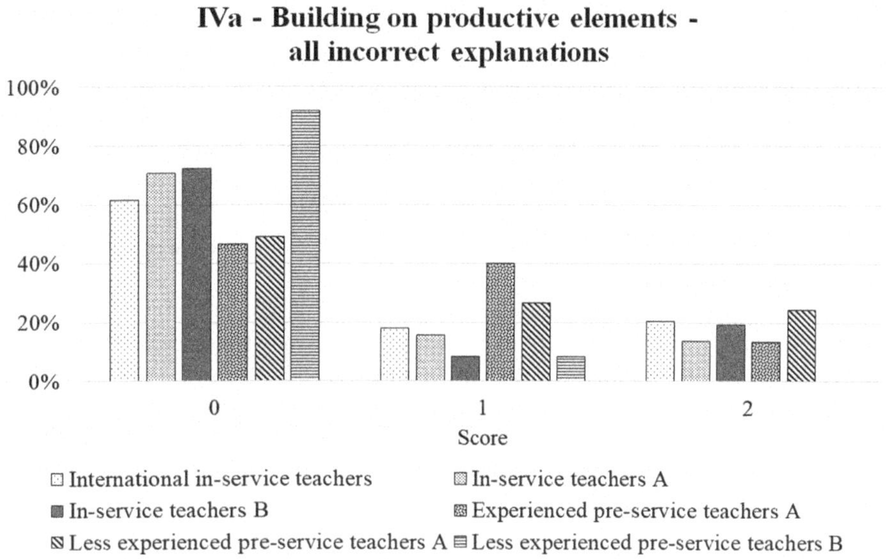

Fig. 4 The scores for building on productive elements (code IVa)

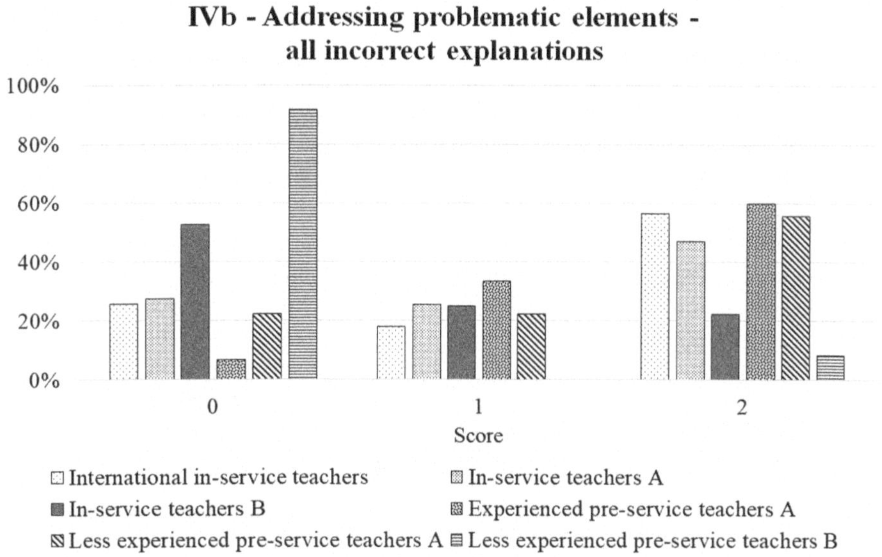

Fig. 5 The scores for addressing problematic elements (code IVb)

When we examined the data further we found another interesting pattern. Most participants did not mention limiting cases at all, whether in strengths or in weaknesses. This finding is especially interesting because the problem itself says "Evaluate the graphs by analysing limiting cases". A possible explanation is that the participants might not be used to solving the problems using limiting case analysis. Therefore, they mostly did not mention limiting cases in their responses. Another possible explanation of this pattern is that the participants realised that this is a "momentum and energy" problem and focused more on those aspects when interpreting students' strengths and weaknesses.

When analysing a situation from a momentum and/or energy point of view it is essential to start by choosing a system of interest. In our study however, we found that most participants did not mention the choice of the system unless it was explicitly mentioned in the student's explanation. In three (out of four) students' explanations the system was implicitly described. This is similar to Seeley et al. (2019) where they found that teachers have trouble identifying the system that is implied in a student's response.

Our findings show that the participants scored higher on addressing apparent problematic elements than on building on productive ones. This means that the participants usually did not build on the students' productive ideas that they found in students' responses but they did address all of the problematic elements they mentioned. A possible explanation for this pattern is that the participants focused too much on the problematic aspects of students' reasoning because they follow misconceptions model of student ideas and they try to help their students get rid of "wrong" ideas. If teachers think that students are carriers of misconceptions then they focus on the problematic things and they try to "fix" what is wrong (Clement 1982; McCloskey 1983; Halloun and Hestenes 1985; Kaltakci Gurel et al. 2016). However, if they think students have knowledge in pieces then they focus on identifying productive things and they build on students' existing knowledge to foster student learning (diSessa 1993; Hammer 2000). Even though the KiP model of student ideas is around thirty years old, our findings indicate that teachers are still struggling with building on students' existing ideas.

7 Limitations of the Study and Implications for Instruction

Our study has some major limitations. Relatively small number of participants does not allow us to make generalisations and all of our participants were a convenience sample. Also, we do not know what teachers would do in their classrooms. We can only say what they did on our survey.

Since our study has significant limitations, we can only infer some implications for instruction. Teacher preparation programs need to help pre-service teachers learn how to build on students' existing ideas. This needs to be done systematically over multiple courses (Etkina et al. 2017). Professional development programs for in-service teachers need to have specific exercises where the participants are provided with examples of student work and engage in activities similar to those in our study with the focus on building on student ideas.

References

Ball DL, Bass H (2003) Toward a practice-based theory of mathematical knowledge for teaching. In: Simmt E, Davis B (eds) Proceedings of the 2002 annual meeting of the Canadian Mathematics Education Study Group, pp 3–14

Ball DL, Thames MH, Phelps G (2008) Content knowledge for teaching: what makes it special? J Teach Educ 59:389–407. https://doi.org/10.1177/0022487108324554

Clement J (1982) Students preconceptions in introductory mechanics. Am J Phys 50:66–71. https://doi.org/10.1119/1.12989

Dehaene S (2021) How we learn: the new science of education and the brain. Penguin Publishing Group

diSessa AA (1993) Toward an epistemology of physics. Cogn Instr 10:105–225. https://doi.org/10.1080/07370008.1985.9649008

Dodlek D, Planinsic G, Etkina E (2024) How to help students learn: an investigation of how in- and pre-service physics teachers respond to students' explanations. Phys Rev Phys Educ Res 20:010120. https://doi.org/10.1103/PhysRevPhysEducRes.20.010120

Etkina E (2015) Millikan award lecture: students of physics—listeners, observers, or collaborative participants in physics scientific practices? Am J Phys 83:669–679. https://doi.org/10.1119/1.4923432

Etkina E, Gregorcic B, Vokos S (2017) Organizing physics teacher professional education around productive habit development: a way to meet reform challenges. Phys Rev Phys Educ Res 13:010107. https://doi.org/10.1103/PhysRevPhysEducRes.13.010107

Etkina E, Gitomer D, Iaconangelo C et al (2018) Design of an assessment to probe teachers' content knowledge for teaching: an example from energy in high school physics. Phys Rev Phys Educ Res 14:010127. https://doi.org/10.1103/PhysRevPhysEducRes.14.010127

Etkina E, Planinsic G, Van Heuvelen A (2019) College physics: explore and apply, 2nd edn. Pearson, New York

Halloun I, Hestenes D (1985) The initial knowledge state of college physics students. Am J Phys 53:1043–1055. https://doi.org/10.1119/1.14030

Hammer D (2000) Student resources for learning introductory physics. Am J Phys 68:S52–S59. https://doi.org/10.1119/1.19520

Hammer D, Elby A (2003) Tapping epistemological resources for learning physics. J Learn Sci 12:53–90. https://doi.org/10.1207/S15327809JLS1201_3

Hammer D, Elby A, Scherr R, Redish E (2005) Resources, framing, and transfer. Transfer of learning from a modern multidisciplinary perspective. Information Age Publishing, Greenwich, Connecticut, pp 89–119

Harlow DB, Bianchini JA (2020) Knowledge-in-pieces. In: diSessa AA, Hammer D (eds) Science education in theory and practice, an introductory guide to learning theory. Springer Texts in Education, pp 389–401

Kaltakci Gurel D, Eryilmaz A, McDermott L (2016) Identifying pre-service physics teachers' misconceptions and conceptual difficulties about geometrical optics. Eur J Phys 37:045705. https://doi.org/10.1088/0143-0807/37/4/045705

McCloskey M (1983) Intuitive physics. Sci Am 248:122–131. https://doi.org/10.1038/scientificamerican0483-122

Otero V, Nathan M (2008) Preservice elementary teachers' views of their students' prior knowledge of science. J Res Sci Teach 45:497–523

Phelps G, Schilling S (2004) Developing measures of content knowledge for teaching reading. Elem Sch J 105:31–48

Posner GJ, Strike KA, Hewson PW, Gertzog WA (1982) Accommodation of a scientific conception: toward a theory of conceptual change. Sci Educ 66:211–227. https://doi.org/10.1002/sce.3730660207

Seeley L, Vokos S, Etkina E (2019) Examining physics teacher understanding of systems and the role it plays in supporting student energy reasoning. Am J Phys 87:510–519. https://doi.org/10.1119/1.5110663

Shepard L (2000) The role of assessment in a learning culture. Educ Res 29:4–14. https://doi.org/10.3102/0013189X029007004

Shulman LS (1986) Those who understand: knowledge growth in teaching. Educ Res 15:4–14. https://doi.org/10.3102/0013189X015002004

Smith JP, diSessa AA, Roschelle J (1993) Misconceptions reconceived: a constructivist analysis of knowledge in transition. J Learn Sci 3:115–163. https://doi.org/10.1207/s15327809jls0302_1

Zull JE (2002) The art of changing the brain: enriching teaching by exploring the biology of learning. Stylus Publishing, Sterling, Virginia

Utilizing Physics Teachers' Epistemological and Pedagogical Conceptions to Develop a Research Instrument for Training Programs

Kalliopi Meli, Dimitrios Koliopoulos, and Konstantinos Lavidas

Abstract In this research, we used a theoretical framework for physics teachers' training as a starting point to develop a new research instrument (online questionnaire) for exploring teachers' conceptions on the epistemology and pedagogy of physics, using thermodynamics as the field of reference. Our goal was to validate the theoretical framework and also use it as a methodological framework to make informed decisions on the design and evaluation of future training programs based on teachers' ideas. Our analysis ($N = 42$ in-service physics teachers in Athens, Greece) indicated four factors that can facilitate a practical use of the theoretical framework. These factors addressed the teachers' prior knowledge, alternative instructional design, traditional framework perception, and intention to change their current instruction. We suggest that further research should address larger populations and make adaptations for additional physics fields to validate and generalize these results.

1 Introduction

Physics teachers' training programs is a common practice for in-service teacher professional development. In this research we support that such programs should consider teachers' prior ideas about the epistemological and pedagogical dimensions of physics as a subject to be taught (Porlán et al. 2004). For the design and evaluation of a webinar series for physics teachers in Greece, we utilized a theoretical framework for physics teachers' training (Gil-Pérez and Pessoa de Carvalho 1997). In this respect, we developed a new research instrument (questionnaire) to identify in-service physics teachers' epistemological and pedagogical conceptions, using thermodynamics as the field of reference. Based on our results, we suggest the utilization of the theoretical framework as a methodological framework as well that can guide the design and evaluation of physics teachers' training programs.

K. Meli (✉) · D. Koliopoulos · K. Lavidas
Department of Educational Sciences and Early Childhood Education, University of Patras, University Campus, 26504 Rio, Greece
e-mail: kmeli@upatras.gr

S. Faletič and J. Pavlin (eds.), *Teaching and Learning Physics Effectively in Challenging Times*, Challenges in Physics Education, https://doi.org/10.1007/978-3-031-72541-8_21

2 Theoretical Framework

Gil-Pérez and Pessoa de Carvalho (1997) suggested a theoretical framework for physics teachers' training that was an appropriate foundation for the development of both research instruments and instructional materials (Gil-Pérez and Pessoa de Carvalho 1997). This framework included four components (or *fundament points*): (a) knowing physics as the subject matter to be taught, (b) knowing teachers' spontaneous ideas on physics and on teaching and learning physics, (c) acquiring theoretical knowledge about the physics teaching and learning process, and (d) teachers' involvement in physics education research and innovation.

2.1 Component 1: Knowing Physics as the Subject Matter to Be Taught

Regarding the first component, the training program should be concerned with teachers' knowledge of physics as a subject matter to be taught. Teachers give prominence to the conceptual aspect of scientific knowledge rather than its methodological and cultural aspects (Fadaei 2012). We consider that teachers' scientific knowledge is bound, in contrast to their epistemological knowledge of physics, since the latter is more "optional" to delve into during physics studies. The main epistemological dimension of physics concerns the history of science, which is usually disconnected from the subject matter to be taught (Apostolou and Koulaidis 2010; Kanderakis et al. 2011). Epistemological knowledge of physics includes conceptual issues, methodological choices, and interactions between science, technology, and society that have driven the construction of physics knowledge (Gil-Pérez et al. 2005). In addition, it is related to the enrichment of scientific knowledge with modern discoveries, to support the dynamic development of physics. Based on the above, teachers are expected to adjust and choose the content they present to students to strengthen an epistemologically correct image of science, but also to be eager to enrich their personal knowledge in this direction.

Fadaei (2012) conducted a survey to 25 physics teachers regarding this first component of the theoretical training framework, asking them to rate its elements in terms of importance. Teachers considered quite important to know how the scientific knowledge has emerged, as well as the interactions between science, technology, and society. For this aspect, more detailed conclusions were added from Irez's (2006) research on the Nature of Science (NOS) in 15 pre-service science teachers. While 86% recognized that society influences science, at the same time only 53% considered science as a cultural product and just 20% believed that scientific knowledge was depended on the social context. Regarding the knowledge of the methodological developments that supported the scientific knowledge, the teachers in Fadaei's (2012) study characterized it as rather unimportant. According to them, the most important element was the acquisition of knowledge related to new scientific discoveries.

Finally, they considered that it was important to be able to choose appropriate content for their teaching, as well as to be predisposed to deepen their own epistemological knowledge.

2.2 *Component 2: Knowing Teachers' Spontaneous Ideas on Physics and on Teaching and Learning Physics*

The second component suggests that for the teachers' training it is important we know their spontaneous ideas on physics and also on the teaching and learning of physics. The most prevailing characteristic is the role of experiments in agreement with an empirical-deductive approach (Aguirre et al. 1990). Educational research highlights a range of physics teachers' conceptions on physics epistemological features, that usually comply to the textbooks and the corresponding teacher instructions. Many of the conceptions that teachers carry are compatible with the positivist tradition of physics. These conceptions have a significant impact on the way students conceive physics as a construct (e.g., Tsai 2007). In this respect, the experimental-inductive approach, during which the formulation of hypotheses and theories is underpinned by free observation and subsequent conclusion drawing, is particularly widespread (Aguirre et al. 1990; Apostolou and Koulaidis 2010). The distorted role of the experiment is consistent with the strict imperatives of an inflexible "scientific method," which predetermines the path to the emergence of established knowledge, mainly through formalistic strategies, therefore obscuring the constant need for creativity that characterizes the construction of scientific knowledge. In the same context belongs the fragmentation of physics into fields, which indeed serve practical purposes, but at the same time distort the unity of science whenever the connections between them are not highlighted. Considering the above, teachers often present physics as a linear scientific achievement, which has not experienced deep crises or reorganizations. Additionally, some general epistemological conceptions that teachers systematically imply, seem to play an important role in the teaching and learning of physics. The reasonings that support scientific knowledge are often treated as "common sense," relying on quick and self-evident answers, which appear as indisputable facts (Ogborn 2008). Another teachers' conception in this respect is the one that projects physics as scientific knowledge "for the few," i.e., mainly for those who can understand the mathematical formalism through which modern science is usually represented. It is therefore not surprising that teachers present the physics construct as the achievement of some genius scientists, who worked individually and acted independently of the socio-economic and technological developments of their time.

Aguirre et al. (1990) studied conceptions of science, teaching and learning held by 74 physics teachers. Concerning their conceptions of physics, 36.2% of the teachers' statements revealed an experimental-inductive conception of science. This outcome was in line with Irez's (2006) results, where 54% of the sample supported the empirical nature of science as a conclusion derived solely from direct evidence. In Aguirre et al. (1990) there was a small percentage (3.8%) who viewed science as a "three-step" process in the sense of the predetermined scientific method. However, the corresponding results in Irez's (2006) sample gave much higher percentages, as 80% supported the existence of a unique scientific method that, according to 54% of the participants, was involving predetermined stages. For the same issues, in the research of Flores et al. (2008) with 12 pre-service physics teachers, 58% of the sample expressed empirical or positivist views concerning conceptual elements, and 62% expressed such views concerning experimental elements. All conceptions mentioned above were also confirmed by Brickhouse (1989) through the interviews she conducted with three teachers (the whole sample), in combination with the lesson observations in their classrooms. In this study, two of the teacher participants agreed with the "linearity" in the construction of knowledge in physics. This finding is also related to the teachers' conceptions in Irez's (2006) study, in which 68% stated that theories and laws are in hierarchical relationships with each other. Finally, one teacher from Brickhouse's (1989) sample intensively formed an image of scientists as isolated and peculiar personalities during his teaching.

2.3 *Component 3: Acquiring Theoretical Knowledge About the Physics Teaching and Learning Process*

It is essential for a training program to engage in the introduction and/or negotiation of the teachers' epistemological conceptions. However, epistemological conceptions directly or indirectly interact with pedagogical conceptions, therefore it is also necessary for a training program to include pedagogical framework elements that concern the teaching and learning of physics in a unified manner (Belo et al. 2014). Regarding the third component, these elements refer to a constructivist view, and especially its socio-cognitive version, which places students' ideas at the center of the teaching process (Mansour 2009). In this context, it should be highlighted that students have already formed conceptions about the scientific subject to be discussed before they attend a formal education setting and any construction of the scientifically accepted knowledge occurs in accordance with their personal pre-existing knowledge. Consequently, learning is meaningful only if each student succeeds in constructing it in a way that meets their individualized cognitive needs and interests (Watts 2013). Yet, only a few teachers appreciate students' prior ideas and the way they construct their knowledge (Gil-Pérez and Pessoa de Carvalho 1997). Furthermore, the utilization of student-cooperative teaching and learning can play a crucial role in highlighting scientific knowledge because of social interactions (Mansour 2009). Finally, it is important that the classroom environment is permeated by the evaluative commitments that also characterize realistic scientific research environments, and not by students' enforcement to engagement with physics (Briscoe 1991).

In Demirci's (2015) study with a sample of 135 pre-service physics teachers who responded in writing to the question "how should physics be taught?", only 15% of them felt that students' prior or alternative ideas should be considered in this respect. In the same sample, 11 interviews were conducted and only 9% of interviewees raised this issue on a constructive basis. At the same time, 27% stated that the "transmission" of knowledge from the teacher to the students is how knowledge constructed, while 64% approached the issue with a conflation of traditional and constructive approaches. Accordingly, in of Flores et al. (2008) only 30% had a constructivist approach to students' prior ideas and the way students construct their knowledge. Finally, from Demirci's (2015) interviews, again only 9% emphasized student-cooperative learning and suggested non-traditional ways of assessment.

2.4 Component 4: Teachers' Involvement in Physics Education Research and Innovation

Finally, the fourth component that concerns teachers' involvement in physics education research and innovation has not been sufficiently explored, indicating the distance that often exists between educational research and practice. According to the examined theoretical framework for physics teacher education, a fruitful strategy for introducing and/or negotiating their epistemological and pedagogical understandings is the teachers' involvement in educational research. This strategy can take on various characteristics. First, teachers should embrace the view that such collaboration could bring solutions to the problems they face in the classroom. Secondly, teachers should be open to the changes that their training may bring when challenging established conceptions and practices. Finally, teachers should actively participate in the investigation, innovation, and construction of physics teaching and learning and to join, as much as possible, the scientific community. Aspects of this axis are particularly unexplored at the empirical research level, which reflects, among other things, the distance between educational research and practice (Saha 2009).

3 Research Questions

Although several elements of the theoretical framework under discussion have been already applied and tested in the various contexts presented in the literature review, so far the framework has not been holistically examined. To do so, we utilize thermodynamics as the field of reference for developing a research instrument and, consequently, a training program based on the components and elements of this theoretical framework (Gil-Pérez and Pessoa de Carvalho 1997). We seek to validate the theoretical framework and explore its potential to work also as a methodological framework to meet physics teachers' needs regarding their training. Therefore, our research questions are the following:

- What is the structure of a methodological framework for a physics teachers' training program that addresses their epistemological and pedagogical conceptions on thermodynamics?
- Which items should be included in a questionnaire for the design and evaluation of a physics teachers' training program on thermodynamics epistemology and pedagogy?

4 Method

Based on the four components of the theoretical framework for physics teachers' training and the elements included in each of these components, we developed a questionnaire that included 22 closed-type questions (with a 5-point Likert scale). With the cooperation of the Regional Centre of Educational Planning for the upper secondary school (15–18 y.o. students) science curriculum (central Athens, Greece), we sent an online questionnaire to 150 physics teachers that taught thermodynamics during that school year (2020–2021). Out of them, 42 teachers answered the questionnaire. The participation was voluntary, the answers were anonymized, and the participants gave their informed consent for the use of their answers for the purpose of this study. All questionnaire items as well as the teachers' responses can be found in Meli et al. (2021).

The questionnaire was quantitatively analysed in SPSS. We started with parallel analysis (O'Connor 2000) to investigate the factorial structure of teachers' responses. Bartlett's Test of Sphericity ($p < 0.01$) indicated that the correlation between the items is adequate for factor analysis as well as Kaiser–Meyer–Olkin (KMO = 0.577) and Measure Sample Adequacy (MSA about 0.6) revealed that the teachers' answers created a satisfactory factorial structure (Tabachnick and Fidell 2007). Subsequently, we used exploratory factor analysis applying the principal component analysis method (Tabachnick and Fidell 2007). Since the factor analysis indicated non-negligible correlations among the factors, we used the oblique rotation to identify a simpler structure pattern. This analysis revealed four factors that presented satisfactory reliability as Cronbach's Alpha coefficient varied from 0.845 to 0.781 (Table 1). The factorial structure of four factors explains 62.7% of the total variation (Factor 1 = 28.4%, Factor 2 = 13.0%, Factor 3 = 11.5%, and Factor 4 = 9.8%). Factor 1 (*Teacher prior knowledge*) was significantly correlated with factor 2 (*Alternative instructional design*) and factor 4 (*Intention to change*). Factors 2 and 4 were additionally correlated with each other. The one factor that had no statistically significant correlation, and in fact had slightly negative relation with the rest factors, was factor 3 (*Traditional framework perception*).

Table 1 Reliability coefficients and product moment Pearson correlations among factors

	1	2	3	4
Factor 1	0.845[c]			
Factor 2	0.414[a]	0.836[c]		
Factor 3	−0.166	−0.173	0.836[c]	
Factor 4	0.314[b]	0.377[b]	−0.183	0.781[c]

[a] Correlation is significant at the 0.01 level (2−tailed)
[b] Correlation is significant at the 0.05 level (2−tailed)
[c] Cronbach's Alpha coefficients

Table 2 Teachers' education level (N = 42)

Degree	Frequency	Percentage frequency (%)	Cumulative percentage
Graduate	18	42.9	42.9
Postgraduate	19	45.2	88.1
Ph.D.	5	11.9	100.0

Table 3 Teachers' experience in teaching thermodynamics (N = 42)

Thermodynamics teaching experience (school years)	Frequency	Percentage frequency (%)	Cumulative percentage
1–3	6	14.3	14.3
4–10	19	45.2	59.5
≥11	17	40.5	100.0

5 Results

5.1 *Research Sample*

In terms of gender, the participants were 73% men and 27% women. More than half participants (57%) were holding a postgraduate degree (Table 2). The vast majority (86%) had already been teaching the thermodynamics course for more than four school years (Table 3).

5.2 *Factor 1: Teacher Prior Knowledge*

The first factor of the final factorial structure included items relating to teachers' prior knowledge, that referred to both epistemological and pedagogical aspects of thermodynamics. Table 4 presents the original theoretical framework elements and the corresponding questionnaire items that included in this factor, in a descending

Table 4 Factor 1: original theoretical framework element (with component and element number) and the respective questionnaire item in descending loading (λ) order

Theoretical framework element	Questionnaire item	λ
Knowing the problems that rose the construction of the knowledge to be taught (1.1)	What is your assessment of your knowledge of the history of thermodynamics concerning the development of its concepts?	0.879
Knowing the methodological orientations employed in the construction of knowledge (1.2)	What is your assessment of your knowledge of the history of thermodynamics concerning the development of its methods?	0.854
Pupils cannot be considered as "tabula rasa" (3.1)	What is your assessment of your knowledge concerning students' conceptions on thermodynamics?	0.769
Acquiring some knowledge of recent scientific developments to transmit a dynamic, non-closed view of physics (1.4)	What is your assessment of your knowledge of the recent scientific developments in thermodynamics (interdisciplinary)?	0.670
A meaningful learning demands that pupils construct their knowledge (3.2)	What is your assessment of your knowledge concerning contemporary constructivist approaches of thermodynamics teaching and learning?	0.639
Knowing the science/technology/society interactions (1.3)	What is your assessment of your knowledge of the history of thermodynamics concerning the cultural context within it was developed?	0.583

loading order. The elements derived from the first (knowing physics as the subject matter to be taught) and the third (acquiring theoretical knowledge about the physics teaching and learning process) component of the original theoretical framework, indicating that teachers' epistemological and pedagogical conceptions were quite interwind.

In the context of the questionnaire, the epistemological conceptions reflected to the teachers' self-assessment of their existing knowledge on the historical development of thermodynamics concerning conceptual, methodological, and cultural aspects, as well as on the recent developments in the field of thermodynamics. The pedagogical conceptions reflected to the teachers' self-assessment of their knowledge on constructivism, with a special focus on the students' prior knowledge, which is fundamental for applying such approaches.

5.3 *Factor 2: Alternative Instructional Design*

The second factor included teachers' ideas for approaching thermodynamics through an alternative instructional design, mostly based on the epistemology of the field and the impact it can have on a broad student audience. Table 5 presents the original theoretical framework elements and the corresponding questionnaire items that included

Table 5 Factor 2: original theoretical framework element (with component and element number) and the respective questionnaire item in descending loading (λ) order

Theoretical framework element	Questionnaire item	λ
A rigid view (algorithmic, exact, infallible... dogmatic) (2.2)	Do you believe that it would be useful to present more historical elements for the development of thermodynamics (e.g., inventions, creative solutions)?	0.909
An individualistic view (2.7)	Do you believe that it would be useful to present the scientists' collaborative work that historically led to the development of thermodynamics (and not just the persons' that were directly connected to the ultimate results)?	0.813
A socially 'neutral' view (2.8)	Do you believe that it would be useful to present the connection of thermodynamics with social issues (historical and recent)?	0.795
A merely accumulative vision (2.4)	Do you believe that it would be useful to present the scientific methods that historically led to the development of thermodynamics (and not just the ultimate results)?	0.792
An exclusively analytical vision (2.3)	Do you believe that it would be useful to present thermodynamics as an interdisciplinary scientific subject?	0.523
A "veiled" and elitist view (2.6)	Do you believe that thermodynamics is a scientific subject that should be it promoted as a meaningful and approachable one for all students?	0.437

in this factor, in a descending loading order. All elements derived from the second component (knowing teachers' spontaneous ideas on physics and on teaching and learning physics) and formed a group of the most prevailing teacher ideas.

In respect to the questionnaire items, teachers were prompted to reveal how deeply they were attached to these ideas and if they would open to overcoming them in the context of an alternative (constructivist) instruction. The focus was shifted toward the most accurate presentation of thermodynamics as a field that has been developing within a cultural and social context throughout a long period and with a variety of interactions among scientists and among relevant fields.

5.4 *Factor 3: Traditional Framework Perception*

The third factor clustered the items that referred to teachers' evaluation of traditional teaching and learning sequences of thermodynamics, including their own self-assessment for implementing such sequences in their classrooms. Table 6 presents the original theoretical framework elements and the corresponding questionnaire items that included in this factor, in a descending loading order. Most elements derived from the third component (acquiring theoretical knowledge about the physics teaching and learning process) and one from the fourth one (teachers' involvement in physics

Table 6 Factor 3: original theoretical framework element (with component and element number) and the respective questionnaire item in descending loading (λ) order

Theoretical framework element	Questionnaire item	λ
The construction of scientific knowledge has axiological commitments (3.5)	What is your assessment of the traditional methods that are used for the introduction of thermodynamics to students?	0.933
To construct knowledge pupils need to deal with problematic situations which may interest them (3.3)	What is your assessment of the traditional methods that are used for the students' penetration of thermodynamics?	0.821
Be conceived in an intimate connection with the teaching practice itself (4.1)	What is your assessment of your teaching of the school knowledge of thermodynamics up to now?	0.812
The construction of scientific knowledge is a social product (3.4)	What is your assessment of the level of group learning during your thermodynamics courses?	0.683

education research and innovation). These refer to the way students approach and construct their knowledge and how teachers act as facilitators in this process.

The respective questionnaire items adhered to the traditional approaches thermodynamics is introduced to students and worked through by the students. It is interesting that the teachers' self-assessment of their instruction, including the level of group learning during their respective courses, made its appearance within this factor. This may indicate that in their everyday practice they found more common ground with the traditional approach of teaching and learning thermodynamics rather the alternative (constructivist) one presented in Factor 2.

5.5 *Factor 4: Intention to Change*

Finally, the fourth factor represented the teachers' intention to deepen their knowledge of the epistemology and pedagogy of thermodynamics and help bridging the gap between educational research and practice. Table 7 presents the original theoretical framework elements and the corresponding questionnaire items that included in this factor, in a descending loading order. The elements derived from almost all components (knowing physics as the subject matter to be taught, teachers' involvement in physics education research and innovation, and knowing teachers' spontaneous ideas on physics and on teaching and learning physics) except for the third (acquiring theoretical knowledge about the physics teaching and learning process). The elements comprising this factor showed teachers' inclination toward new, research-informed, knowledge on teaching and learning. The emergence of the teachers' "extreme inductivism" within this factor was exceptionally intriguing.

As for the respective questionnaire items, teachers appeared to interwind their potential training with the development of new teaching and learning sequences

Table 7 Factor 4: original theoretical framework element (with component and element number) and the respective questionnaire item in descending loading (λ) order

Theoretical framework element	Questionnaire item	λ
Oriented to favor the experiencing of innovating proposals and explicit teaching reflection (4.2)	Are you interested in approaching the school knowledge of thermodynamics with a new teaching and learning sequence?	0.867
Designed to incorporate teachers to the investigation and innovation and involve them in the construction of the specific knowledge body of science teaching (4.3)	Are you interested in approaching the school knowledge of thermodynamics with a new teaching and learning sequence that will be implemented to contribute to the pertinent educational research?	0.845
Knowing how to choose appropriate content (1.5)	What is your assessment of your potential to restructure, enrich, or alternate the standard suggested school knowledge of thermodynamics?	0.705
Being prepared to deepen the knowledge (1.6)	What is your assessment of your readiness for the enrichment of your knowledge of thermodynamics as a subject to be taught?	0.575
Extreme inductivism (2.1)	Do you believe that it would be useful to further utilize experiments for the school knowledge of thermodynamics?	0.434

on thermodynamics, that would require a shift from the traditional approach. The teachers' inductivism may require a different interpretation in the context of this factor, as it seemed that teachers included the role and extent of experimental work in class subjected to change in agreement to their alternated instruction.

6 Discussion

In this study, we utilized a theoretical framework for physics teachers' training (Gil-Pérez and Pessoa de Carvalho 1997) to develop an online questionnaire as a research instrument that would facilitate physics teachers' training programs. Our aim was to examine in-service physics teachers' conceptions on physics epistemology and pedagogy to validate the theoretical framework and use it as a methodological framework as well. Instead of approaching physics as a "generic" subject matter to be taught, we utilized thermodynamics as a field of reference. At a great extent, we validated the elements of the initial framework (Aguirre et al. 1990; Flores et al. 2008; Fadaei 2012); however, our exploratory analysis indicated a different structure for the included elements when the theoretical framework was put into practice. We used the results for the design and evaluation of a training program on thermodynamics teaching and learning in upper secondary school in Greece (Meli et al. 2021).

The first factor we extracted (*Teacher prior knowledge*) included elements from both the epistemological and pedagogical aspects of thermodynamics as a subject matter to be taught, indicating that teachers may need to deal with these aspects in a unified manner. This factor combined teachers' knowledge of thermodynamics as a history-dependent scientific field and of students' initial ideas as they start the course. As for the epistemological aspect, previous studies (e.g., Irez 2006; Fadaei 2012), indicated that teachers thought higher of the conceptual development of thermodynamics in comparison to its methodological and cultural one. This finding agreed with their interest in any new developments in the field, suggesting that they may were more comfortable dealing with the scientific content rather than its epistemology. However, in our research these two elements (conceptual development and new developments) were grouped along with the methodological and cultural development, and also with the science-technology-society interactions. As for the pedagogical aspect, our literature review showed teachers' low engagement to constructivist approaches (e.g., Flores et al. 2008; Demirci 2015), but our results indicated that the teachers focused on their awareness about two crucial parameters: students' prior knowledge and the way students construct their knowledge.

The second factor (*Alternative instructional design*) formed a group of the most persistent "traditional" teacher ideas when they come to instruction. The teachers gave prominence to the rigid, individualistic, socially "neutral", accumulative, and analytical views of thermodynamics that may characterize them as scientists and interfere with their instruction. These views were sporadically identified in the literature (e.g., Brickhouse 1989; Irez 2006), yet they had not been correlated with each other. The teachers in our sample unified these views under one factor that reflects a very narrow perspective of science as a culturally isolated product which progresses almost randomly, depending on the immediate success of solitary scientists, and concerns only "gifted" individuals (Apostolou and Koulaidis 2010). However, the respective questionnaire items were formulated in a way that prompted teachers to identify and revise them in order to move toward a constructivist approach of thermodynamics teaching and learning.

The third factor that was formed (*Traditional framework perception*) reflected teachers' conceptions on the traditional they commonly introduce and elaborate thermodynamics in their respective courses. In this context, students are prompted to construct knowledge ignoring the attached axiological commitments or problematic situations relevant to that knowledge that may interest them. In addition, they mostly work individually in the classroom. These elements clearly interacted with teachers' self-assessment of their usual instruction and therefore they were grouped together. This finding was in-line with the results of previous studies (e.g., Flores et al. 2008; Demirci 2015) and highlight how teachers conceive the learning obstacles students encounter in a thermodynamic course, but also the teaching difficulties that teachers face in the context of a traditional approach.

Finally, the fourth extracted factor (*Intention to change*) indicated the connections that physics education research can have with the everyday teaching and learning practices. The teachers saw the development of their instruction in terms of the content they incorporate in relation to the deepening of their thermodynamics epistemological and pedagogical knowledge. An element that interestingly made its appearance within this factor is the teachers' "extreme inductivism" as a prominent idea they bring to their instruction and is mostly presented through an intense focus to experiments, as it has often presented in the literature (e.g., Aguirre et al. 1990; Irez 2006; Flores et al. 2008). In the context of this factor, however, teachers probably demonstrated their need to inform their experimental instruction with education research-based ideas (e.g., replace the "recipe-like" approach). The findings that fall under this Factor may imply that the "gap" between educational research and practice is not as wide as previously perceived (Saha 2009); at least from the teachers' point of view.

Future studies should test the questionnaire to larger populations of in-service physics teachers and, additionally, to pre-service physics teachers to indicate possible similarities and differences between these samples. Considering that the choice of thermodynamics as a field of reference may have impacted the teacher's answers (in comparison to *physics* in general or any other field, e.g., *mechanics*) it would be interesting to adapt it for different physics fields (beyond thermodynamics) to explore its generalization potential. Consequently, the reframed theoretical framework as a methodological one can be used for the design and evaluation of future physics teachers' training programs. A limitation inherited in all questionnaires that uses self-reports is that they have a risk of socially desired responses and measurement bias (Lavidas et al. 2022). A comparison of the instrument presented in this study with ones with similar content can define the criterion validity for such questionnaires that explore teacher conceptions.

References

Aguirre JM, Haggerty SM, Linder CJ (1990) Student-teachers' conceptions of science teaching and learning: a case study in preservice science education. Int J Sci Educ 12:381–390

Apostolou A, Koulaidis V (2010) Epistemology and science education: a study of epistemological views of teachers. Res Sci Technol Educ 28:149–166

Belo NAH, van Driel JH, van Veen K, Verloop N (2014) Beyond the dichotomy of teacher-versus student-focused education: a survey study on physics teachers' beliefs about the goals and pedagogy of physics education. Teach Teach Educ 39:89–101

Brickhouse NW (1989) The teaching of the philosophy of science in secondary classrooms: case studies of teachers' personal theories. Int J Sci Educ 11:437–449

Briscoe C (1991) The dynamic interactions among beliefs, role metaphors, and teaching practices: a case study of teacher change. Sci Educ 75:185–199

Demirci N (2015) Prospective high school physics teachers' beliefs about teaching practices: from traditionalist to constructivist. Eurasia J Math Sci Technol Educ 11:693–711

Fadaei AS (2012) Investigating the effects of teacher training on learning physics. Lat Am J Phys Educ 6:348–351

Flores F, Lopez A, Gallegos L, Barojas J (2008) Transforming science and learning concepts of physics teachers. Int J Sci Educ 22:197–208

Gil-Pérez D, Pessoa de Carvalho AM (1997) Physics teacher training: analysis and proposals. Int Comm Phys Educ 1, chapter D4

Gil-Pérez D, Vilches A, Fernández IC et al (2005) Technology as 'applied science.' Sci Educ 14:309–320

Irez S (2006) Are we prepared?: an assessment of preservice science teacher educators' beliefs about nature of science. Sci Educ 90:1113–1143

Kanderakis N, Dossis S, Koliopoulos D (2011) Teachers' conceptions about the implementation of a HPS sequence concerning the movement of a simple pendulum. In: Proceedings of the 11th international IHPST and 6th Greek history philosophy and science teaching joint conference, Aristotle University of Thesssaloniki, Thessaloniki, 1–5 July 2011, pp 687–696

Lavidas K, Papadakis S, Manesis D et al (2022) The effects of social desirability on students' self-reports in two social contexts: lectures vs. lectures and lab classes. Inf 13:491

Mansour N (2009) Science teachers' beliefs and practices: issues, implications, and research agenda. Int J Environ Sci Educ 4:25–48

Meli K, Koliopoulos D, Lavidas K (2021) Physics teachers' training webinars for teaching and learning introductory thermodynamics in upper secondary school in Greece. In: ESERA 2021 "Fostering scientific citizenship in an uncertain world", University of Minho, Braga, 30 August–3 September 2021, pp 1126–1135

O'Connor BP (2000) SPSS and SAS programs for determining the number of components using parallel analysis and Velicer's MAP test. Behav Res Methods Instrum Comput 32:396–402

Ogborn J (2008) Science and commonsense. In: Vicentini M, Sassi E (eds) Connecting research in physics education with teacher education, vol 2. International Commission of Physics Education. https://web.phys.ksu.edu/icpe/Publications/teach2/index.html

Porlán R, Martín R, Pozo D (2004) The conceptions of in-service and prospective primary school teachers about the teaching and learning of science. J Sci Teach Educ 15:39–62

Saha LJ (2009) The dissemination of knowledge about research on teachers to the teachers. Int Handb Res Teach Teach 71–79

Tabachnick BG, Fidell LS (2007) Using multivariate statistics, 5th edn. Allyn and Bacon/Pearson Education

Tsai CC (2007) Teachers' scientific epistemological views: the coherence with instruction and students' views. Sci Educ 91:222–243

Watts M (2013) Constructivism, re-constructivism, and task-oriented problem solving. In: Fensham PJ, Gunstone RF, White RT (eds) The content of science: a constructivist approach to its teaching and learning. Routledge, pp 51–70